ENGINEERING PRACTICES FOR MILK PRODUCTS

Dairyceuticals, Novel Technologies, and Quality

Innovations in Agricultural and Biological Engineering

ENGINEERING PRACTICES FOR MILK PRODUCTS

Dairyceuticals, Novel Technologies, and Quality

Edited by
Megh R. Goyal, PhD, PE
Subrota Hati, PhD

Apple Academic Press Inc.
3333 Mistwell Crescent
Oakville, ON L6L 0A2, Canada

Apple Academic Press Inc.
1265 Goldenrod Circle NE
Palm Bay, Florida 32905, USA

First issued in paperback 2021

Exclusive worldwide distribution by CRC Press, a member of Taylor & Francis Group
No claim to original U.S. Government works

ISBN 13: 978-1-77463-476-9 (pbk)
ISBN 13: 978-1-77188-801-1 (hbk)

Library and Archives Canada Cataloguing in Publication

Title: Engineering practices for milk products : dairyceuticals, novel technologies, and quality / edited by Megh R. Goyal, Subrota Hati.

Names: Goyal, Megh Raj, editor. | Hati, Subrota, editor.

Series: Innovations in agricultural and biological engineering.

Description: Series statement: Innovations in agricultural and biological engineering | Includes bibliographical references and index.

Identifiers: Canadiana (print) 20190164247 | Canadiana (ebook) 20190164271 | ISBN 9781771888011 (hardcover) | ISBN 9780429264559 (ebook)

Subjects: LCSH: Dairy products industry—Technological innovations. | LCSH: Dairy products.

Classification: LCC SF250.5 .E54 2019 | DDC 637—dc23

Library of Congress Cataloging-in-Publication Data

Names: Goyal, Megh Raj, editor. | Hati, Subrota, editor.

Title: Engineering practices for milk products : dairyceuticals, novel technologies, and quality / edited by Megh R. Goyal, Subrota Hati.

Other titles: Innovations in agricultural and biological engineering.

Description: Oakville, ON : Palm Bay, Florida : Apple Academic Press, [2020]. | Series: Innovations in agricultural and biological engineering | Includes bibliographical references and index. | Summary: "While also addressing the need for more effective processing technologies for increased safety and quantity, the dairy industry needs to address the growing customer demand for new and innovative dairy foods with enhanced nutritional value. This volume, Engineering Practices for Milk Products: Dairyceuticals, Novel Technologies, and Quality, looks at new research, technology, and applications in the engineering of milk products, specifically covering functional bioactivities to add value while increasing the quality and safety of milk and fermented milk products. Chapters in the book look at the functional properties of milk proteins and cheese, functional fermented milk-based beverages, biofunctional yoghurt, antibiotic resistant pathogens, and other probiotics in dairy food products. The information found here will be of value to the growing dairy industry in India, globally the largest milk producing country since 1997, and the research will also be useful for dairy food engineering professionals wherever milk and dairy products are produced and consumed"-- Provided by publisher.

Identifiers: LCCN 2019035119 (print) | LCCN 2019035120 (ebook) | ISBN 9781771888011 (hardcover) | ISBN 9780429264559 (ebook)

Subjects: LCSH: Dairy products--Quality--Research. | Dairy products industry--Technological innovations--Research.

Classification: LCC SF250.5 .E545 2020 (print) | LCC SF250.5 (ebook) | DDC 338.4/7637--dc23

LC record available at https://lccn.loc.gov/2019035119

LC ebook record available at https://lccn.loc.gov/2019035120

Apple Academic Press also publishes its books in a variety of electronic formats. Some content that appears in print may not be available in electronic format. For information about Apple Academic Press products, visit our website at **www.appleacademicpress.com** and the CRC Press website at **www.crcpress.com**

THE BOOK SERIES: INNOVATIONS IN AGRICULTURAL AND BIOLOGICAL ENGINEERING, APPLE ACADEMIC PRESS INC.

Under the book series titled *Innovations in Agricultural and Biological Engineering*, Apple Academic Press Inc. is publishing subsequent volumes in the specialty areas defined by the American Society of Agricultural and Biological Engineers (<asabe.org>) over a span of 8–10 years. Academic Press Inc. wants to be principal source of books in agricultural biological engineering. We seek book proposals from readers in areas of their expertise.

The mission of this series is to provide knowledge and techniques for agricultural and biological engineers (ABEs). The series offers high-quality reference and academic content in agricultural and biological engineering (ABE) that is accessible to academicians, researchers, scientists, university faculty and university-level students, and professionals around the world.

Agricultural and biological engineers ensure that the world has the necessities of life, including safe and plentiful food, clean air and water, renewable fuel and energy, safe working conditions, and a healthy environment by employing knowledge and expertise of the sciences, both pure and applied, and engineering principles. Biological engineering applies engineering practices to problems and opportunities presented by living things and the natural environment in agriculture.

ABE embraces a variety of the following specialty areas (asabe.org): aquacultural engineering, biological engineering, energy, farm machinery and power engineering, food and process engineering, forest engineering, information & electrical technologies engineering, natural resources, nursery and greenhouse engineering, safety and health, and structures and environment.

For this book series, we welcome chapters on the following specialty areas (but not limited to):

1. Academia-to-industry-to-end-user loop in agricultural engineering
2. Agricultural mechanization
3. Aquaculture engineering
4. Biological engineering in agriculture

5. Biotechnology applications in agricultural engineering
6. Energy source engineering
7. Food and bioprocess engineering
8. Forest engineering
9. Hill land agriculture
10. Human factors in engineering
11. Information and electrical technologies
12. Irrigation and drainage engineering
13. Nanotechnology applications in agricultural engineering
14. Natural resources engineering
15. Nursery and greenhouse engineering
16. Potential of phytochemicals from agricultural and wild plants for human health
17. Power systems and machinery design
18. GPS and remote sensing potential in agricultural engineering
19. Robot engineering in agriculture
20. Simulation and computer modeling
21. Smart engineering applications in agriculture
22. Soil and water engineering
23. Structures and environment engineering
24. Waste management and recycling
25. Any other focus area

For more information on this series, readers may contact:

Megh R. Goyal, PhD, PE
Book Series Senior Editor-in-Chief
Innovations in Agricultural and
Biological Engineering
E-mail: goyalmegh@gmail.com

BOOKS ON AGRICULTURAL & BIOLOGICAL ENGINEERING BY APPLE ACADEMIC PRESS, INC.

Evapotranspiration: Principles and Applications for Water Management

Flood Assessment Modeling and Parameterization

Management of Drip/Trickle or Micro Irrigation

Book Series: Research Advances in Sustainable Micro Irrigation
Senior Editor-in-Chief: Megh R. Goyal, PhD, PE

Book Series: Innovations and Challenges in Micro Irrigation
Senior Editor-in-Chief: Megh R. Goyal, PhD, PE

Book Series: Innovations in Agricultural & Biological Engineering
Senior Editor-in-Chief: Megh R. Goyal, PhD, PE

- Novel Dairy Processing Technologies: Techniques, Management, and Energy Conservation
- Sustainable Biological Systems for Agriculture: Emerging Issues in Nanotechnology, Biofertilizers, Wastewater, and Farm Machines
- State-of-the-Art Technologies in Food Science: Human Health, Emerging Issues and Specialty Topics
- Scientific and Technical Terms in Bioengineering and Biological Engineering
- Engineering Practices for Management of Soil Salinity: Agricultural, Physiological, and Adaptive Approaches
- Processing of Fruits and Vegetables: From Farm to Fork
- Technological Processes for Marine Foods, from Water to Fork: Bioactive Compounds, Industrial Applications, and Genomics
- Engineering Practices for Milk Products: Dairyceuticals, Novel Technologies, and Quality
- Nanotechnology and Nanomaterial Applications in Food, Health, and Biomedical Sciences
- Nanotechnology Applications in Dairy Science: Packaging, Processing, and Preservation

ABOUT THE SENIOR EDITOR-IN-CHIEF

Megh R. Goyal, PhD, PE

Retired Professor in Agricultural and Biomedical Engineering, University of Puerto Rico, Mayaguez Campus;
Senior Acquisitions Editor,
Biomedical Engineering and Agricultural Science, Apple Academic Press, Inc.

Megh R. Goyal, PhD, PE, is a Retired Professor in Agricultural and Biomedical Engineering from the General Engineering Department in the College of Engineering at the University of Puerto Rico–Mayaguez Campus; and Senior Acquisitions Editor and Senior Technical Editor-in-Chief in Agriculture and Biomedical Engineering for Apple Academic Press, Inc. He has worked as a Soil Conservation Inspector and as a Research Assistant at Haryana Agricultural University and Ohio State University.

During his professional career of 50 years, Dr. Goyal has received many prestigious awards and honors. He was the first agricultural engineer to receive the professional license in Agricultural Engineering in 1986 from the College of Engineers and Surveyors of Puerto Rico. In 2005, he was proclaimed as "Father of Irrigation Engineering in Puerto Rico for the Twentieth Century" by the American Society of Agricultural and Biological Engineers (ASABE), Puerto Rico Section, for his pioneering work on micro irrigation, evapotranspiration, agroclimatology, and soil and water engineering. The Water Technology Centre of Tamil Nadu Agricultural University in Coimbatore, India, recognized Dr. Goyal as one of the experts "who rendered meritorious service for the development of micro irrigation sector in India" by bestowing the Award of Outstanding Contribution in Micro Irrigation. This award was presented to Dr. Goyal during the inaugural session of the National Congress on "New Challenges and Advances in Sustainable Micro Irrigation" held at Tamil Nadu Agricultural University. Dr. Goyal received the Netafim Award for Advancements in Microirrigation: 2018 from the American Society of Agricultural Engineers at the ASABE International Meeting in August 2018.

A prolific author and editor, he has written more than 200 journal articles and textbooks and has edited over 70 books. He is the editor of three book series published by Apple Academic Press: Innovations in Agricultural & Biological Engineering, Innovations and Challenges in Micro Irrigation, and Research Advances in Sustainable Micro Irrigation. He is also instrumental in the development of the new book series Innovations in Plant Science for Better Health: From Soil to Fork.

Dr. Goyal received his BSc degree in engineering from Punjab Agricultural University, Ludhiana, India; his MSc and PhD degrees from Ohio State University, Columbus; and his Master of Divinity degree from Puerto Rico Evangelical Seminary, Hato Rey, Puerto Rico, USA.

ABOUT THE EDITOR

Subrota Hati, PhD

Assistant Professor, Dairy Microbiology Department, SMC College of Dairy Science, Anand Agricultural University, Anand, Gujarat, India

Subrota Hati, PhD, is currently working as an Assistant Professor in the Dairy Microbiology Department at the SMC College of Dairy Science, Anand Agricultural University, Anand, Gujarat, India. He has expertise in purification and characterization of food-derived bioactive peptides and probiotic fermented dairy foods. He has also worked at Mother Dairy, India, as a Quality Assurance Executive.

His BTech degree was in Dairy Technology from West Bengal University of Animal and Fishery Sciences, India. With ICAR-NDRI fellowship, he received Master and the PhD degree in Dairy Microbiology from National Dairy Research Institute, Karnal, India.

He is working on biofunctional properties of fermented milks produced by lactic acid bacteria, particularly probiotics. He has developed acidophilus milk, soy-based fermented curd, and peptides rich dahi. He is also working on isolation and purification of ACE inhibitory bioactive peptides derived from cow milk, goat milk, buffalo milk, camel milk, and soy milk.

He isolated and characterized 45 lactic acid bacteria and yeasts from fermented foods of the North Eastern Region (NER) of India, particularly Meghalaya, and developed the soy-based fermented foods for the Khasi Tribals of the Tura region. He has also isolated and characterized 25 lactic acid bacteria from fermented foods of the western part of India and deposited eight well-characterized dairy starter cultures in the Microbial Type Culture Collection and Gene Bank (MTCC), Chandigarh, for an International Depositary Authority (IDA) patent.

He has purified and characterized novel ACE inhibitory peptides enriched with IPP and VPP amino acids from different fermented milks of cow, goat, camel and soymilk and submitted to the National Center for

Biotechnology Information as a reference database. His work on purification and characterization of antihypertensive peptides derived from fermented milk-produced by indigenous well-characterized lactic acid bacteria helps to combat the hypertension without having any side effects. His work on different milks derived functional peptide adds novel sequences of antihypertensive peptides in AHTPDB databases from indigenous breeds of our country.

He was the recipient of various awards: Urmilabala Gold Medal by the Indian Dairy Association, New Delhi; Prof. Sukumer De Gold Medal as a Topper by the West Bengal University of Animal & Fishery Sciences, Kolkata (Vice Chancellor Gold Medal); three best paper awards by the Indian Dairy Association, New Delhi; 15 best poster awards from the National and International Seminars or Conferences; Silver Medal by All Bengal Teachers' Association, Govt. West Bengal; Young Scientist Award by Bioved, Allahabad; Best Young Scientist Award by Network on Fermented Foods (SASNET-FF), AAU, Anand and Lund University, Sweden and Best Oral Presentation Award by International Institute of Food and Nutritional Sciences (IIFANS), New Delhi, etc.

He has published 52 research papers, 11 review articles, 15 technical articles in various national and international peer-reviewed journals and also published 20 book chapters and presented 35 abstracts in various national and international seminars/conferences. He also edited two books, namely *Recent Advancement of Dairy Product Technology* and *Dairy Product Technology.*

Recently he was awarded with a best teacher award (2018) by Anand Agricultural University, Anand; an Indo-Australia Career Boosting Gold Fellowship (2018) by DBT, New Delhi; an Indian National Science Academy Visiting Scientist Award (2018); and several others.

For more details, the reader may contact him at: subrota_dt@yahoo.com.

CONTENTS

CONTRIBUTORS

Divyasree Arepally
Department of Agricultural and Food Engineering, Indian Institute of Technology, Kharagpur, 721302, India, Mobile: +91-9966995212, E-mail: divyasreearepally@gmail.com

Narender Kumar Chandla
Assistant Professor, College of Dairy Science and Technology,
Guru Angad Dev Veterinary and Animal Sciences University, Ludhiana–141004, India,
Mobile: +91-9464116738, E-mail: chandla84@gmail.com

Rekha Chawla
Assistant Professor, Department of Dairy Technology, College of Dairy Science and Technology,
Guru Angad Dev Veterinary and Animal Sciences University (GADVASU), Ludhiana 141004,
Punjab, India, Mobile: +91-7589145459, E-mail: mails4rekha@gmail.com

Kunal M. Gawai
Assistant Professor, Dairy Microbiology Department, Sheth MC College of Dairy Science,
Anand Agricultural University (AAU), Anand 388110, Gujarat, India, Mobile: +91-8488034122,
E-mail: kunalgawai@gmail.com

Gurlal Singh Gill
School of Public Health & Epidemiology, Guru Angad Dev Veterinary and Animal Sciences University,
Ludhiana–141004, India, Mobile: +91-7986214854, E-mail: gill.kangar@gmail.com

Apurba Giri
Assistant Professor and Head, Department of Nutrition, Mugberia Gangadhar Mahavidyalaya
(Affiliated with Vidyasagar University), P.O. Bhupatinagar 721425, District Purba Medinipur,
West Bengal, India, Mobile: +91-9564289290, E-mail: apurbandri@gmail.com

Tridib Kumar Goswami
Professor, Department of Agricultural and Food Engineering, Indian Institute of Technology,
Kharagpur, 721302, India, Mobile: +91-9647485515, E-mail: tkg@agfe.iitkgp.ac.in

Megh R. Goyal
Retired Faculty in Agricultural and Biomedical Engineering from the College of Engineering at the
University of Puerto Rico–Mayaguez Campus; and Senior Acquisitions Editor and Senior Technical
Editor-in-Chief in Agricultural and Biomedical Engineering for Apple Academic Press Inc.;
PO Box 86, Rincon–PR–006770086, USA, E-mail: goyalmegh@gmail.com

Lopamudra Haldar
Associate Professor, Department of Dairy Microbiology, Faculty of Dairy Technology,
West Bengal University of Animal and Fishery Sciences, Mohanpur-741252, Nadia, India,
Mobile: +91-9862714273, E-mail: mohor7@gmail.com

Subrota Hati
Assistant Professor, Dairy Microbiology Department, Anand Agricultural University (AAU),
Anand–388110, Gujarat, India, Mobile: +91-8617630431, E-mail: max037.ndri@gmail.com

S. K. Kanawjia
Emeritus Scientist, Cheese, and Fermented Foods Laboratory, Dairy Technology Division,
ICAR-National Dairy Research Institute (NDRI), Karnal132001, Haryana, India,
Mobile: +91-9896782850, E-mail: skkanawjia@rediffmail.com

Mital R. Kathiriya
Assistant Professor, Dairy Microbiology Department, Anand Agricultural University (AAU),
Anand–388110, Gujarat, India, Mobile: +91-9913468082, E-mail: mital@aau.in

Jashandeep Kaur
Junior Research Fellow, College of Dairy Science and Technology,
Guru Angad Dev Veterinary and Animal Sciences University, Ludhiana–141004, India,
Mobile: +91-7508099339, E-mail: jashan99339@gmail.com

Anju B. Khatkar
Scholar, Department of Food Science and Technology, Punjab Agricultural University,
Ludhiana–141004, India, Mobile: +91-9876870672, E-mail: abkhatkar@gmail.com

Sunil Kumar Khatkar
Assistant Professor (Dairy Technology), College of Dairy Science and Technology,
Guru Angad Dev Veterinary and Animal Sciences University, Ludhiana–141004, India,
Mobile: +91-9876870671, E-mails: absuneelkhatkar@gmail.com; suneel_khatkar@yahoo.co.in

Narendra Kumar
Former Research Scholar, Dairy Microbiology Division, ICAR-National Dairy Research Institute,
Karnal-132001, Haryana, India, Mobile: +91-9468376537, E-mail: narendra289186@gmail.com

Vandna Kumari
Former Research Scholar, Dairy Microbiology Division, ICAR-National Dairy Research Institute,
Karnal-132001, Haryana, India, Mobile: +91-9416193538, E-mail: vandna92@gmail.com

Jagrani Minj
Functional Fermented Foods and Bioactive Peptides Lab, Dairy Microbiology Division,
ICAR-National Dairy Research Institute (NDRI), Karnal–132001, Haryana, India,
Mobile: +91-9991103386, E-mail: jagrani.ndri@gmail.com

Krishan Kumar Mishra
Associate Professor, Department of Medicine, College of Veterinary Science and Animal Science
(NDVSU), Rewa 486001, MP, India, Mobile: +91-8966888486, E-mail: drmishra79@gmail.com

Santosh Kumar Mishra
Assistant Professor, Department of Dairy Microbiology, College of Dairy Science and Technology,
Guru Angad Dev Veterinary and Animal Sciences University (GADVASU), Ludhiana 141004,
Punjab, India, Mobile: +91-9464995049, E-mail: skmishra84@gmail.com

Veena Nagarajappa
Assistant Professor, Department of Dairy Chemistry, College of Dairy Science and Technology,
Guru Angad Dev Veterinary and Animal Sciences University (GADVASU), Ludhiana 141004,
Punjab, India, Mobile: +91-9855886831, E-mail: veena.ndri@gmail.com

B. Surendra Nath
Principal Scientist, Dairy Chemistry Section, Southern Regional Station (SRS),
ICAR-National Dairy Research Institute, Adugodi, Hosur Road, Bengaluru-560030, Karnataka, India,
Mobile: +91-9449028628, E-mail: bsn_ndri@yahoo.com

Anil Kumar Panghal
Research & Consultancy Coordinator & Associate Professor, School of Agriculture,
Lovely Professional University, Jalandhar-Delhi G.T. Road (NH-1), Phagwara 144034, Punjab, India,
Mobile: +91-9988049760, E-mail: anil.20785@lpu.co.in

Ami Patel
Senior Assistant Professor, Mansinhbhai Institute of Dairy & Food Technology (MIDFT),
Dudhsagar Dairy campus, Mehsana-384002, Gujarat, India, Mobile: +91-9825067311,
E-mails: ami@midft.com; amiamipatel@yahoo.co.in

Falguni Patra
Assistant Professor, Mansinhbhai Institute of Dairy & Food Technology (MIDFT),
Dudhsagar Dairy Campus, Mehsana-384002, Gujarat, India,
Mobile: +91-9106710890, E-mail: falguni@midft.com

J.B. Prajapati
Professor, and Head, Dairy Microbiology Department, Anand Agricultural University (AAU),
Anand–388110, Gujarat, India, Mobile: +91-9879105948, E-mail: jbprajapati@aau.in

Sudharshan Reddy Ravula
Department of Agricultural and Food Engineering, Indian Institute of Technology,
Kharagpur, 721302, India, Mobile: +91-9492883137, E-mail::r.sudharshanreddy@gmail.com

Pinaki Ranjan Ray
Professor, Department of Dairy Chemistry, Faculty of Dairy Technology,
West Bengal University of Animal and Fishery Sciences, Mohanpur Campus,
Nadia-741252, WB, India, Mobile: +91-9903040914, E-mail: pinakirray@gmail.com

Soumyashree Saha
Research Scholar, Dairy Microbiology Division, National Dairy Research Institute (NDRI),
Karnal-132001, Haryana, India, Mobile: +91-8902654189, E-mail: soumyashreesaha22@gmail.com

Nihir Shah
Assistant Professor, Mansinhbhai Institute of Dairy & Food Technology (MIDFT),
Dudhsagar Dairy campus, Mehsana-384002, Gujarat, India, Mobile: +91-9925605480,
E-mail:nihirshah13@yahoo.co.in

Brij Pal Singh
Assistant Professor, Department of Microbiology, School of Science, R.K. University,
Rajkot–360020, Gujarat, India, Mobile: +91-8950465147, E-mail: bpsingh03@gmail.com

V. Sreeja
Assistant Professor, Dairy Microbiology Department, Sheth MC College of Dairy Science,
Anand Agricultural University (AAU), Anand 388110, Gujarat, India,
Mobile: +91-9998326542, E-mail: sreeja_p70@rediffmail.com

Neelam Upadhyay
Scientist (Food Technology), Dairy Technology Division, ICAR- National Dairy Research Institute,
Karnal-132001, Haryana, India, Mobile: +91-9255772587, E-mail: neelam_2912@yahoo.co.in

Yogesh V. Vekariya
Assistant Professor, Department of Dairy Engineering, Anand Agricultural University (AAU),
Anand–388110, Gujarat, India, Mobile: +91-7405508077, E-mail: yogeshvekariya@aau.in

Shilpa Vij
Principal Scientist, Functional Fermented Foods and Bioactive Peptides Lab,
Dairy Microbiology Division, ICAR-National Dairy Research Institute (NDRI),
Karnal–132001, Haryana, India, Mobile: +91-9996262863,
E-mail: shilpavijn@yahoo.co.in

ABBREVIATIONS

Å	angstrom
AAB	acetic acid bacteria
ABTS	2,2'-azino-di-3-ethylbenzthiazoline sulphonate
ACE	angiotensin converting enzyme
ACE-I	angiotensin-I converting enzyme
ACE-II	angiotensin-II converting enzyme
AFLP	amplified fragment length polymorphism
ALA	α-linolenic acid
ALAD	alanine dehydrogenase
ALP	alkaline phosphatase
ALS	acetolactate synthase
AMP	antimicrobial packaging
AMPs	antimicrobial peptides
AMR	antimicrobial resistance
AP-PCR	arbitrarily primed polymerase chain reaction
APC	antigen-presenting cells
APD	antimicrobial peptides database
APEDA	Agricultural and Processed Food Exports Development Authority
ARDRA	amplified rDNA restriction analysis
AS	aggregation substance
AST	aspartate transaminase
ATCC	American Type of Culture Collection
ATP	adenosine triphosphate
BCM5	beta-casomorphin-5
BCM7	beta-casomorphin-7
BDI	borculo domo ingredients
BHA	butylated hydroxy anisole
BHT	butylated hydroxyl toluene
BLG	β-lactoglobulin
BSA	bovine serum albumin
BTM	bulk tank milk
CAC	Codex Alimentarius Commission
$CaCl_2$	calcium chloride

cas	CRISPR associated gene
CAT	catalase
CCP	critical control point
CD	celiac disease
CFU	colony forming unit
CFU/g	colony forming units per gram
CHD	coronary heart disease
CLA	conjugated linoleic acid
CNS	central nervous system
CO_2	carbon dioxide
COD	chemical oxygen demand value
CPP	calcium phosphopeptidase
CPV	conical process vat
CRISPR	clustered regularly interspaced short palindromic repeats
CS	carrot-shred
CVD	cardiovascular diseases
CVS	computer vision syndrome
DCs	dendritic cells
DEC	diarrheogenic *E. coli*
DF	diafiltration
DH	degree of hydrolysis
DHA	docosahexaenoic acid
DMBA	dimehylbenza anthracene
DNA	deoxyribonucleic acid
DPA	docosapentaenoic acid
DPPH	α,α-diphenyl-β-picrylhydrazyl
DVS	direct vat set
EAI	emulsion activity index
EDTA	ethylene diamine tetra acetic acid
EIA	Export Inspection Agency
EMS	environmental management system
EPA	eicosapentaenoic acid
EPS	exopolysaccharides
ERIC	enterobacterial repetitive intergenic consensus
ESBL	extended-spectrum β-lactamase
ESI	emulsion stability index
ESβL	extended spectrum β-lactamase
FAD	flavin adenine dinucleotide

FAME	fatty acid methyl ester
FDA	Food and Drug Administration
FFA	free fatty acid
FM-D	foam-mat drying
FMN	flavin mononucleotide
FMS-D	filtermat spray drying
FOSHU	food for specified health uses
FSMS	Food Safety Management System
FuFoSE	Functional Food Science in Europe
GABA	gamma amino butyric acid
GAP	good agricultural practices
GHP	good handling practices
GI	gastrointestinal
GIT	gastrointestinal tract
GM	genetically modified
GMP	good manufacturing practice
GMP	good manufacturing procedures
GPx	glutathione peroxidase
GRAS	generally recognized/regarded as safe
GSMM	genome scale metabolic model
HACCP	hazard analysis and critical control point
HDL	high density lipoproteins
HGT	horizontal gene transfer
HHP	high pressure processing
HIV	human immune-deficiency virus
HP	high pressure
HPLC	high performance liquid chromatography
HPP	high pressure processing
HP-USD	high-power ultrasound drying
HTST	high temperature short time
HY-D	hybrid dryer
IBD	inflammatory bowel disease
ICTV	International Committee on Taxonomy of Viruses
IDF	International Dairy Federation
IFN	interferon
IG	immunoglobulin
IgA	immunoglobulin A
IgG	immunoglobulin G
IL	interleukin

ILSI	International Life Science Institute
inlA	internalins A
inlB	internalins B
inlC	internalins C
IS	insertion sequence
ISO	International Organization for Standardization
ISSHE	inclined scraped surface heat exchanger
ITS	intergenic transcribed sequence
KCl	potassium chloride
KNOS	kinin-nitric oxide system
LA	lactic acid
LAB	lactic acid bacteria
LCFA	long chain fatty acids
LC-MS/MS	liquid chromatography–mass spectrometry/mass spectrometry
LDH	lactate dehydrogenase
LDL	low density lipoprotein
LDL-C	low density lipoprotein cholesterol
LGG	Lactobacillus rhamnosus GG
LGü	lactoglobulin ü
LLO	listeriolysin O
LPS	lipopolysaccharides
m/m	mass by mass
MAP	modified atmospheric packaging
MDR	multidrug resistance
MF	microfiltration
Mg	magnesium
MGE	mobile genetic elements
MIC	minimum inhibitory concentration
Micro LC-TOF-MS	micro liquid chromatography coupled to time-of-flight mass spectrometry
MICs	minimum inhibition concentration
MLN	mesenteric lymph nodes
MLST	multi-locus sequence typing
MP	metalized polyester
MPCs	milk protein concentrates
MPI	milk protein isolate
MRS	DeMan, Rogosa, and Sharpe agar
MRSA	methicillin-resistant *Staphylococcus aureus*

MSSA	methicillin-susceptible *Staphylococcus aureus*
MSS-D	multi-stage spray dryer
MW-D	microwave-related drying
NaCl	sodium chloride
NADH	nicotinamide adenine dinucleotide
NASA	National Aeronautics and Space Administration
NCDC	National Collection of Dairy Cultures
NDDB	National Dairy Development Board
NDRI	National Dairy Research Institute
NF	nanofiltration
NICE	nisin-controlled expression
NK	natural killer
NK-cell	natural killer cell
OHSAS	occupational health and safety management system
OM	outer membrane
ORAC	oxygen radical absorbance capacity
OTR	oxygen transmission rate
pABA	para-aminobenzoicacid
PCR	polymerase chain reaction
PCS	processed cheese spread
PEG	polyethylene glycol
PFA	prevention of food adulteration
PFGE	pulsed-field gel electrophoresis
PGM	phosphoglucomutase
pI	isoelectric point
PI-PLC	phosphatidylinositol-specific phospholipase C
PP	polypropylene
PPAR	peroxisome proliferators activated receptors
PPAR-γ	proliferators activated receptors -γ
PPI	proton pump inhibitor
PUFA	polyunsaturated fatty acids
PVC	poly vinyl chloride
PVL	Panton-Valentine leukocidin
QMS	quality management system
RAPD	randomly amplified polymorphic DNA
RAS	rennin-angiotensin system
RCR	rolling circle replication
rDNA	recombinant DNA
REP	repetitive extragenic palindromic

RFLP	restriction fragment length polymorphism
RH	relative humidity
RNS	reactive nitrogen species
RO	reverse osmosis
ROS	reactive oxygen species
RPE	reverse pathway engineering
RP-HPLC	reverse phase high performance liquid chromatography
RT	reverse transcriptase
RVA	*rapid* visco analyzer
SCC*mec*	staphylococcal cassette chromosome *mec*
SDS–PAGE	sodium dodecyl sulfate–polyacrylamide gel electrophoresis
SF-D	spin flash dryer
SGOT	serum glutamic oxaloacetic transaminase
SGPT	serum glutamic pyruvic transaminase
SHMT	serine hydroxymethyltransferase
SNF	solid-not-fat
SOD	superoxide dismutase
ssDNA	single stranded deoxyribonucleic acid
SSHE	scraped surface heat exchanger
TAP	triplet arbitrary primed
TBHQ	tertiary butylated hydroquinone
TC	total cholesterol
tetM	*gene for tetracycline resistance*
TFSSHE	thin film scraped surface heat exchanger
TLRs	toll-like receptors
TNF	tumor necrosis factor
TQM	total quality management
TSSHE	thin film scraped surface heat exchanger
UF	ultrafiltration
UHT	ultra high temperature
US-FDA	U.S. Food and Drug Administration
UTIs	urinary tract infections
VB-D	vacuum belt dryer
VLDL	very-low-density lipoprotein
VRE	vancomycin-resistant enterococci
VREF	vancomycin-resistant *Ent. faecium*
WHO	World Health Organization
WPC	whey protein concentrate

WPH	whey protein hydrolysate
WPIs	whey protein isolates
Zn	zinc
α-LA	α-lactalbumin
β-CN-f	beta casein fragment
β-LG	β-lactoglobulin
ω-3 FAs	omega-3 fatty acids

BOOK ENDORSEMENT

Dairy science and technology plays a significant role in the advancement of dairy processing, quality, and safety of dairy foods and also explores the innovative ideas on novel product development for combating modern lifestyle diseases. This book has covered all the aspects of recent happenings in dairy and food industry. This edited book has been compiled with chapters from scientists working in different disciplines in dairy science and technology on the functional milk proteins, functional cheese, probiotics, and functional dairy beverages, mechanization of dairy processing, non-thermal technology, antibiotic-resistant pathogens, and quality and safety of dairy foods. This book will be a useful guide for undergraduate and postgraduate students as well as dairy and food personnel for upgrading their knowledge and concepts on recent happenings in dairy science and technology.

Yogesh Khetra, PhD
Professor, Dairy Technology Division
National Dairy Research Institute
Karnal–132001, Haryana, India
E-mail: yogesh.khetra@gmail.com

PREFACE 1 BY MEGH R. GOYAL

At the 49th annual meeting of the Indian Society of Agricultural Engineers at Punjab Agricultural University during February 22–25 of 2015, a group of ABEs convinced me that there is a dire need to publish book volumes on the focus areas of agricultural and biological engineering (ABE). This is how the idea was born on a new book series titled, Innovations in Agricultural and Biological Engineering.

The contributions by the cooperating authors to this book volume have been most valuable in the compilation. Their names are mentioned in each chapter and in the list of contributors. This book would not have been written without the valuable cooperation of these investigators; many of whom are renowned scientists who have worked in the field of ABE throughout their professional careers. Subrota Hati joins me as editor of this book volume. He is a frequent contributor to my book series and a staunch supporter of my profession. His contribution to the contents and quality of this book has been invaluable.

The goal of this book volume, *Engineering Practices for Milk Products: Dairyceuticals, Novel Technologies, and Quality,* is to guide the world science community on how engineering practices for milk products has evolved from dairy barn to fork.

We thank editorial and production staff and Ashish Kumar, Publisher and President, at Apple Academic Press, Inc., for making every effort to publish the book when the diminishing resources are a major issue worldwide.

I express my deep admiration to our families for their understanding and collaboration during the preparation of this book. As an educator, I wish to give a piece of advice to one and all in the world: *"Permit that our Almighty God, our Creator, allow us to inherit new technologies for a better life at our planet. I invite my community in agricultural engineering to contribute book chapters to the book series by getting married to my profession."* I am in total love with our profession by length, width, height, and depth. Are you?

—**Megh R. Goyal, PhD, PE**
Senior Editor-in-Chief

PREFACE 2 BY SUBROTA HATI

Milk is an almost complete food and nourishes all age groups by providing proper nutrition since long ago. Milk and milk products meet the daily demands of consumers. Milk procurement and processing are increasing at a faster pace in India with the advancement of dairy science and technology. We are on the way to spread the knowledge and share the recent technological advancements in the dairy and food sector.

As we know that India is producing 140 million tonnes of milk every year approximately, and India is the largest milk-producing country in the world, even though only 16% of the total milk is being processed in the organized sector. Raw milk is a perishable product, and it is an expensive product to ship long distances. This fact provides the economic force that has contributed to keeping milk production near major population centers. Economies of scale have increased dramatically in milk processing, brought on by improved processing technology, biological engineering, and improved milk quality. This increase in economies of scale has reduced values of many milk products and has resulted in several milk products becoming a generic commodity. Fluid milk, many cheese products, milk powder, condensed milk, and even some ice cream are all products that have developed a commodity pricing structure throughout the dairy food industry.

The dairy industry has a golden opportunity to handle a huge amount of milk in the organized sector. However, value-added products, as well as the automation system in processing plants, have given a new insight to the dairy industry. Most of the dairy players have set up their R&D to invent or discover different aspects of dairy science. Our motto is to value addition of the dairy science and technology, incorporating the new concepts, value-added product developments, and recent advancements; however, consumers demand healthy foods with proper nutrition and that contribute to a balanced diet. To meet this challenge, novel value-added dairy foods are developed and sold to the market. Nowadays, dairy and food industries are looking forward to the novel technology that could be applied for developing innovative foods with sufficient nutrition. Fermented dairy foods have occupied the market share in foods and are growing steeply as probiotics-added foods are the first choice for modern consumers for maintaining the balance of gut flora. Probiotics provide various health benefits, including antimicrobial, anticancer, antitumor, immunomodulatory effects, etc. But antibiotic-resistant pathogens are now a huge threat to the food industries and are causing disastrous health problems. Probiotic-fermented dairy foods could be an alternative solution for combating this problem. Value-added dairy foods like cheese, fermented dairy beverages, and composite dairy foods are the focused interest for the new generation foods. These foods will help to solve the long-standing problems of anemia or malnutrition, etc. Food industries are demanding for the development of non-thermal technology, which will process foods at low temperature, preserving their nutrition and sensory properties as well as heat-sensitive biomolecules present in foods.

Therefore, this book is edited to scatter more ideas and knowledge on novel value-added dairy foods and probiotic dairy foods with their health benefits. Recent advancements of food processing technology, particularly non-thermal technology and the potential of dairy foods in human health, need attention by scientists. This book will be a helpful guide to undergraduates, postgraduates, and personnel from the food industries.

I dedicate this book to my wife, Priyanka, and my daughter, Samprity, for continuously supporting and inspiring me. I am also grateful to Dr. Megh R. Goyal for his continuous technical guidance and motivation as a Senior Editor.

—Subrota Hati, PhD
Editor

PART I
Functional Bioactivities of Milk Products

PART I

CHAPTER 1

FUNCTIONAL PROPERTIES OF MILK PROTEINS

VEENA NAGARAJAPPA, NEELAM UPADHYAY, REKHA CHAWLA,
SANTOSH KUMAR MISHRA, and B. SURENDRA NATH

ABSTRACT

Milk proteins not only have high nutritive value in terms of high biological value, protein digestibility corrected amino acid score, protein efficiency ratio, and bioactive peptides, but also have desirable sensory characteristics and great techno-functional or value-adding functional properties. Due to these reasons, concentrates, and isolates of milk proteins are finding numerous applications in the food industry. The demand for milk-derived proteins will continue to increase as the field of food applications widens. The improvement and development in novel functional properties of both casein and whey proteins have potential to drive their growth even more in the years to come, because the consumer now prefers to consume more of natural ingredients as part of the daily diet.

1.1 INTRODUCTION

Milk is a multi-component, nutrient-dense food; and nourishes and provides immunological protection to the offspring of mammals. It is a complex liquid wherein different types of molecules are present in different forms, such as fats as emulsions, proteins along with some minerals as colloidal particles and lactose, some soluble proteins and few minerals as true solutions. The important protective and physiological functions are being impacted by oligosaccharides and proteins and peptides, such as metal-binding proteins, immunoglobulins (IGs), hormones, enzymes, growth factors, and antibacterial agents present in milk [41, 131]. Fluid milk from the bovine species hasabout 12.7% of solids consisting of 87.3, 3.7, 3.4, 4.8, and 0.7% of water, fat, protein, lactose, and ash, respectively [40].

Proteins are major constituents of our diet required for proper growth and supporting the general well-being. Out of all known sources of dietary proteins, milk proteins have been best described and researched [41]. Traditionally, milk proteins are classified into two major categories: casein accounting for around 80% of total proteins, while the rest 20% being the whey proteins. Casein is reported to be precipitated at its isoelectric point (~pH 4.6 at a temperature above 8°C). It consists of mainly four fractions, namely: α_{s1}-, α_{s2}-, and β- and k-casein @ 10, 2.6, 9.3 and 3.3 g/L, respectively [135]. Most of these fractions exist as colloidal particles called casein micelles ranging in size from 30 to 600 nm having an average molecular weight of 10^8 Da.

Physiologically, caseins are essential for the neonates since this act as a source of amino acids needed for proper growth of the infant. However, averting pathological calcification of mammary glands is suggested to be the major physiological feature of the casein micelle system [58]. Caseins are reported to be a robust source of biologically active peptides [71]. The favorable health effects can be ascribed to a number of peptide sequences that exhibit antioxidative, antibacterial, antihypertensive, antithrombotic, opioid, and immunomodulatory bioactivities, etc. [38].

The portion of proteins remains soluble after precipitation (isoelectric) of caseins is referred to as whey proteins that are globular and heat-sensitive proteins [43]. The major components among whey proteins include β-lactoglobulin (β-LG), α-lactalbumin (α-LA), bovine serum albumin (BSA) and IG having a corresponding concentration of 3.2 g/L, 1.2 g/L, 0.4 g/L, and 0.7 g/Land represent a respective 50, 20, 10 and 10% of total whey protein fraction, respectively [135]. The other minor constituents of this fraction include lactoferrin, lactolin, blood transferrin, and proteose-peptones. Due to the presence of varied amino acid composition [91], whey proteins show different globular structures. Being high-quality proteins, they are extremely important for digestibility, bioavailability, amino acid profile, and biological value, besides functional properties and sensory characteristics. Furthermore, the bioactive peptides of whey proteins origin are reported to possess physiological benefits leading to improvement in health and reduction in the risks of occurrence of diseases [86].

Caseins are immensely heat-stable and thus do not coagulate even when subjected to a temperature of 100°C for 24 h or 140°C for 20–25 min [41]. Caseins are phosphoproteins, and degree of phosphorylation depends on casein fractions. A number of phosphate residues in the individual casein imparts molecular charges on them and thereby metal binding (especially calcium ions) properties, heat stability, hydration, and solubility. Due to inter-micellar steric stability and electrostatic repulsion, casein micelles are not aggregated under natural conditions due to the hairy structure of C-terminal region of κ–casein from the surface of the micelle [25, 59].

Whey proteins, on the other hand, are not phosphorylated, but are richer in sulfur content (1.7%) in comparison to caseins containing around0.8% Sulfur. The high content of proline in caseins leads to prohibition of formation of secondary structure, resulting in relatively open 'rheomorphic' structure of amphipathic, small, and randomly coiled protein molecules. These possess a tendency to undergo self-association [129]. On the contrary, whey proteins possess globular conformations showing a high proportion of ordered structure in their sequences. These structures exhibit pronounced hydrophilicity, smaller amphipathicity and a minute tendency for self-association. These show little sensitive to changes in pH and ionic strength over caseins [66].

Milk proteins are the important functional components that provide favorable characteristics to food systems into which they are added [137]. Generally, the functionality of proteins is due to their physicochemical properties. It is majorly classified into two main groups: hydrodynamic or hydration-related, which includes water absorption, viscosity, solubility, and gelation; while surface-active properties, which include foaming, emulsification, and film formation [65, 124].

Surface-active properties are dependent on the hydrophobicity and hydrophilicity of protein domains, whereas the hydrodynamic properties are related to size, conformation, and flexibility of a protein molecule. Functional properties indicate natural features of protein molecules, i.e., size, shape, flexibility, susceptibility to denaturation, composition, and sequence of amino acid, charge, and its distribution, hydrophobicity/hydrophilicity, characteristics of micro-domain structures, capability of a domain or the whole molecule to adapt to environmental changes, specificity of interactions with other food components, and the environmental characteristics like temperature, pH, pressure, and ionic strength [28].

The design of functionality by means of enzymatic, physical or chemical methods can permit a protein to fulfill a variety of purposes in food products. Many different categories and kinds of milk protein-enriched products like milk protein concentrates (MPCs), caseins, and caseinates, whey protein concentrates (WPCs), and isolates (WPIs) are prepared; and specific protein blends designed especially for specific applications are also fabricated from milk proteins.

The chapter discusses the functional properties of casein and whey proteins in a comprehensive manner.

1.2   MANUFACTURE OF MILK PROTEIN-BASED PRODUCTS

Skim milk when treated with diluted mineral acid (sulfuric acid or hydrochloric acid) results in precipitation of casein proteins at its isoelectric pH (4.6) leading to formation of acid casein [127], while the action of chymosin or synthetic rennet on skim milk results in cleavage of κ-casein, which disrupts the micellar casein leading to formation of rennet casein curd in the presence of released calcium. Casein, after precipitation by acid (referred to as curd), is heated to remove the whey entrapped in it. It is washed with water several times, followed by mechanical centrifugation or pressing [127]. The washed and dried acid casein curd is then neutralized (pH 6.6). The neutralizers used for this purpose may be sodium, potassium, or calcium hydroxide leading to the dissolution of acid casein, which on spray-drying produces caseinates of sodium, potassium, or calcium, respectively. Caseinates possess higher solubility, and thus functional attributes, in comparison to purified casein powder.

Advancement in the membrane processing technology (like microfiltration (MF), ultrafiltration (UF), diafiltration (DF), reverse osmosis (RO) and ion-exchange) have led to the development of various whey protein derived products, e.g., WPC, and WPI that contains 50–85% and 90–98% protein (dry matter basis), respectively. The other finished whey products are hydrolyzed whey, demineralized whey, and reduced lactose whey [60]. The final composition of each of these whey products contains a varied amount of proteins, lactose, fat, and minerals, besides having different functional properties [5, 82].

The UF permits selective retention of proteins and selective permeation of water, minerals, lactose, and other molecules of low molecular weight compounds, making this technology industrially viable to produce WPC [60]. However, DF approach is adopted for further increasing the protein content of WPC and to produce purification grade of WPC. The credit for greater economic value (i.e., 3 to 40 times high) of WPC over whey powder is attributed to its outstanding nutritional and functional characteristics [7].

The ion-exchange chromatographic method is used for the manufacture of WPI from whey. In this process, the pH of whey is adjusted to <4.6 using acid, followed by reacting with a cation-exchange resin so as to allow the adsorption of proteins. The non-adsorbed material, including lactose, is eluted out with the whey. The proteins are desorbed from the ion-exchanger by first washing the resin with water followed by adjusting the pH to >5.5 with alkali. The pH of the eluted solution of proteins is adjusted, concentrated by UF, evaporated (optional) and finally spray- dried to obtain WPI [14, 99].

The WPI products, thus obtained, possess good functionality, besides providing a high concentration of protein and low level of lactose and lipid. The functionality of WPI is superior to WPC when compared on an equal protein basis, but its production is restricted owing to the huge cost involved in its production. The limited hydrolysis under strict conditions results in the formation of whey protein hydrolysates (WPH), which can be used for the particular functional and nutritive purpose. They are generally used in infant formulae on account of being pre-digested, which enable ready absorption in the gut [41, 61].

Predominantly, three methods (namely dry blending, precipitation, and membrane filtration) are employed to produce MPCs [64]. However, membrane processing is the most used method for domestic production of MPC. MPCs are obtained directly from skim milk by combining UF/DF that lead to the removal of most of soluble minerals and lactose while retaining milk protein. The retentate, thus obtained, is concentrated via evaporation followed by spray drying [93]. MPCs can have protein contents ranging from 42 to 85%, retaining casein and whey proteins in their native form, similar to that found in milk. MPC manufactured with UF prevents both the extrinsic adjustment of pH as well as calcium chloride addition. This enables the casein micelle to be intact and offers different functionality to MPC produced by UF treatment than MPC manufactured by dry blending or precipitation [54, 64].

Common MPC products include: MPC85, MPC80, MPC70, MPC42 (in order of decreasing protein content), etc. Milk Protein Isolates (MPI), on the other hand, are milk protein powders having a very high content of protein, i.e., ≥90% and very low concentration of minerals and lactose [64].

Milk proteins can be modified by several methods and have the potential for optimizing functional properties. The specialty ingredients are produced by altering the physicochemical properties of milk proteins by employing physical, chemical, and/or enzymatic modification. Primary structure of proteins can be chemically modified for improving the functional properties.

Main classes of reactions used to modify the side chain of amino acids chemically are: acylation, alkylation, oxidation, and reduction. The chemical modification should be avoided for food application due to harsh conditions adopted during the reaction, the non-specific chemical reagents, and solvents used for the reactions and the amount of residual chemical that remains in the final product due to incomplete removal of the same. However by employing enzymatic hydrolysis, the functional properties of proteins are altered as enzymatic treatment leads to disruption of the tertiary structure of proteins that leads to a reduction in molecular weight, besides enhancing the interaction of peptides generated with themselves as well as with the environment [24, 76]. The enzymatic hydrolysis, thus, has the potential to modify and optimize the biological, techno-functional, and nutritional characteristics of proteins to be used in food products [39, 56, 119].

1.3   FUNCTIONAL PROPERTIES OF MILK PROTEINS

The functional properties of milk proteins are basically physical and chemical properties that influence the behavior of proteins during processing, preparation, storage, and consumption [68]. They also contribute to the quality and organoleptic characteristics of food systems [123]. An overview of the major functional properties of milk proteins is presented in this section.

1.3.1   SOLUBILITY

Solubility determines the use of protein in the food system. In other words, it is an essential prerequisite that manifests several other functional properties. Proteins are required to possess higher solubility in order to exhibit superior emulsifying, gelation, foaming, and whipping properties [94].

Micellar casein was hydrolyzed by Grufferty and Fox [44] using alkaline milk proteinase. The results revealed increased soluble nitrogen content to release the proteose peptones and peptides with good surface activity. Abert and Kneifel [1] observed an increase in the nitrogen solubility index, owing to an increase in the degree of acid casein. The solubility of casein hydrolysates is reported to vary slightly as a function of the degree of hydrolysis and pH. Sodium caseinate hydrolysates with varying degree of hydrolysis are suggested to be highly soluble over a wide range of pH (from 2 to 8) with more than 90% solubility and high solubility values at pH 4 [108].

A comparative study using three different proteases (i.e., papain, trypsin, and pancreatin) revealed that the solubility of sodium caseinate hydrolysates varies with the enzyme type, presenting an interesting correlation with the enzymes specificities. Out of three enzymes, the hydrolysates obtained with papain exhibited the higher ability of enhancing the solubility of sodium caseinate, followed by trypsin and pancreatin [79]. This could be because papain is a cysteine endopeptidase having broad substrate specificity. It hydrolyzes specifically the bonds, including arginine, lysine, and phenylalanine [128].

Whey proteins possess a unique property of being soluble in water (over the pH range of 2 to 9), which makes them very useful ingredient to be used in the preparation of beverage [133]. However, the solubility of whey proteins decreases with increasing concentration of salt due to the salting-out effect [97]. Also, the whey proteins denature at a temperature higher than 70°C resulting in a reduction in whey protein-solubility on thermal denaturation [91].

Pelegrine and Gasparetto [103] indicated that the solubility of the proteins was decreased at pH 4.5. Also, the increase in temperature resulted in a decrease in solubility, indicating the occurrence of protein denaturation [103]. Mont and Jost [89] reported that the treatment with neutrase or papain and trypsin action resulted in 80 and 100% of solubility of heat-denatured whey proteins, respectively due to their smaller molecular sizes and a lower degree of secondary structure than that present in native proteins.

MPC has been reported to retain good solubility for around 6–8 months when stored at a low relative humidity (RH) and at ambient temperature or under refrigerated conditions. However, MPC with elevated moisture content and water activity, when stored above ambient temperature, can suffer from poor solubility [4, 8, 36]. The reduction in solubility of MPCs could be attributed to a higher degree of protein-protein interactions on the surface of the powder particle.

The insoluble protein material comprises primarily of caseins, majorly α- and β-caseins with less or no whey protein [4, 50]. This insoluble material was formed as a result of feeble, non-covalent (hydrophobic) protein-protein interaction, with few disulfide-linked protein aggregates comprising of κ-casein, β-LG, and α_{s2}-casein. The instability of proteins, in turn, affecting the solubility of MPC has been attributed to modifications in protein conformation and interaction between water and protein [46].

Schuck et al., [118] reported that a slower rate of water transfer to the interior of the grain could be the reason for the increase in the insoluble matter in MPC and not the denaturation. It is suggested that the contact with water creates a high surface viscosity leading to poor transfer of water to the interior and thus can slow the process of internal hydration. The high protein content of the particles may be the reason for higher surface viscosity. The findings of another research indicate that an increasing proportion of calcium in mineral fraction leads to a decrease in the solubility of MPC [6].

Pierre et al. [106] noted that on increasing the reconstitution temperature from 24 to 50°C, the solubility index was increased from 74 to 97.2%. Several studies have reported that the addition of monovalent salts like chloride salts of sodium and potassium (NaCl and KCl) can improve the solubility MPC and micellar casein. The improved reconstitution of micellar casein could be attributed to the hygroscopic nature of NaCl. Moreover, these minerals may be associated with modified protein-protein interactions [117] or modification in the casein micelle structure [9] leading to improved rehydration.

The significant improvement in the solubility of MPC80/ MPC85 on treatment with monovalent salt over untreated samples has been reported by several researchers [11, 17, 122]. Almost higher than 90% solubility in sodium- and potassium-treated MPC is reported to be retained even after 12 months of storage at 21°C. This has been suggested to be due to firstly, increase in electrostatic repulsive forces between casein micelles [50]; and secondly, to release of calcium and phosphate from the casein micelles [121, 122]. Besides electrostatic forces and hydrophobic forces are also reported to play a crucial role in the solubility of high-protein milk powders by modifying protein-protein interactions leading to restricted protein aggregate formation [4, 50, 81].

1.3.2 GELATION PROPERTIES

The functional groups within the protein are required to be exposed for the formation of a gel. The exposed group's make the interaction easier that finally leads to the formation of a three-dimensional network. However, gel formation is a complicated process that is affected by the concentration of protein, the quantity of water, pH, ionic strength, time, and temperature along with the interaction with other components in the food system [110]. The heat treatment makes the native protein to denature, which leads to the formation of disulfide bonds and exposure of hydrophobic amino acid residues [120]. When heating is continued during post-denaturation, the proteins aggregate and interact with other proteins to form a gel or coagulum.

When raw milk or caseinate solutions are acidified, the moderate loss of electrostatic repulsions and other physical interactions like Van der Waals and hydrophobic forces, as well as entropic and steric effects, lead to protein aggregation [114]. The gel formation, in case of raw milk, initiates at approximately pH 4.9. However, a higher acidification temperature would result in the inception of gelation at a higher pH [77]. Therefore in case of heated milk, gel formation begins at higher pH values (around pH 5.4 depending on the pre-heating conditions and the acidification temperature). The composition of milk gel is reportedly different for heated and unheated milk. For instance, whey proteins and disulfide bridges are important parts of the gel network for heated milk gels; while weak casein-casein interactions result in the formation of a gel network for unheated milk. However, the formation of caseinate gel is a less complicated phenomenon. In a solution having protein concentration comparable to that of milk, the sodium caseinates are found to be present as self-assembled aggregates at the neutral pH [115]. The limited proteolysis is resulting in the hydrolysis of κ-casein, which stabilizes the casein micelles to produce para-κ-casein, leads to coagulation of casein in the presence of Ca^{2+}. Casein micelles can also be destabilized, resulting in the formation of gel or precipitate on adding an equal volume of ~60% (v/v) ethanol to milk [55].

The consumer acceptance of various foods depends on the capability of whey proteins to form a gel that not only can hold water, lipids, and other components, but also to provide textural characteristics. Majorly two types of gel are formed from whey protein have been extensively studied: (1) firstly, gel induced at room temperature [34, 95, 113, 116]; and (2) secondly, heat-induced gel [80, 92]. The first type is the consequence of cold-set gelation. These gels are produced by first unfolding and aggregating the whey proteins by heating whey proteins solutions, then rapidly cooling to room temperature, followed by obtaining gelation with the addition of NaCl or $CaCl_2$ [113]. WPC products, at a concentration of 80–120 g/L, can gel at around 60–90°C. The gelling process depends on the temperature and duration of heating, ionic strength, pH and concentration of protein, lipids, salt, and sugar. However, the gelation time of rennet-induced gels was reported not to be affected by hydration treatment or MPC type (MPC70/ MPC85) [72].

The rheological properties of rennet-induced gels obtained from MPC 56/70/90 over reconstituted skim milk have been studied by Ferrer et al., [37]. MPC dispersions have relatively lower gelation time and higher storage modulus than that of reconstituted skim milk having an equal concentration of protein. In subsequent work, rennet gelation behavior of reconstituted MPC and skim milk was studied [83]. It was observed that the reconstituted MPC coagulated only after being supplemented with around 2 mmol of $CaCl_2$.

The desirable coagulation properties could be obtained using high-protein MPCs (e.g., MPC80) only after addition of $CaCl_2$ [37, 72, 83], recommending that the ionic equilibrium is important for ensuring the functionality of MPC. This could be due to excessive removal of soluble calcium during the production of MPCs by UF and DF processes. The study suggests that rennet coagulation kinetics and gel strength of reconstituted MPC could be like the raw skim milk on the addition of an appropriate amount of calcium.

1.3.3 EMULSIFYING PROPERTIES

Proteins have the ability to be used as emulsifiers on account of having amphiphilic nature and also of their ability to unfold and re-orientate at the interface [53]. The stability of the film formed by protein at the oil-water interface depends on equilibria between protein molecules and for both the phases. The emulsifying capacity of a protein is ascertained from the particle size of emulsion droplets obtained under set-conditions of homogenization from a given concentration of protein. The emulsifying properties of proteins are better at smaller droplet size due to the larger surface area [137].

The emulsifying properties of proteins may be improved on limited denaturation, which does not decrease the solubility drastically but increases the surface hydrophobicity [134]. These proteins are replaced by some common emulsifiers that grant weak stabilization but are also reported to have toxic effects on long-term use [53, 73]. Milk proteins possess the good emulsifying ability, and the groups buried inside the tertiary structure could be exposed on hydrolysis permitting fresh interactions at the emulsion droplet interface, e.g., disulfide, and nonpolar bonds for improving the emulsification capacity [52, 53].

Casein is an excellent emulsifier because it can easily unfold at the interface. β-Casein is the best emulsifying agent among caseins [25]. The emulsifying ability and capacity of sodium caseinate are reported to be increased on heat treatment, i.e., 50–100°C for 5 min near the isoelectric point [62]. This increase in functionality of the soluble protein fractions from sodium caseinate was suggested to be due to exposure of the hidden hydrophobic domains located on protein backbone on account of the application of heat. Haque and Mazaffar [47] reported an increase in the emulsifying activity index (EAI) and a decrease in emulsion stability index (ESI) after proteolysis of acid casein. Casein was hydrolyzed up to 5% degree of hydrolysis by trypsin resulting in the formation of β-casein (48–63 and 129–184) and α_{s1}-casein (167–208) peptides possessing excellent emulsion properties [132].

Luo et al., [79] reported a reduced surface hydrophobicity of casein on treatment with papain and trypsin. However, surface hydrophobicity of casein was increased during the first 4 h of treatment with pancreatin followed by a significant reduction after 24 h. The EAI and ESI were increased for three proteases, except for the papain treatment after 1 h and pancreatin treatment after 24 h. At the same DH (around 20%), EAI, and ESI are reported to be higher on trypsin treatment of casein after 24 h in comparison to treatment with pancreatin and papain, suggesting the high substrate specificity of trypsin over pancreatin and papain.

Whey proteins are good emulsifiers in the food system and thus find wide applications in food systems. Many studies have been on whey proteins individually and in combination for identification of conditions, where these act as emulsifiers [29, 30, 70, 85]. WPI, a natural emulsifier, is used in foods and beverage industry for the formation of stabilized o/w emulsion [2]. WPI lead to the formation of comparatively thin interfacial layer, which stabilizes emulsion forms relatively thin interfacial layers that mainly stabilize emulsions for flocculation by electrostatic repulsion. The emulsion, thus formed, is sensitive to ionic strength and pH [84]. WPI denatures at a higher temperature making the WPI-stabilized emulsions highly sensitive to thermal treatment. Thus, WPI is suitable only for the products, where the composition of the product and environmental conditions favor the formation of a stable product [18, 19].

The emulsifying ability of whey proteins and sodium caseinate is superior to the aggregated milk protein products, like MPC [35, 123]. The emulsifying ability of MPCs is associated with the extent of casein aggregation. Furthermore, emulsions formed with sodium caseinate are less stable to creaming than the ones prepared with large casein aggregates of MPC 85 and non-fat dried milk. This could be due to the larger size of protein aggregates of MPC and skim milk powder that may not be effective in causing depletion in flocculation and result in the formation of very viscous emulsion and reduce the creaming [35]. The report suggests that the emulsion stability of MPC-stabilized emulsion was decreased on increasing the concentration of protein [33]. The authors reported that unheated emulsion showed viscoplastic behavior, and on increasing the content of protein, the behavior was changed from shear-thinning to shear thickening. The emulsion showed pseudoplastic behavior on heating, and the higher preheating temperature resulted in more conspicuous shear-thinning character and higher emulsion viscosity.

Ye [137] revealed that the emulsion formed with low calcium MPC formed a finer emulsion having lower total surface protein concentration. The aggregation state of casein is reported to influence the emulsifying and absorption properties of MPC. The increase in emulsion droplet size decreases the stability of the emulsions at low-protein concentrations in low-calcium MPC-stabilized emulsions. The increase in the concentration of protein beyond a maximum level results in a decrease in emulsion stability, suggesting depletion in flocculation of the emulsions caused by the protein state in low-calcium MPCs [137]. The presence of few non-micellar casein fractions confers a protective stabilizing effect on whey protein aggregates, in both continuous as well as a dispersed phase during secondary heat treatment in emulsions stabilized with low calcium MPC and results in better emulsification [74].

1.3.4 FOAMING PROPERTIES

Foams can be defined as dispersions of gas in the liquid, stabilized by a variety of solid particles and/or low or high molecular weight surface-active substances, the physicochemical properties (structure, rate of adsorption, interactions, etc.) of which determine properties of the foam itself [136]. It has been suggested that the polypeptide chain of an ideal foam-forming protein should unfold relatively with ease, besides the protein having good solubility, high surface hydrophobicity and low total charge at the pH value of food product [107].

The foamability (or foaming capacity) of a given solution can be defined in terms of the volume of the foam formed under fixed conditions like temperature, foaming time, stirring intensity, etc.), or in terms of time required to obtain a pre-set volume. After the formation of foam, several processes occur each leading to the destruction of foam, i.e., drainage of liquid, the coalescence of the bubble, and disproportionation of the bubble [31, 32]. Foam stability generally refers to the ability of the foam to maintain some of its properties constant with time (e.g., size of the bubble and/or volume of foam and/or content of liquid).

The foams with higher overrun are generally produced from caseinates. However, foams produced from WPC or WPI have higher stability over that of the foams formed by caseinates. Foam forming ability of hydrolysates obtained by hydrolysis of casein is generally high in comparison to parental proteins [12, 78]. The enzyme specificity and DH, however, influence the foam stability of hydrolysates [78, 125].

Patel [100] hydrolyzed casein and identified three β-casein peptides, namely: β-casein (101–145), (107–145) and (107–135) in the mixture after hydrolysis. The 10% DH shows a reduction in foam-forming ability in comparison to DH being 5%. The hydrolysis of β-casein by plasmin leading to the formation of hydrophobic peptides results in improved foam formation and foam stabilization at a pH of 4.0 [15]. Foam capacity of enzymatically hydrolyzed sodium caseinate showed an increase initially followed by a decrease with hydrolysis time. Similar results have been observed for foam stability for the samples treated with α-chymotrypsin, showing the least value of foam stability after hydrolysis for 80 min [108]. It is suggested that foaming is increased on limited hydrolysis, but it has a deteriorative effect on foam stability.

Variations in foaming properties of whey products are reported to be influenced by the degree of denaturation of whey proteins and to residual lipid, ash, non-protein nitrogen, glycomacropeptide, and lipoprotein contents [51]. These characteristics of whey products may be influenced by cheese making and whey processing conditions. In one of the studies, regulated heating of whey proteins at <3.9 pH led to a limited denaturation with no aggregation [63, 88] and therefore, it was concluded that the foaming properties were enhanced strongly [90].

Phillips et al., [105] showed that heat treatment beyond that used for pasteurization was necessary to increase the foam stability of WPI. The micro-filtered WPCs with a fat content of 2 to 4% on a dry weight basis are reported to have better foaming characteristics than that whey that have not been micro-filtered [51, 102]. The poor foaming properties of WPI have been attributed to residual fat content [104].

The good foaming properties are obtained by MPC having high protein concentrations. Sodium caseinate is reported to have lesser foaming characteristics in comparison to co-precipitated MPC [75]. Also, acid co-precipitated MPC has lower whipping and foaming characteristics than co-precipitated MPC containing high-calcium concentrations [68].

1.4 FOOD APPLICATIONS

1.4.1 CASEIN-BASED PRODUCTS

Casein and its hydrolysates are important ingredients for the food industry on account of excellent functional properties, besides being naturally abundant. They not only improve and impart color and flavor to food commodities, but also their surface-related and hydrodynamic properties result in suitable functionalities that are employed for the manufacture of a number of food products [27, 68].

Caseins contain a higher amount of lysine, making them outstanding nutritional supplements for cereals, which lack in lysine. Therefore, casein/caseinates are added to breakfast cereals, milk biscuits, protein-enriched bread, and biscuits, high-protein cookies as nutritional supplements and as emulsifiers for improving texture for frozen baked cakes and cookies. However, sodium caseinate is not an accepted ingredient for controlling texture and uniformity in baked products like waffles, biscuits, doughnuts, and raised doughs [23]. Sodium caseinate is used for its emulsifying and foaming characteristics in milkshakes, while it is added to yogurts for increasing gel firmness and reducing syneresis. Also, it is used in powdered coffee creams and whiteners as fat encapsulator/emulsifier; imparting body, whitening, improved flavor, and resistance to feathering to the product [13, 22].

Milk protein products are added for the manufacture of conventionally processed and imitation milk products (such as imitation Mozzarella cheese and imitation cream cheese) for supplementing the protein content. Imitation cheese, commonly referred to as cheese analog, is prepared using caseins, salts, and vegetable oil, which lead to a remarkable saving in cost in comparison to the naturally manufactured cheese. The principal functional properties of casein used in such application are fat and water binding, stringiness, texture enhancing, melting properties, and shredding ability [97]. The most commonly used ingredient for cheese analog includes acid casein, rennet casein, and caseinates.

Casein products are also used in candy for the development of chewy-firm body [67]. These also contribute to color and flavor of confectionary. The whipping, foaming, and stabilizing properties of casein products are exploited in effervescent drinks, drinking chocolate, cream liqueurs, and beverages. Sodium casein is also used in frozen desserts and ice cream substitutes for improving the body, texture, whipping properties, besides acting as a stabilizer. Similarly, it is used in whipped toppings and infant puddings, and also it acts as film former and emulsifier [97]. Casein products are also used as fining agents, for decreasing stringency and color and for aiding in the clarification of wine and beer industries.

1.4.2 WHEY PROTEIN-BASED PRODUCTS

Whey protein products, such as WPC and WPI, possess good functional and nutritive value and thus find a way in the food industry. WPC or WPI products can be added for the production of soft, semi-hard, and hard cheese. The yield, consistency, and nutritional value of yogurt, fermented products, and cheese are improved by the addition of whey protein products. Banks et al., [10] reported an increase in yield of Cheddar cheese with the addition of WPC when compared with the traditional process.

Hinrichs [57] reported an increase in the yield, nutritive value, and the sensory properties of low-fat cheese on the addition of whey proteins into the casein matrix via the microparticulation process. Incorporation of WPC in yogurt results in higher water holding capacity and firmness value than the yogurt prepared by adding skim milk powder [21, 109, 112, 126], due to more cross-linkage of the network with WPC added yogurts.

Inclusion of whey protein in bread dough is reported to increase water absorption [23, 69]; and improvement in the functionality during the manu-facture of bread has been associated with the degree of denaturation of whey protein [48]. These days protein-fortified beverages like fruits juices, milk-based beverage, soft drinks are reported to contain whey protein products to have nutritious products often commercialized as 'sports' beverage, having high clarity over a wide range of pH. The β-LG have high foam overrun capacity, whipping ability, and heat stability identical to egg white (even after the addition of sugar) [20].

Milk-based flavored beverages are reported to contain WPCs and WPIs, which impart with the body and colloidal stability, viscosity, besides acting as protein supplements in frozen juice concentrate and flavored beverages [97]. WPI is an important natural emulsifier commonly used in foods and beverages [2]. Recently Ozturk et al., [98] reported that WPI was better than gum Arabic to produce smaller droplets at low concentration of cmulsifier used for emulsion-based delivery of vitamin E acetate. WPI shows a typical aggregation behavior on heating, which has been used for the manufacturing of food foams like whipped toppings, mousses, soufflés, angel food cakes, and meringue [96].

1.4.3 MPC-BASED PRODUCTS

The use of MPCs in the food and beverage industry is due to a range of func-tional properties, such as water binding, gelling, viscosity, emulsification, foaming/whipping, and heat stability. MPC production and its demand have increased over the period of years due to its use as an ingredient in a variety of products. For example, MPCs containing lower amount of protein (i.e., from 42 to 50%) are effective as ingredients in yogurt, cheese, and soup, while MPCs containing a higher amount of protein (i.e., 70% or more) are used in pediatric and geriatric foods, beverages, medical foods, enteral foods and preparation of protein bar.

MPCs are one of the ingredients for non-standard cheeses like Ricotta, Baker's cheese, Hispanic, and Feta cheese, processed cheese, processed cheese spread (PCS) products, and other fresh cheeses. The research studies have been carried out on the application of MPCs for the manufacture of Feta [49, 72], Mozzarella [42, 49] and Cheddar cheeses [49, 111]. Consistent cheese can be produced round the year by using MPCs or ultra-filtered milk for standardizing cheese milk [111].

Caro et al., [16] reported an increase in cheese yield on the addition of commercial MPC to whole milk for the preparation of Mexican-style Oaxaca cheese compared with the use of skim milk, which decreased the yield. Francolino et al., [42] reported an increase in the cheese yield from 13.8 to 16.7%, when MPC was used for standardization of cheese milk. The researchers attributed it to higher recovery of proteins and total milk solids in MPC cheese and also to the slightly higher cheese moisture.

In yogurts, non-fat dried milk, skim milk powder, and whole milk powders are replaced by MPCs on an equal protein or milk solids nonfat basis for improving texture and stability, with a concomitant decrease in whey separation of yogurt. It has been reported that nonfat dried milk replaced with MPC showed no negative effect on the textural properties of yogurt [45, 87].

A similar study conducted on ice cream mix by using MPC 56 or 80 showed no negative effect on the physical properties of ice cream [3]. The protein content of ice cream can be increased by using MPCs without affecting its lactose content [101]. Thus, the protein content of foods and beverages can be increased by utilizing MPCs having higher protein content, which also imparts a clean dairy flavor to the product without significantly increasing the lactose and hence causing no browning. MPCs can also be used for the production of lactose-free fermented milk [130]. Therefore, these high-protein MPCs are finding applications in low-lactose, high-protein products such as cheese sauces, and UHT beverages.

1.5   SUMMARY

The chapter focuses on the functional properties of both casein and whey proteins in a comprehensive manner. The chapter discusses: Classification of milk proteins and their heterogeneity, physicochemical, functional, and nutritional attributes of milk proteins (i.e., casein, and whey proteins); Manufacture of casein-, whey protein-, and milk protein-based products; functional properties of principal milk proteins like casein, whey protein and milk protein-based products; and various food applications of different milk proteins and their uses in food industry.

KEYWORDS

- **bovine serum albumin**
- **casein micelles**
- **emulsion activity index**
- **enzymatic hydrolysis**
- **flocculation**
- **milk protein**
- **nitrogen solubility index**
- **solubility index**
- **whey proteins**
- **α-lactalbumin**
- **β-lactoglobulin**

REFERENCES

1. Abert, T., & Kneifel, W., (1992). Physicochemical and functional properties of protein hydrolysates as obtained by treatment with different enzymes. In: *Protein and Fat Globule Modification by Heat Treatment, Homogenization and Other Technological Means for High Quality Dairy Products* (pp. 125–131). International Dairy Federation, Brussels, Belgium.
2. Adjonu, R., Doran, G., Torley, P., & Agboola, S., (2014). Whey protein peptides as components of nanoemulsions: A review of emulsifying and biologicalfunctionalities. *Journal of Food Engineering, 122*(1), 15–27.
3. Alvarez, V. B., Wolters, C. L., Vodovotz, Y., & Ji, T., (2005). Physical properties of ice cream containing milk protein concentrates. *Journal of Dairy Science, 88*(3), 862–871.
4. Anema, S. G., Pinder, D. N., Hunter, R. J., & Hemar, Y., (2006). Effects of storage temperature on the solubility of milk protein concentrate (MPC85). *Food Hydrocolloids, 20*(2/3), 386–393.
5. Atra, R., Vatai, G., Bekassy-Molnar, E., & Balint, A., (2005). Investigation of ultra-and nanofiltration for utilization of whey protein and lactose. *Journal of Food Engineering, 67*(3), 325–332.
6. Babella, G., (1989). Scientific and practical results with use of ultrafiltration in Hungary. *Bulletin of the International Dairy Federation, 244*, 7–25.
7. Baldasso, C., Barros, T. C., & Tessaro, I. C., (2011). Concentration and purification of whey proteins by ultrafiltration. *Desalination, 278*(1–3), 381–386.
8. Baldwin, A. J., & Truong, G. N. T., (2007). Development of insolubility in dehydration of dairy milk powders. *Food and Byproducts Processing, 85*(3), 202–208.
9. Baldwin, A. J., (2010). Insolubility of milk powder products–a mini-review. *Dairy Science & Technology, 90*(2/3), 169–179.

10. Banks, J. M., Stewart, G., Muir, D. D., & West, I. G., (1987). Increasing the yield of Cheddar cheese by the acidification of milk containing heat denatured whey protein. *Milchwissenschaft*, *42*(4), 212–215.

11. Bhaskar, G. V., Singh, H., & Blazey, N. D., (2001). *Milk Protein Concentrate Products and Process* (p. 45). International Patent Specification WO01/41578, Dairy Research Institute, Palmerstone North, New Zealand.

12. Britten, M., Giroux, H. J., & Gaudin, V., (1994). Effect of pH during heat processing of partially hydrolyzed whey protein. *Journal of Dairy Science*, *77*(3), 676–684.

13. Buchheim, W., (1983). *Proceedings of IDF Symposium on Physicochemical Aspects of Dehydrated Protein-Rich Milk Products* (pp. 319–330). Statens Forsogsmejeri, Hillerod, Denmark.

14. Burgess, K. J., & Kelly, J., (1979). Technical note: Selected functional properties of a whey protein isolate. *Journal of Food Technology*, *14*(3), 325–329.

15. Caessens, P. W. J. R., Visser, S., Gruppen, H., Van Aken, G. A., & Voragen, A. G. J., (1999). Emulsion and foam properties of plasmin derived β-casein peptides. *International Dairy Journal*, *9*(3–6), 347–351.

16. Caro, I., Soto, S., Franco, M. J., Meza-Nieto, M., Alfaro-Rodriguez, R. H., & Mateo, J., (2011). Composition, yield, and functionality of reduced-fat Oaxaca cheese: Effects of using skim milk or a dry milk protein concentrate. *Journal of Dairy Science*, *94*(2), 580–588.

17. Carr, A. J., Bhaskar, G. V., & Ram, S., (2007). *Monovalent Salt Enhances Solubility of Milk Protein Concentrate* (p. 56). Patent number EP-1–553–843-B1 from European Patent Specification, New Zealand Dairy Board, Wellington.

18. Chanamai, R., & McClements, D. J., (2002). Comparison of gum Arabic, modified starch, and whey protein isolate as emulsifiers: Influence of pH, $CaCl_2$ and temperature. *Journal of Food Science,* *67*(1), 120–125.

19. Charoen, R., Jangchud, A., Jangchud, K., Harnsilawat, T., Naivikul, O., & McClements, D. J., (2011). Influence of biopolymer emulsifier type on formation and stability of rice bran oil-in-water emulsions: Whey protein, gum Arabic, and modified starch. *Journal of Food Science*, *76*(1), E165–E172.

20. Chattertona, D. E. W., Smithersb, G., Roupasb, P., & Brodkorb, A., (2006). Bioactivity of β-lactoglobulin and α-lactalbumin-Technological implications for processing. *International Dairy Journal*, *16*(11), 1229–1240.

21. Cheng, L. J., Augustin, M. A., & Clarke, P. T., (2000). Yogurts from skim milk-whey protein concentrate blends. *Australian Journal of Dairy Technology*, *55*(2), 110–110.

22. Clarke, R. J., & Love, G., (1974). *Chemistry & Industry* (p. 151). John Wiley & Sons, New York.

23. Cocup, R. O., & Sanderson, W. B., (1987). Functionality of dairy ingredients in bakery products. *Food Technology*, *41*(10), 86–91.

24. Corredig, M., & Dalgleish, D. G., (1997). Studies on the susceptibility of membrane derived proteins to proteolysis as related to changes in their emulsifying properties. *Food Research International*, *30*(9), 689–697.

25. Dalgleish, D. G., (1997). Adsorption of protein and the stability of emulsions. *Trends in Food Science & Technology*, *8*(1), 1–6.

26. Dalgleish, D. G., (2011). On the structural models of bovine casein micelles-review and possible improvements. *Soft Matter*, *7*(6), 2265–2272.

27. Damodaran, S., (1994). Structure-function relationship of food proteins. In: *Protein Functionality in Food Systems* (pp. 1–37). Marcel Dekker, New York.

28. Darewicz, M., Dziuba, J., & Dziuba, M., (2006). Functional properties and biological activities of bovine casein proteins and peptides. *Polish Journal of Food and Nutrition Science, 56*(1), 79–86.

29. Demetriades, K., Coupland, J. N., & McClements, D. J., (1997). Physicochemical properties of whey protein-stabilized emulsions as affected by heating and ionic strength. *Journal of Food Science, 62*(3), 342–347.

30. Demetriades, K., Coupland, J. N., & McClements, D. J., (1997). Physical properties of whey protein-stabilized emulsions as related to pH and NaCl. *Journal of Food Science, 62*(3), 462–467.

31. Denkov, N. D., & Marinova, K. G., (2006). Antifoam effects of solid particles, oil drops and oil-solid compounds in aqueous foams. In: *Colloidal Particles at Liquid Interfaces* (pp. 383–444). Cambridge University Press, Cambridge.

32. Denkov, N. D., (2004). Mechanisms of foam destruction by oil-based antifoams. *Langmuir, 20*(22), 9463–9505.

33. Dybowska, B. E., (2007). Influence of protein concentration and heating conditions on milk protein-stabilized oil-in-water emulsions. *Milchwissenschaft, 62*(2), 139–142.

34. Elofsson, C., Dejmek, P., Paulson, M., & Burling, H., (1998). Characterization of cold/gelling whey protein concentrate. *International Dairy Journal, 7*(8/9), 601–608.

35. Euston, S. R., & Hirst, R. L., (1999). Comparison of the concentration-dependent emulsifying properties of protein products containing aggregated and non-aggregated milk protein. *International Dairy Journal, 9*(10), 693–701.

36. Fang, Y., Selomulya, C., & Chen, X. D., (2010). Characterization of milk protein concentrate solubility using focused beam reflectance measurement. *Dairy Science & Technology, 90*(2/3), 253–270.

37. Ferrer, M. A., Hill, A. R., & Corredig, M., (2008). Rheological properties of rennet gels containing milk protein concentrates. *Journal of Dairy Science, 91*(3), 959–969.

38. FitzGerald, R. J., & Meisel, H., (2003). Milk protein hydrolysates and bioactive peptides. In: *Advanced Dairy Chemistry: Proteins 1A* (pp. 675–698). Kluwer Academic/Plenum Press, New York.

39. Foegeding, E. A., Davis, J. P., Doucet, D., & McGuffey, M. K., (2002). Advances in modifying and understanding whey protein functionality. *Trends in Food Science and Technology, 13*(5), 151–159.

40. Fox, P. F., & McSweeney, P. L. H., (1998). *Dairy Chemistry and Biochemistry* (1st edn., p. 478). Blackie Academic & Professional, London–United Kingdom.

41. Fox, P. F., & McSweeney, P. L. H., (2003). *Advanced Dairy Chemistry–Proteins* (3rd edn., p. 1323). Part A, Kluwer Academic/Plenum Publishers, New York, USA.

42. Francolino, S., Locci, F., Ghiglietti, R., Lezzi, R., & Mucchetti, G., (2010). Use of milk protein concentrate to standardize milk composition in Italian citric mozzarella cheese making. *LWT-Food Science and Technology, 43*(2), 310–314.

43. Goff, H. D., & Hill, A. R., (1993). Chemistry and physics. In: *Dairy Science and Technology Handbook. Principles and Properties* (pp. 1–82). VCH Publishers, New York.

44. Grufferty, M. B., & Fox, P. F., (1988). Functional properties of casein hydrolyzed by alkaline milk proteinase. *New Zealand Journal of Dairy Science & Technology, 23*(2), 95–108.

45. Guzmán-González, M., Morais, F., Ramos, M., & Amigo, L., (1999). Influence of skimmed milk concentrate replacement by dry dairy products in a low fat set-type yogurt model system. I: Use of whey protein concentrates, milk protein concentrates and skimmed milk powder. *Journal of the Science of Food and Agriculture, 79*(8), 1117–1122.

46. Haque, E., Bhandari, B. R., Gidley, M. J., Deeth, H. C., Moller, S. M., & Whittaker, A. K., (2010). Protein conformational modifications and kinetics of water-protein interactions in milk protein concentrate powder upon aging: Effect on solubility. *Journal of Agricultural and Food Chemistry, 58*(13), 7748–7755.

47. Haque, Z. U., & Mazaffar, Z., (1992). Casein hydrolysates, II: Functional properties of peptides. *Food Hydrocolloids, 5*(6), 559–571.

48. Harper, W. J., & Zadow, J. G., (1984). Heat induced changes in whey protein concentrates as related to bread manufacture. *New Zealand Journal of Dairy Science & Technology, 19*(3), 229–237.

49. Harvey, J., (2006). Protein fortification of cheese milk using milk protein concentrate: Yield improvement and product quality. *Australian Journal of Dairy Technology, 61*(2), 183–185.

50. Havea, P., (2006). Protein interactions in milk protein concentrate powders. *International Dairy Journal, 16*(5), 415–422.

51. Hawks, S. E., Phillips, L. G., Rasmussen, R. R., Barbano, D. M., & Kinsella, J. E., (1993). Effects of processing treatment and cheese-making parameters on foaming properties of whey protein isolates. *Journal of Dairy Science, 76*(9), 2468–2477.

52. He Tan, Y., Tian, Z., Chen, L., Hu, F., & Wu, W., (2011). Food protein-stabilized nano-emulsions as potential delivery systems for poorly water-soluble drugs: Preparation, *in vitro* characterization, and pharmacokinetics in rats. *International Journal of Nano-medicine, 6*, 521–533.

53. He, W., Lu, Y., Qi, J., Chen, L., Hu, F., & Wu, W., (2013). Nanoemulsion-templated shell-crosslinked nanocapsules as drug delivery systems. *International Journal of Pharmaceutics, 445*(1/2), 69–78.

54. Henning, D. R., Baer, R. J., Hassan, A. N., & Dave, R., (2006). Major advances in concentrated and dry milk products, cheese, and milk fat-based spreads. *Journal of Dairy Science, 89*(4), 1179–1188.

55. Hewedi, M. M., Mulvihill, D. M., & Fox, P. F., (1985). Recovery of milk protein by ethanol precipitation. *Irish Journal of Food Science and Technology, 9*(1), 11–23.

56. Hidalgo, M. E., Correa, A. P. F., Canales, M. M., Daroit, D. J., Brandelli, A., & Risso, P., (2015). Biological and physicochemical properties of bovine sodium caseinate hydrolysates obtained by a bacterial protease preparation. *Food Hydrocolloids, 43*(1), 510–520.

57. Hinrichs, J., (2001). Incorporation of whey proteins in cheese. *International Dairy Journal, 11*(4–7), 495–503.

58. Holt, C., (1997). The milk salts and their interaction with casein. In: *Advanced Dairy Chemistry* (pp. 233–256). Chapman & Hall, London.

59. Horne, D. S., (1986). Steric stabilization and casein micelle stability. *Journal of Colloid and Interface Science, 111*(1), 250–260.

60. Huffman, L. M., (1996). Processing whey protein for use as a food ingredient. *Food Technology, 50*(2), 4952.

61. Huppertz, T., Fox, P. F., & Kelly, A. L., (2006). High pressure-induced changes in bovine milk proteins: A review. *Biochimica Biophysica Acta (BBA)–Proteins and Proteomics, 1764*(3), 593–598.

62. Jahaniaval, F., Kakuda, Y., Abraham, V., & Marcone, M. F., (2000). Soluble protein fractions from pH and heat treated sodium caseinate: Physicochemical and functional properties. *Food Research International, 33*(8), 637–647.

63. Jelen, P., & Buchheim, W., (1984). Stability of whey protein upon heating in acidic conditions. *Milchwissenschaft, 39*(4), 215–218.

64. Kelly, P., (2011). Milk protein products: Milk protein concentrate. In: *Encyclopedia of Dairy Sciences* (pp. 848–854). Academic Press, San Diego.
65. Kilara, A., & Panyam, D., (2003). Peptides from milk proteins and their properties. *Critical Reviews in Food Science and Nutrition*, *43*(6), 607–633.
66. Kinsella, J. E., & Whitehead, D. M., (1989). Proteins in whey: Chemical, physical and functional properties. *Advances in Food and Nutrition Research*, *33*(1), 343–438.
67. Kinsella, J. E., (1970). Functional chemistry of milk products in candy and chocolate manufacture. *Manufacturing Confectioner*, *50*(10), 45–48.
68. Kinsella, J. E., (1984). Milk proteins: Physicochemical and functional properties. *Critical Reviews in Food Science and Nutrition*, *21*(3), 197–262.
69. Kinsella, J. E., (1982). Relationships between structure and functional properties of food proteins. In: *Food Proteins* (pp. 51–103). Applied Science Publishers, London.
70. Klemaszewski, J. L., Das, K. P., & Kinsella, J. E. J., (1992). Formation and coalescence stability of emulsions stabilized by different milk proteins. *Food Science*, *57*(2), 366–371.
71. Korhonen, H., & Pihlanto, A., (2003). Food-derived bioactive peptides-opportunities for designing future foods. *Current Pharmaceutical Design*, *9*(16), 1297–1308.
72. Kuo, C. J., & Harper, W. J., (2003). Effect of hydration time of milk protein concentrate on cast Feta cheese texture. *Milchwissenschaft*, *58*(5), 283–286.
73. Lam, R. S. H., & Nickerson, M. T., (2013). Food proteins: A review on their emulsifying properties using a structure-function approach. *Food Chemistry*, *141*(2), 975–984.
74. Liang, Y., Patel, H., Matia-Merino, L., Ye, A., & Golding, M., (2013). Effect of pre- and post-heat treatments on the physicochemical, microstructural and rheological properties of milk protein concentrate-stabilized oil-in-water emulsions. *International Dairy Journal*, *32*(2), 184–191.
75. Linden, G., & Lorient, D., (1999). *New Ingredients in Food Processing: Biochemistry and Agriculture* (p. 360). CRC Press, Boca Raton, USA.
76. Liu, Q., Kong, B., Xiong, Y. L., & Xia, X., (2010). Antioxidant activity and functional properties of porcine plasma protein hydrolysate as influenced by the degree of hydrolysis. *Food Chemistry*, *118*(2), 403–410.
77. Lucey, J. A., (2007). Microstructural approaches to the study and improvement of cheese and yogurt products. In: *Understanding and Controlling the Microstructure of Complex Foods* (pp. 600–621). Woodhead Publishing, Cambridge.
78. Ludwig, I., Krause, W., & Hajos, G., (1995). Functional properties of enzymatically modified milk proteins. *Acta Alimentaria*, *24*(3), 289–296.
79. Luo, Y., Pan, K., & Zhong, Q., (2014). Physical, chemical and biochemical properties of casein hydrolyzed by three proteases: Partial characterizations. *Food Chemistry*, *155*(11), 146–154.
80. Manigo, M. E., (1992). Gelation of whey protein concentrates. *Food Technology*, *46*(1), 114–117.
81. Mao, X., Tong, P. S., Gualco, S., & Vink, S., (2012). Effect of NaCl addition during diafiltration on the solubility, hydrophobicity, and disulfide bonds of 80% milk protein concentrate powder. *Journal of Dairy Science*, *95*(7), 3481–3488.
82. Marshall, K., (2004). Therapeutic applications of whey protein. *Alternative Medicine Review*, *9*(2), 136–156.
83. Martin, G. J. O., Williams, R. P. W., & Dunstan, D. E., (2010). Effect of manufacture and reconstitution of milk protein concentrate powder on the size and rennet gelation behavior of casein micelles. *International Dairy Journal*, *20*(2), 128–131.

84. McClements, D. J., (2004). Protein-stabilized emulsions. *Current Opinion in Colloid & Interface Science, 9*(5), 305–313.

85. McCrae, C. H., Law, A. J. R., & Leaver, J., (1999). Emulsification properties of whey proteins in their natural environment: Effect of whey protein concentration at 4 and 18% milk fat. *Food Hydrocolloids, 13*(5), 389–399.

86. Mclntosh, G. H., Royle, P. J., Le Leu, R. K., Regester, G. O., Johnson, M. A., Grinsted, R. L., Kenward, R. S., & Smithers, G. W., (1998). Whey proteins as functional food ingredients? *International Dairy Journal, 8*(5/6), 425–434.

87. Mistry, V. V., & Hassan, H. N., (1992). Manufacture of nonfat yogurt from a high milk protein powder. *Journal of Dairy Science, 75*(4), 947–957.

88. Modler, H. W., & Harwalker, V. H., (1981). Whey protein concentrate prepared by heating under acidic conditions, I: Recovery by ultrafiltration and functional properties. *Milchwissenschaft, 36*(9), 537–542.

89. Monti, J. C., & Jost, R., (1978). Enzymatic solubilization of heat-denatured cheese whey protein. *Journal of Dairy Science, 61*(9), 1233–1237.

90. Morr, C. V., & Foegeding, E. A., (1990). Composition and functionality of commercial whey and milk protein concentrates and isolates: A status report. *Food Technology, 44*(4), 100–112.

91. Mullvihill, D. M., & Donovan, M., (1987). Whey proteins and their thermal denaturation-A review. *Irish Journal of Food Science and Technology, 11*(1), 43–75.

92. Mullvihill, D. M., & Kinsella, J. E., (1987). Gelation characteristics of whey proteins and β-lactoglobuin. *Food Technology, 41*(9), 102–111.

93. Mulvihill, D. M., (1992). Production, functional properties and utilization of milk protein products. In: *Proteins. Advanced Dairy Chemistry* (pp. 369–405). Elsevier Applied Science, London- UK.

94. Nakai, S., & Li-Chan, E., (1985). Structure modification and functionality of whey proteins: Quantitative structure-activity relationship approach. *Journal of Dairy Science, 68*(10), 2763–2772.

95. Nakamura, M., Sato, K., Koisumi, S., Kawachi, K., & Nakajima, I., (1995). Preparation and properties of salt- induced gel of whey protein. *Nippon Shokuhin Kagaku Kaishi, 42*(1), 1–6.

96. Nicorescu, I., Loisel, C., Riaublanc, A., Vial, C., Djelveh, G., Cuvelier, G., & Legrand, J., (2009). Effect of dynamic heat treatment on the physical properties of whey protein foams. *Food Hydrocolloids, 23*(4), 1209–1219.

97. O'Regan, J., Ennis, M. P., & Mulvihill, D. M., (2009). Milk proteins. In: *Handbook of Hydrocolloids* (2nd edn, pp. 298–358.). Woodhead Publishing, New York.

98. Ozturk, B., Argin, S., Ozilgen, M., & McClements, D. J., (2015). Formation and stabilization of nanoemulsion-based vitamin E delivery systems using natural biopolymers: Whey protein isolate and gum Arabic. *Food Chemistry, 188*(12), 256–263.

99. Palmer, D. E., (1982). Recovery of proteins from food factory waste by ion exchange. In: *Food Proteins* (pp. 341–352). Applied Science Publishers, London.

100. Patel, G. C., (1994). *Structure and Foaming Properties of Peptides Isolated from Casein Hydrolysates* (p. 232). Unpublished PhD Thesis, The Pennsylvania State University, Harrisburg–PA, USA.

101. Patel, M. R., Baer, R. J., & Acharya, M. R., (2006). Increasing the protein content of ice cream. *Journal of Dairy Science, 89*(5), 1400–1406.

102. Pearce, R. J., Marshall, S. C., & Dunkerley, J. A., (1992). Reduction of lipids in whey protein concentrates by microfiltration-effect on functional properties. In: *New*

Applications of Membrane Processes, IDF Publ. No. 9201 (p. 118). International Dairy Federation, Brussels, Belgium.

103. Pelegrine, D. H. G., & Gasparetto, C. A., (2005). Whey proteins solubility as function of temperature and pH. *LWT–Food Science and Technology, 38*(1), 77–80.

104. Peltonen-Shalaby, R., & Mangino, M. E., (1986). Compositional factors that affect the emulsifying and foaming properties of whey protein concentrates. *Journal of Food Science, 51*(1), 91–95.

105. Phillips, L. G., Schulman, W., & Kinsella, J. E., (1990). pH and heat treatment effects on foaming of whey protein isolate. *Journal of Food Science, 55*(4), 1116–1119.

106. Pierre, A., Fauquant, J., Le Graet, Y., Piot, M., & Maubois, J. L., (1992). *Preparition de phosphocasienpar microfiltration sur membrane* (Preparation of phosphocasien by microfiltration membrane). *LeLait, 72*(5), 461–474.

107. Poole, S., & Fry, J., (1987). *Developments in Food Proteins* (pp. 257–298). Elsevier Applied Science, New York.

108. Pralea, D., Dumitrascu, L., Borda, D., & Stănciuc, N., (2011). Functional properties of sodium caseinate hydrolysates as affected by the extent of chymotrypsinolysis. *Journal of Agroalimentary Processes and Technology, 17*(3), 308–314.

109. Puvanenthiran, A., Williams, R. P. W., & Augustin, M. A., (2002). Structure and viscoelastic properties of set yogurt with altered casein to whey protein ratios. *International Dairy Journal, 12*(4), 383–391.

110. Raikos, V., Campbell, L., & Euston, S. R., (2007). Rheology and texture of hen's egg protein heat-set gels as affected by ph and the addition of sugar and/or salt. *Food Hydrocolloids, 21*(2), 237–244.

111. Rehman, S. U., Farkye, N. Y., Considine, T., Schaffner, A., & Drake, M. A., (2003). Effects of standardization of whole milk with dry milk protein concentrate on the yield and ripening of reduced-fat cheddar cheese. *Journal of Dairy Science, 86*(5), 1608–1615.

112. Remeuf, F., Mohammed, S., Sodini, I., & Tissier, J. P., (2003). Preliminary observations on the effects of milk fortification and heating on microstructure and physical properties of stirred yogurt. *International Dairy Journal, 13*(9), 773–782.

113. Resch, M., & Danbert, I., (2002). Rheological and physicochemical properties of derivatized whey protein concentrate powders. *International Journal of Food Properties, 5*(2), 419–434.

114. Roefs, S. P. F. M., & Van Vliet, T., (1990). Structure of acid casein gels, II: Dynamic measurements and type of interaction forces. *Colloids and Surfaces, 50*(1), 161–175.

115. Ruis, H. G. M., Venema, P., & Van Der Linden, E., (2007). Relation between pH induced stickiness and gelation behavior of sodium caseinate aggregates as determined by light scattering and rheology. *Food Hydrocolloids, 21*(4), 545–554.

116. Sato, K., Nakamura, M., Koisumi, S., Kawachi, K., & Nakajima, I., (1995). Changes in hydrophobicity and SH content on salt-induced gel of whey protein. *Nippon Shokuhin Kagaku Kogaku Kaishi, 42*(1), 7–13.

117. Schuck, P., Briard, V., Mejean, S., Piot, M., Famelart, M. H., & Maubois, J. L., (1999). Dehydration by desorption and by spray drying of dairy proteins: Influence of the mineral environment. *Drying Technology, 17*(7/8), 1347–1357.

118. Schuck, P., Piot, M., Mejean, S., Le Graet, Y., Fauquant, J., Brule, G., & Maubois, J. L., (1994). *Deshydration par atomization de phosphor caseinatenatifobtenu par microfiltration sur membrane* (Spray dehydration of alternating phosphor casein obtained by membrane microfiltration.). *LeLait, 74*(5), 375–388.

119. Shanmugam, V. P., Kapila, S., & Sonfack, T. K., (2015). Antioxidative peptide derived from enzymatic digestion of buffalo casein. *International Dairy Journal*, *42*, 1–5.

120. Shimada, K., & Matsushita, S., (1980). Thermal coagulation of egg albumin. *Journal of Agricultural and Food Chemistry*, *28*(2), 409–412.

121. Sikand, V., Tong, P. S., & Walker, J., (2013). Effect of adding salt during the diafiltration step of milk protein concentrate powder manufacture on mineral and soluble protein composition. *Dairy Science & Technology*, *93*(4/5), 401–413.

122. Sikand, V., Tong, P. S., Vink, S., & Walker, J., (2012). Effect of powder source and processing conditions on the solubility of milk protein concentrates. *Milchwissenschaft*, *67*(3), 300–303.

123. Singh, H., & Ye, A., (2009). Interaction and functionality of milk proteins in food emulsions. In: *Milk Proteins: From Expression to Food* (pp. 321–346). Elsevier Applied Science, London, UK.

124. Singh, H., (2011). Milk protein products: Functional properties of milk proteins. In: *Encyclopedia of Dairy Sciences* (pp. 887–893). Academic Press, New York.

125. Slattery, H., & Fitzgerald, R. J., (1998). Functional properties and bitterness of sodium caseinate hydrolysates prepared with a *Bacillus* proteinase. *Journal of Food Science*, *63*(2), 418–422.

126. Sodini, I., Montella, J., & Tong, P. S., (2005). Physical properties of yogurt fortified with various commercial whey protein concentrates. *Journal of the Science of Food and Agriculture*, *85*(5), 853–859.

127. Southward, C. R., (2012). *Casein Products*. New Zealand Dairy Research Institute, http://nzic.org.nz/ChemProcesses/dairy/3E.pdf (Accessed on 20 July 2019).

128. Storer, A. C., & Menard, R., (2013). Papain. In: *Handbook of Proteolytic Enzymes* (pp. 1858–1861). Academic Press, London.

129. Swaisgood, H. E., (2003). Chemistry of the caseins. In: *Advanced Dairy Chemistry-1: Proteins* (3rd edn., pp. 140–201). Kluwer Academic/Plenum Publishers, London.

130. Szigeti, J., Krasz, A., & Varga, L., (2006). A novel technology for production of lactose-free fermented milks. *Milchwissenshcaft*, *61*(2), 177–180.

131. Thompson, A., Boland, M., & Singh, H., (2009). *Milk Proteins from Expression to Food* (p. 533). Elsevier Inc, San Diego, CA, USA.

132. Turgeon, S. L., Sanchez, C., Gauthier, S. F., & Paquin, P., (1996). Stability and rheological properties of salad dressing containing peptide fraction of whey proteins. *International Dairy Journal*, *6*(6), 645–658.

133. Vardhanabhuti, B., & Foegeding, E. A., (2008). Effects of dextran sulfate, NaCl, and initial protein concentration on thermal stability of β-lactoglobulin and α-lactalbumin at neutral pH. *Food Hydrocolloids*, *22*(5), 752–762.

134. Walstra, P., De Roos, A. L., (1993). Proteins at air-water and oil-water interfaces: Static and dynamic aspects. *Food Review International*, *9*(4), 503–525.

135. Walstra, P., & Jenness, R., (1984). *Dairy Chemistry and Physics* (p. 467). John Wiley and Sons, New York.

136. Weaire, D., & Hutzler, S., (1999). *The Physics of Foams* (p. 264). Clarendon Press, Oxford.

137. Ye, A., (2011). Functional properties of milk protein concentrates: Emulsifying properties, adsorption and stability of emulsions. *International Dairy Journal*, *21*(1), 14–20.

CHAPTER 2

COMPOSITE DAIRY FOODS: SCOPE, APPLICATIONS, AND BENEFITS

REKHA CHAWLA, SANTOSH KUMAR MISHRA, and
VEENA NAGARAJAPPA

ABSTRACT

The unique combination of industrial partnership and researchers with different nutritional profile usually results in new "value-added product" with enhanced health benefits. These products being nutritious and healthy can be tried by all age and income groups. Therefore, further initiatives in the production of such products at research levels and institutional-industrial partnerships are required for further proliferation of this concept and acceptability at a wider scale.

2.1 INTRODUCTION

India has been ranked on the forefront in developing national food and nutrition records, undertaking research studies/surveys detailing the ongoing agriculture, food, and nutrition and health changeovers. However, despite all these efforts, a major concern to combat micronutrient deficiency is a burning issue. Therefore, during the past decade, deficiencies due to micronutrient or macronutrient have been attracting the attention of both academicians and administrators. Apart from this, 30–40% of produce is being lost during postharvest operations till today. By reduction in these losses, population with malnutrition among children and women can be provided with an adequate supply of daily diet.

Nutrition delivery via means of common food items is the most significant step to fight hidden deprivation, undernutrition, and ill-health. There is a need to combine the ingredients from other sources with milk-based ingredients to lower down the rate of malnutrition, and hidden hunger. Therefore, to address the demand of foods with improved and enhanced dietary status and to further deliver food with inherent bioactive components along with the benefits of certain therapeutic attributes, the concept of composite foods has come up.

Composite foods can be defined as complementation and combination of different food categories so as to increase the nutrition value of a resultant product. Therefore, it leads to the development of a new product with improved nutritional status, improved functionality, and health virtues. Although there does not exist precise class or group as "composite food," yet the terminology can be employed for any formulation, which contains considerable quantities of two or more unlike food clusters. Among various groups of composite foods, composite dairy foods, particularly milk-cereal or milk- fruits and vegetable–blends remain in demand and are gaining popularity day by day.

Milk (being a perfect medium and source of various valued macronutrients like fat, protein, and milk sugar -lactose, vitamins, and micronutrients; available in the form of minerals) is considered as a complete food or a 'nutritious food,' as it contains all vital nutrients required by the body for its proper functioning. But the milk lacks certain essential components such as micronutrients (iron, copper, etc.) and fiber. And, milk in itself has certain limiting factors such as it is a cause of allergenicity for many, the inability of lactose digestion for lactose intolerance population, presence of cholesterol and saturated fat content, etc. Still to combat the lacunae of the milk, with the help of advancements in expertise and tools available, numerous components obtained through milk are being used in combination with other non-milk components for preparing versatile range of food products with enhanced nutritional output in the form of 'value-added products' with better assimilability and superior functionality along with benefits of wholesomeness. Also, such combinations always create nutritionally of superior category, which otherwise single ingredient cannot provide.

For a population with milk allergenicity or lactose intolerance problems, various other additives can be added to milk so that a complete food can be prepared with supplementation and complementation of various micronutrients by adding cereals, millets, fruits, vegetables, and legumes, etc. Also, non-dairy components play an important role in increasing the acceptability of composite foods by participating in various chemical reactions and mingling well with elements of milk products. This approach not only yields improved sensory and nutritional profile of the resultant product but also leads to decreased cost [32].

A sharp increase in the cost of milk or milk-based ingredients has also dramatically affected the cost-returns of milk and non-dairy food products, when used in conjunction with composite foods. Cost of milk and milk products has also increased to such an extent that these have become inaccessible for many weaker groups of society. Therefore, all these facts and data on rising costs, demands for the development of composite dairy products at a lower cost is increasing due to health benefits [58].

Therefore with this strategy, a variety of functional and health-promoting foods can be developed. However, these composite foods are not very popular in India, and some of them are even regional specific. Therefore, due to lack of technical know-how and scientific technologies to prepare the same, commercialization at a wider scale of these products is a challenging task.

This chapter focuses on: (1) Various possible categories of composite dairy foods prepared from a combination of milk, fruits, vegetables, cereals, and millets; and (2) The issues and precautions related to process of their combination.

2.2 COMPOSITE DAIRY FOODS

The Codex standard of 1999 defines the composite dairy product as a food product from milk and/or its constituents, milk products, with or without the addition of milk processing by-products and non-dairy components included in as individual ingredients, added for a purpose other than the substitution of milk constituents [4].

Earlier the aim of substitution of non-dairy ingredients in composite foods was dependent on the improvement in sensory characteristics and on the product integrity, but nowadays it also takes into account the targeted benefits to be derived from the resultant product. Also, as few persons are conscious and doubtful with respect to consumption of milk and milk products either due to certain health-related issues or considering the presence of saturated fats, cholesterol, lactose in the milk, etc. Therefore, an entire range of innovative and novel products have come up to suit the palate of the diverse customers, keeping in mind the demands laid by today's consumers as per their needs of lifestyle, statistics, social, financial, and developmental background. The concept of composite foods is well appreciated by researchers, academicians, and dairy technologists, as combining non-dairy functional entities into milk-based foods increases the consumption and consequently production of such foods [10].

After milk, whey is an excellent choice for preparation of composite foods. Due to its high protein quality score and presence of amino acids of branched nature (BCAAs) in a high percentage, whey has extensively been predominant in the health industry and especially bodybuilding as a supplement. Research has shown that whey protein helps in body healing and repair injured tissues [12]. Similarly, helping in boosting the immune system and fight against infections and anti-aging effects of whey protein have also been listed by various researchers. However, they also advocate the application of whey protein in various other segments such as a functional food or a nutraceutical and its administration in cases of cancer, hepatitis B, HIV, CVD, osteoporosis, and even enduring stress [8]. These have also practically proven in lowering the occurrence of GI tumors in the test animals [39].

2.2.1   CHOICE FOR COMPOSITE DAIRY FOODS

A variety of options can be opted for the preparation of composite dairy foods. However, the major choice of manufacturer remains the supplementation of major cereals like wheat and rice, being easily available. Apart from this, these sources not only being a good source of carbohydrates and energy but also provide bulk and taste to the composite foods.

2.2.2   CEREAL-BASED COMPOSITE FOODS

Combination of milk or milk-based foods with cereals provides a complete nutritional profile. Thus, the deficiencies borne by the milk can be compensated well by the introduction of cereals wherein fiber, lysine (the amino acid absent in dairy foods) and other nutrients absent in milk are abundantly available. The classic examples of dairy- cereal combination composite foods are *kheer, dhoda burfi*, porridge (*dalia*), and malted milk foods.

Various cereal milk-based products have been developed like groundnut *burfi* [30]. Use of sorbic acid imparted good stability and acceptability up to 6 months in polypropylene (PP) and over a period of eight months in metalized polyester (MP) pouches when stored under ambient temperatures (15–34°C). The control samples got spoiled within one month of storage, exhibited microbial growth particularly mold growth and also produced fermented smell; whereas the samples with sorbic acid as preservative remained acceptable and did not show any signs of spoilage due to microbial growth up to 6 months. Likewise, supplementation with various types of flour also results in good quality products. Examples are the use of soy flour in *Gulabjamun* [52], and the use of pearl millet in instant *kheer* [13].

2.2.3 GLUTEN-FREE COMPOSITE FOODS

Apart from this, a variety of millets can be utilized for the preparation of composite dairy foods. The alarming increase in wheat allergy (celiac disease (CD)) has led to the development of such products and has forced the consumer to divert from normal wheat-based products to other minor or coarse cereals (as millets do not contain gluten protein, which otherwise is the abundant and main protein of wheat). The CD is an autoimmune disease featured by an unusual response to dietary gluten in vulnerable individuals that results in the injury of the small intestine and is associated with diverse systemic outcomes [48]. It is due to the presence of protein gluten in wheat (gliadin fraction) and another protein, which is soluble in alcohol (prolamines) commonly found in barley, and rye. The study has indicated that 1 in 310 persons suffer from CD [55] and 1 in 133 Americans (about 2% of the population) is reportedly to have CD [3].

In another study, 8–25% prevalence of CD has been seen amongst first-degree relatives. Thus, the increasing celiac population has forced food manufacturers to focus on non-wheat products or gluten-free products. Considering the rise in allergic populations of celiac, use of millets is an essential and excellent approach due to health priority. Also, its use in the preparation of snack foods is commendable due to other health benefits of millets, such as gluten-free, least allergenic, slower release of glucose in the body, and a good reservoir of energy.

Also, millets are high in minerals, particularly Fe, Ca, Mg, P, Zn, and K. Packed with such qualities, now these are referred to as *nutria-millets* or *nutria-cereals*. Millets provide essential components such as protein, fatty acids, minerals, vitamins, dietary fiber, and polyphenols [17]. Protein obtained from millets contain a good amount of sulfur containing amino acids (methionine and cysteine), which are a good source of antioxidants, such as phenolic acids and glycated flavonoids. Also, foods prepared from millets have enhanced viability or functionality of probiotics with substantial health benefits. When these millets can be combined with other essential ingredients, rich in other vital nutrients, such as milk, the quality of the resultant product can be increased further beyond limits.

It is high in fat content compared with other cereals such as maize, rice, and sorghum [44]. Nutritionally, millets also contain magnesium and phosphorus in sufficient quantity. Magnesium has the ability to lower the risk of migraine and heart attacks, while phosphorus is a known precursor of energy with adenosine triphosphate (ATP) as an essential component [5, 17, 37].

Among millets, barley carries a versatile nature to be further used in various culinary preparations and is one of the oldest known cereal grains with a smooth nutty flavor and an attractive chewy, pasta-like body. Barley has also been used in breweries. Also, barley plays a vital role in the formulation of various health foods as well as used in animal fodder. Considering the various health benefits of barley, it has long been associated with preparations of composite foods. It is also effective in reducing plasma total cholesterol (TC) and LDL cholesterol when incorporated in healthy diets [9]. Barley functional dry soup mix powder was developed by Kaur and Das [29]. Similarly, malted wheat flour was used for the preparation of *Doda burfi* by Jha [26], whereas Gajbhiye et al. [22] standardized the manufacturing process of cereal milk-based confection *Doda burfi*.

2.2.4   FRUITS–BASED COMPOSITE FOODS

There is an emerging need to reduce postharvest losses in fruits and vegetables especially in developing countries like India, where these losses of fresh produce range from 20 to 50% due to prevalent poor infrastructure and lack of marketing facilities [21].

Therefore, the use of fruits and vegetables in the preparation of composite dairy foods is an affirmative choice for manufacturers and consumers. Combination of these two products also results in mouthwatering and nutritionally superior products, balanced in all proportions. The acceptability of such products relies on the fact that the addition of a variety of ingredients does not support the particular flavor and aroma of one commodity and is liked by people of all ages. Furthermore, the presence of active ingredients in fruits and vegetables makes the product more likable by the nutritionists as well. Various research studies have been conducted in this regard and variety of innovative products such as Sapota added *Kalakand* and mango-based dairy products. These studies are briefly discussed in the next section.

2.3   RESEARCH STUDIES: FRUIT-BASED COMPOSITE DAIRY FOODS

Sawant et al. [51] studied the preparation of *Kalakand* using sapota fruit. In this study, *kalakand* was prepared wherein sapota pulp was used to replace milk at 10% in treatment T1 and 20% sapota pulp in treatment T2. They found that the fat content and acidity of the product decreased with the addition of pulp, whereas TS, carbohydrate, and mineral content increased on an increasing proportion of sapota pulp. The highest overall acceptability was recorded (8.94) for treatment T1 with 10% replacement. Similarly, the authors also tried the same product using mango fruit pulp. *Kalakand* was prepared by replacing milk with mango pulp at two different concentrations of 10 and 20% whereas control (T0) was without replacement to milk. Results indicated that there was an increase in total solids content, carbohydrate, and ash content, whereas fat and protein decreased proportionally with the increase in mango pulp. Acidity of the product was also affected negatively with the addition of mango pulp, and the overall acceptability was highest for treatment T1.

Kamble et al., [28] evaluated the effects of incorporation of varying levels of pineapple pulp in *burfi* on sensory and chemical properties. They revealed that the *burfi* prepared using 15% pineapple pulp had higher overall acceptability scores and was superior among other treatments. Also, the chemical composition of *burfi* with respect to different constituents like fat, protein, total solids, moisture, and ash varied significantly due to variable levels of pineapple pulp. A similar attempt was made by Banker et al., [7], who reported an acceptable quality of *burfi* (prepared using *khoa* from standardized buffalo milk with 6% fat and 9% SNF). Results advocated the addition of 10 parts of pineapple pulp and 90 parts of *khoa* by weight basis with the addition of 30% sugar for the preparation of this *burfi*.

Effect of variable levels of fruit additives on certain physicochemical properties of yogurt during storage was studied by Yousef et al., [61], whereas Wanik et al., [60] analyzed the optimum level of orange pulp and its effect on physicochemical and textural properties of *Santraburfi*. To prepare an acceptable quality *Santraburfi*, the researchers reported that the amount of orange pulp should not be less than 10%. Value-added *Kalakand* using papaya fruit (*papita*) pulp at different levels; 5, 10, 15, and 20% in buffalo milk was studied by Patel and Roy [46], who reported that 5% addition of papaya pulp with 95% buffalo milk is optimum to get a fruit-based *kalakand* burfi.

Apart from this, a variety of foods can be added in dairy ingredients to produce value-added products with milk and fruits as a supplement. Table 2.1 depicts characteristics of selected fruits for development of composite dairy products to fetch maximum health benefits.

TABLE 2.1   Bioactivities of Selected Fruits for Development of Food Products

Fruit	Characteristics of Selected Fruits	References
Banana	It is rich in potassium (2440 mg per 100 g) and low in sodium (1mg per 100g). It is a calming food due to the large content of tryptophan (an amino acid), which is converted to serotonin (an inhibitory neurotransmitter of the brain). It is a good lubricator for intestine, beneficial in the treatment of constipation and ulcers. It is also a good source of vitamin C, B- group vitamins (such as folate, niacin, riboflavin, and pantothenic acid).	[33, 35]
Blueberry	It is a source of ample amount of vitamin C and K, along with Manganese and several other antioxidants. It contains a good amount of anthocyanins. The fruit is known to boost enzymes that grow new nerve cells in the brain (important for memory). It also contains tannins, which are astringent to the digestive system and can reduce inflammation.	[14]
Cherry	It contains a high level of antioxidants and is a rich source of ß-carotene (20.0% more than blueberries and strawberries), vitamin C and vitamin K. It is a good source of minerals like Fe, Mg, and K.	[24]
Concentrated grape juice	Concentrated grape juice has 82.0% solids, 0.63% of protein, having a high percentage of minerals especially calcium and iron (5–10 mg/100 g). High iron content is beneficial for patients suffering from anemia.	[45]
Mango	It is highly rich in ß- carotene (445µg/100 g: an antioxidant), vitamin C (28 mg/100 g), and vitamin B (134 mg/100 g). It provides minerals such as calcium, magnesium, iron, and zinc. The tartaric acid, malic acid, and a trace of citric acid help to maintain the alkali reserve of the body.	[27]
Pomegranate	It is high in vitamin C, potassium, folate, thiamine, and many minerals like potassium, copper; and manganese providing 234 Kcal per 100 g. It is low in saturated fats and sodium. Pomegranate is a potent source of several antioxidants and has antiviral activity within. It also reduces intestinal disorders, dysentery, worms, and diarrhea. It has been claimed to lower cholesterol and reduce certain life-threatening cancers such as risks of breast, skin, prostate, lungs, etc.	[31]

2.4   RESEARCH STUDIES: VEGETABLE-BASED COMPOSITE DAIRY FOODS

Vegetables not only provide taste to the plate, but are also a source of various minerals, vitamins, fiber, and phytochemicals [20, 59]. Studies have proven that diets rich in vegetables are certainly associated with lowering the menace of cardiovascular diseases [42]. The 2007 report by WHO confirmed that a large number of deaths per year (2.7 million) were accountable to diets less in vegetables or diets with less than the required quantity of complex carbohydrates and fiber content. Therefore, these stands at the top of 10 risk factors, which highly contribute towards mortality [18].

Researchers have conducted studies to supplement the deficiencies borne by an entity with milk to get a blend, which has richer taste and texture. For example, carrots being a rich source of carotenoids help in vision, and also contain a good amount of polyphenols and vitamins, which act as anti-oxidants, anticarcinogens, and immune enhancers [19]; therefore blend of carrots and milk has been used by various researchers. Study has also been conducted to optimize variables to prepare carrot-milk cake to check the effects of different carrot-shred (CS) lengths, cooking methods, levels of *khoa*, sugar, fat; and stages of fat addition on product preparation yielded that CS lengths of 1 cm, cooked in open pan with 30% *khoa* and 22.5% sugar along with addition of 5% fat after cooking of CS resulted in carrot-milk cake with unmatched resultant product [23].

Bandyopadhay et al. [6] studied the effect of incorporation of carrot paste (0, 10, 20, 30, 40, and 50%) on quality improvement of sweet *rasogolla*. The addition of carrot to chhana resulted in decreased acid and free fatty acid (FFA) development in *Rasogolla* syrup, decreased absorptivity, and also provided color stability. Therefore, incorporation of carrot up to a concen-tration of 30% improves the quality of *rasogolla*. Similarly, *kheer* prepared from carrots added @ 30% has been standardized, and the burfi from the same was prepared by Mathur [38].

Srivastva and Saxena [56] conducted studies on preparation and analysis of novel bitter gourd (*karela*)-based milk product. Sirsat et al. [54] conducted studies on preparation of ash gourd *peda* from different proportions of buffalo milk blended with different proportions (5, 10, and 15%) of ash gourd pulp and they concluded that increasing the proportion of ash gourd decreased the protein and fat content of the product, however the sensory attributes remained unaffected. Acceptable quality *peda* can be prepared by incorpora-tion of 5% ash gourd pulp.

Similarly, Dadge et al. [16] prepared sweet potato kheer from buffalo milk with sweet potato added at various levels (@ 2.5, 5, and 7.5% in milk) along with the addition of sugar, rice, and cardamom. Bhutkar et al. [11] studied the effect of various levels of ash Gourd (*petha*) pulp for preparation of *kalakand*. The results directed that adding 10 parts of ash gourd pulp and 90 parts of chhana leads to an acceptable quality of *kalakand*. The cost of production of the final product was in the range of 356, 343, 331 and 318 Rs/Kg for T1, T2, T3, and T4 treatments, respectively.

2.5   ISSUES ASSOCIATED WITH FORMULATION OF COMPOSITE FOODS

Although the formulation with the aim to fortify and complement deficient nutrients in composite foods looks simple, yet it is a complex task. One should be aware of following items in terms of its phenomenon occurring inside the foods when mixed in a combination:

- Addition of ingredients that are colored in nature due to naturally occurring pigment in them, either lose their color due to processing to a higher temperature or could not retain their activity potential (due to pigment degradation) when used in combination for composite foods.
- Combination of millets or other non-dairy ingredients, when mixed with milk, can cause destabilization of internal pH system of milk at a higher temperature; thereby can lead to coagulation of proteins.
- Presence of some polyphenols, anti-nutritional factors, and their presence even after processing, should be taken into consideration while formulating composite products. It has been observed in various research studies that natural antinutrients present in plant foods are purely accountable for malabsorption of nutrients and various minerals and trace elements. Some of these plant chemicals have also shown harmful effects on human health and undoubtedly advantageous to human health when taken in suitable amounts [57].
- Some rich sources of micronutrients (like iron: a known pro-oxidant) can oxidize fat-rich dairy products and can adversely affect its organoleptic properties in terms of flavor and color changes and may affect the storage stability of foods.
- The biopolymers possess complex physicochemical properties and these properties are highly dependent on many factors such as the chemical composition of the biopolymer support, their structure (2°, 3°, 4°), their molecular weight and the reactivity of any side chains or groups.
- The physicochemical properties of two dissimilar systems (i.e., starch, and milk) depend chiefly on their comparative concentrations and physicochemical properties. Therefore, a systematic approach and understanding of the two dissimilar systems can lead towards the option for altering the composition and processing conditions of foodstuffs to modify their textural behavior and sensory attributes [15].

Thus, the effect of all these changes reflected in terms of sensorial properties does not match the expectations of the consumer and are not accepted by the mass as a whole. Therefore, a complete strategic methodology and scientific intervention should be studied before undertaking such projects at commercial level.

2.6   COMMERCIAL-BASED PERSPECTIVE COMPOSITE DAIRY PRODUCTS

- **Gluten Free Bakery Products:** Such products are still available on the shelves of supermarkets and will be boon for celiac patients if further introductions in the same arena can be made. Combination of such gluten-free flour was used to attempt success in this category was made by Rai et al., [47] and Murugkar et al., [43].
- **Malted Milk Foods:** They offer good product profile wherein a variety of ingredients can be mixed together to complement and supplement the nutrients. Thus, the end product further can be mixed with milk to get substantial benefits. In this context Modi [41], developed a low-cost complementary infant food employing skim milk and whey in combination with added millets such as germinated pearl millet (*bajra*), barley malt (*Jau*), and cornflour (*makai*).
- **Milk and Coarse Cereal-based Convenience Foods:** With the boom in newer innovations in the arena of Science and Technology, and an increasing focus on the development of functionally active and value-added products, the drift in development of functional or value-added cereal products has shaken the baking industry. Cereal Technology is a broad-spectrum arena, where a number of specialized products are coming day after day in the market wherein cookies, or biscuits occupy a special space and are liked by children and kids alike. Variants of value-added cookies have been tried and prepared by researchers using a blend of various flours such as oligofructose, dietary fiber, and lower calorie fat [36] or added extruded orange pulp [34]. Similarly, different cereals in a blend like a soybean, corn, and carrot meal have also been tried for the production of cookies to improve the functionality of the cookies [2]; and blend of soybean and maize flour with improved nutritional demands [1].
- **Milk Cereal-based Fermented Foods:** The best example in this category is *Rabadi*, in which coarse cereals like millets, maize can be added to buttermilk, and the mixture is fermented using lactic acid

bacteria (LAB). The product is boiled, salted as per taste, and can be served chilled during hot summer months. It has been reported that the process reduces the level of anti-nutrients, and improves anti-oxidant potential.

Along with this, various extruded products are also becoming the choice of the consumer being lighter and tastier in their appeal. Extrusion Technology and the products out of it is another popular category to create a variety of RTE based snacks. The extruded products possess enormous advantages with respect to the nutritional value of the final product, such as inactivation of antinutrients, destruction of aflatoxins along with benefits of increasing the digestibility of fiber of such products [49, 53]. These can be prepared from corn, rice, soy, millets, etc. Meena [40] developed milk, millet-based extruded snacks.

- **Mixture of Cereals and Millets**: It can be added for the preparation of various probiotic drinks, and other fermented drinks.

Though the list of such products is numerous, yet the wide acceptability of such products and mass awareness regarding the nutritional profile of such products and their commercial undertaking by entrepreneurs is still lacking. Therefore, concerted effects should be made on these lines for profitability.

2.7   SUMMARY

We are in the era of novel and health-promoting foods, which contain certain bioactive components that confer therapeutic properties. The increasing inclination of functional foods has paved the way towards many innovative products in the market, and this has renewed interest in exploring medicinal benefits from fruits and vegetables. Although consumers want healthy foods, yet they also do not want to compromise with taste. In this context, fruits and vegetables are an excellent choice. They are natural, health-promoting, and possess several medicinal properties.

Although milk and milk-based products contain essential minerals, yet these are deficient in fiber and other micronutrients like iron, copper, and certain vitamins. Also, there are some limitations of milk constituents like causing allergic reactions, lactose intolerance, cholesterol, saturated fat content, etc. For example, milk fat can cause health problems if consumed in excess quantities. Therefore, there is a need to overcome these deficiencies through supplementation of milk and milk products with required nutrients and health-promoting components through proper processes. One such process involves blending of dairy products with fruits, vegetables, herbs, millets, legumes, and cereals to form composite dairy foods to increase its nutritional value and to provide numerous health benefits.

Most common options for such formulations are fruit-based, vegetable-based, and cereal-based composite dairy foods. Numerous technological issues may arise on mingling different components such as cereals, millets, or fruits and vegetables with milk, because of the difference in inherent chemical constituents and nature of different entities. For example, mixing of fruit's pulp in milk can hinder the stability of milk proteins and damage the inherent salt balance of the milk system. It also aids in lowering the pH of the milk system by imparting acid content in the product that can cause coagulation of milk proteins during heating. Apart from this, there are certain micronutrients (Fe), which enhances, and aggravates the oxidative reactions in fat-rich dairy products and thus affect the flavor adversely along with decreasing the storage stability of such foods.

Some of the key limitations of such formulations are unstable products (heat stability affected), fat oxidation in case of fat-rich dairy products, uneven textural attributes, bitter/off-flavor and increased acidity of the final product. The incorporation of non-dairy components in milk is mainly focused towards the enhancement in sensory profile and formulations with enhanced organoleptic improvement with improved health attributes.

KEYWORDS

- **allergenicity**
- **carotenoids**
- **celiac disease**
- **flavonoids**
- **lactose intolerance**
- **malted milk foods**
- **neurotransmitter**
- **organoleptic**
- **polyphenols**
- **ready to eat**
- **whey protein**

REFERENCES

1. Akubor, P. I., & Onimawo, I. A., (2003). Functional properties and performance of cowpea/plantain/wheat flour blends in biscuits. *Journal Plant Food Human Nutrition, 58*, 1–8.
2. Akubor, P. I., & Ukwuru, M. U., (2005). Functional properties and biscuit making potential of soybean and cassava flour blends. *Plant Foods for Human Nutrition, 58*(3), 1–12.
3. Anonymous, (2015). *Celiac Disease: Fast Facts.* http://www.celiaccentral.org/celiac-disease/facts-and-figures/ (Assessed on July 20 2018).
4. Anonymous, (2015). Codex general standard for the use of dairy terms. http://www.fao.org/docrep/015/i2085e/i2085e00.pdf (Assessed on July 20 2018).
5. Badau, M., Nkama, H. I., & Jideani, I. A., (2005). Phytic acid content and hydrochloric acid extractability of minerals in pearl millet as affected by germination time and cultivar. *Food Chemistry, 92*(3), 425–435.
6. Bandyopadhyay, M., Mukherjee, R. S., Chakraborty, R., & Raychaudhuri, U., (2006). A survey on formulations and process techniques of some special Indian traditional sweets and herbal sweets. *Indian Dairyman, 58*, 23–35.
7. Bankar, S. N., Korake, R. L., Gaikwad, S. V., & Bhutkar, S. S., (2013). Studies on preparation of pineapple *burfi*. *Asian Journal of Dairy & Food Research, 32*(1), 40–45.
8. Bayford, C., (2010). Whey protein: A functional food. *Journal of Clinical Nutricant, 3*(4), 471–478.
9. Behall, K. M., Scholfield, D. J., & Hallfrisch, J., (2004). Diets containing barley significantly reduce lipids in mildly hypercholesteromic men and women. *American Journal of Clinical Nutrition, 80*, 1185–1193.
10. Berry, D., (2002). Healthful ingredients sell dairy foods. *Dairy Foods, 103*, 54–56.
11. Bhutkar, S. S., Nimbalkar, S. S., & Kumbhar, T. V., (2015). Effect of different levels of ash gourd pulp for manufacturing *kalakand*. *Journal of Agriculture & Veterinary Science (IOSR-JAVS), 8*(3), 4–6.
12. Bucci, L., (1995). Nutrition applied to injury rehabilitation and sports medicine. *Nutrition in Exercise & Sport, 5,* S39–S61.
13. Bunkar, D. S., Jha, A., & Mahajan, A., (2014). Optimization of the formulation and technology of pearl millet based 'ready-to-reconstitute' kheer mix powder. *Journal of Food Science & Technology, 51*(10), 2404–2414.
14. Camire, M. E., & Dougherty, M. P., (2006). Frozen wild blueberry-tofu-soy-milk desserts. *Journal Food Science, 71*, S119–S123.
15. Considine, T., Noisuwan, A., Hemar, Y., Wilkinson, B., Bronlund, J., & Kasapis, S., (2011). Rheological investigations of the interactions between starch and milk proteins in model dairy systems: A review. *Food Hydrocolloids, 25*(8), 1–10.
16. Dadge, A. V., Thorat, B. N., Londhe, G. K., & Awaz, H. B., (2014). Effect of different levels of sweet potato paste on physico-chemical properties of *kheer*. *Journal of Veterinary Science and Technology, 5*(3), 21.
17. Devi, N., Lakshmi, S., Sajid, A., Kalpana, K., & Soumya, M., (2014). Utilization of extrusion technology for the development of millet based complementary foods. *Journal of Food Science and Technology, 51*(10), 2845–2850.
18. Dias, J. S., (2011). World importance, marketing and trading of vegetables. *Acta Horticulturae, 921*, 153–169.

19. Dias, J. S., (2014). Nutritional and health benefits of carrots and their seed extracts. *Food and Nutrition Sciences, 5*, 2147–2156.
20. Dias, J. S., & Ryder, E., (2011). World vegetable industry: Production, breeding, trends. *Horticulture Review, 38*, 299–356.
21. FFTC, (2015). *Postharvest Losses of Fruit and Vegetables in Asia*, http://www.fftc. agnet.org/library.php?func=view&style=type&id=20110630151214 (Assessed on 20 July 2018).
22. Gajbhiye, S. R., Goel, B. K., Uprit, S., Asgar, S., & Singh, K. C. P., (2005). Standardization of *Doda Burfi* (Sprouted wheat based milk product). *Poster Paper Presented at National Seminar on Value Added Dairy Products* (p. 1). National Dairy Research Institute, Karnal–India.
23. Gupta, M., Bajwa, U., & Sandhu, K. S., (2005). Optimization of variables associated with processing of carrot-milk cake. *Journal of Food Science and Technology, 42*(2), 16–22.
24. Hedrick, T., Markakis, P., & Wagnitz, S., (1969). Cherries in spreads and cottage cheese sauce. *Journal of Dairy Science, 52*(12), 2057–2059.
25. Itagi, H. B. N., & Singh, V., (2012). Preparation, nutritional composition, functional properties and antioxidant activities of multigrain composite mixes. *Journal of Food Science and Technology, 49*(1), 74–81.
26. Jha, A., (2003). *Development and Evaluation of Malted Barley Flour Supplemented Doda Burfi* (p. 118). M. Sc. Thesis, Bundelkhand University, Jhansi, India.
27. Kadam, R. M., Bhambure, C. V., Burte, R. G., & Joshi, S. V., (2009). Process standardization for manufacture of Mango Burfi. In: *Souvenir of the National Seminar on 'Novel Dairy and Food Products of the Future* (pp. 177–183). DSC Alumni Association and SMC College of Dairy Science, Anand Agricultural University, Anand, Gujarat, India.
28. Kamble, K., Kahate, P. A., Chavan, S. D., & Thakare, V. M., (2010). Effect of pine-apple pulp on sensory and chemical properties of *burfi*. *Veterinary World, 3*(7), 329–331.
29. Kaur, S., & Das, M., (2015). Nutritional and functional characterization of barley flaxseed based functional dry soup mix. *Journal of Food Science & Technology, 52*(9), 5510–5521.
30. Khan, M. A., Semwal, A. D., Sharma, G. K., Yadav, D. N., & Srihari, K. A., (2008). Studies on the development and storage stability of groundnut (*Arachis Hypogea*) Burfi. *Journal of Food Quality, 31*(5), 612–626.
31. Konowalchuk, J., & Speirs, J. I., (1976). Yogurt preparation using pomegranate. *Journal of Food Science, 41*, 103–117.
32. Krupa, H., Atanu, J., & Patel, H. G., (2011). Synergy of dairy with non-dairy ingredients or product: A review. *African Journal of Food Science, 5*(16), 817–832.
33. Kumar, P., Tyagi, S. M., Chauhan, G. S., & Sharma, H. K., (2001). Physicochemical changes during fermentation of banana-whey blended beverages. *Egyptian Journal of Dairy Science, 29*, 53–61.
34. Larrea, M. A., Chang, Y. K., & Martinez-Bustos, F., (2005). Some functional properties of extruded orange pulp and its effect on the quality of cookies. *LWT–Food Science and Technology, 38*, 213–220.
35. Laxminarayana, G. S., Ghosh, B. C., & Kulkarni, S., (1997). Technology of ready-to-use banana milk shake powder. *Journal of Food Science and Technology, 34*, 41–45.

36. Lee, S. A., George, E., Inglett, B., Palmquist, D. B., & Warner, K., (2009). Flavor and texture attributes of foods containing β-glucan-rich hydrocolloids from oats. *LWT–Food Science and Technology, 42*, 350–357.

37. Liang, S., & Yang, G. M. Y., (2010). Chemical characteristics and fatty acid profile of foxtail millet bran oil. *Journal of American Oil Chemists Society, 87*, 63–67.

38. Mathur, S., (2008). Carrot burfi. In: *Indian Sweets* (pp. 35, 36). Ocean Book Ltd. Pub., New Delhi.

39. Mcluntosh, G. H., (1995). Dairy proteins protect against dimethylhydrazine-induced intestinal cancers in rats. *Journal of Nutrition*, 809–816.

40. Meena, R., (2010). *Process Development for Milk Protein Rich Extrude Snacks Based on Pearl Millet* (pp. 58–64). M.Tech. Thesis, National Diary Research Institute, Karnal-India.

41. Modi, R., (2009). *Process Development for Complementary Food Based on Whey-Milk-Cereal Blend* (pp. 80–89). M.Tech Thesis, ICAR- National Diary Research Institute, Karnal-India.

42. Mullie, P., & Clarys, P., (2011). Association between cardio- vascular disease risk factor knowledge and lifestyle. *Food and Nutrition Science, 2*(10), 1048–1053.

43. Murugkar, D. A., Gulati, K. N., & Gupta, C., (2015). Evaluation of nutritional, textural and particle size characteristics of dough and biscuits made from composite flours containing sprouted and malted ingredients. *Journal of Food Science and Technology, 52*(8), 5129–5137.

44. Obilana, A. B., & Manyasa, E., (2002). Millets. In: Belton, P. S., & Taylor, J. R. N., (eds.), *Pseudo Cereals and Less Common Cereals: Grain Properties and Utilization Potential* (pp. 177–217). Springer-Verlag: New York.

45. Ozturk, B. A., & Oner, M. D., (1999). Production and evaluation of yogurt with concentrated grape juice. *Journal of Food Science, 64*, 530–532.

46. Patel, K. M., & Roy, S. K., (2015). Development of value added *kalakand* using papaya fruit. *Indian Journal of Dairy Science, 68*(1), 91–94.

47. Rai, S., Kaur, A., & Singh, B., (2014). Quality characteristics of gluten free cookies prepared from different flour combinations. *Journal of Food Science and Technology, 51*(4), 785–789.

48. Reilly, N., & Green, P. R., (2014). Presentation of celiac disease in children and adults. In: Rampertab, S. D., & Mullin, G. E., (eds.), *Celiac Disease* (pp. 95–105). Springer, New York.

49. Saalia, F. K., & Phillips, R. D., (2011). Degradation of aflatoxins by extrusion cooking: Effects on nutritional quality of extrudates. *LWT–Food Science and Technology, 44*(6), 1496–1501.

50. Sawant, V. Y., Chauhan, D. S., Padghan, P. V., & Thombre, B. M., (2007). Formulation and evaluation of mango fruit *kalakand*. *Journal of Food Science and Technology, 31*(5), 389–394.

51. Sawant, V. Y., Thombre, B. M., Chauhan, D. S., & Padghan, P. V., (2006). Preparation of *kalakand* with sapota fruit. *Journal of Dairying, Foods & Home Science, 25*(3/4), 186–189.

52. Singh, A. K., Kadam, D. M., Saxena, M., & Singh, R. P., (2011). Effect of soy flour supplementation on the quality and shelf life of gulabjamun. *International Journal of Food Science and Nutrition Engineering, 1*(1), 11–17.

53. Singh, S., Gamlath, S., & Wakeling, L., (2007). Nutritional aspects of food extrusion: A review. *International Journal of Food Science and Technology, 42*(8), 916–929.

54. Sirsat, A. B., Shinde, A. T., & Korake, R. L., (2013). Studies on preparation of ash gourd *peda. Indian J. Dairy Science, 66*(3), 213–217.

55. Sood, A., Midha, V., Sood, N., Avasthi, G., & Sehgal, A., (2006). Prevalence of celiac disease among school children in Punjab, North India. *Journal of Gastroenterolology and Hepatology, 21,* 1622–1625.

56. Srivastava, T., & Saxena, D. C., (2012). Optimization of total polyphenol content and antioxidant activity on preparation of novel bitter gourd sweet. *IRACST- Engineering Science and Technology: An International Journal, 2*(5), 861–874.

57. Ugwu, F. M., & Oranye, N. A., (2006). Effects of some processing methods on the toxic components of African breadfruit (*Treculia Africana*). *African Journal of Biotechnology, 5,* 2329–2333.

58. Wang, H., (1980). The importance of traditional fermented foods. *Journal of Biosciences, 30,* 402–404.

59. Wargovich, M. J., (2000). Anticancer properties of fruits and vegetables. *Horticulture Science, 35,* 573–575.

60. Wasnik, P. G., Nikam, P. B., Dhotre, A. V., Waseem, M., Khodwe, N. M., & Meshram, B. D., (2013). Physicochemical and textural properties of Santra *burfi* as influenced by orange pulp content. *Journal of Food Science and Technology, 52*(2), 1158–1163.

61. Yousef, M., Nateghi, L., & Azadi, E., (2013). Effect of different concentration of fruit additives on some physicochemical properties of yogurt during storage. *Annals of Biological Research, 4*(4), 244–249.

CHAPTER 3

FUNCTIONALITY ENHANCEMENT IN CHEESE

APURBA GIRI and S. K. KANAWJIA

ABSTRACT

Cheese is a well-known milk product that is valued for its nutritional superiority. It contains bioactive peptides, which is a good source of calcium having good bioavailability, high amounts of conjugated linoleic acid (CLA) (in processed cheese), etc. In addition to all these bioactive compounds, the milk fat in cheese does not have negative effects on serum cholesterol level. Several studies have reported successful addition of different functional ingredients such as different dietary fibers (inulin, oat fiber, and soy-fiber), phytosterols, omega-3 fatty acids, whey protein concentrate (WPCs), cocoa, probiotics, polyphenolic compounds, etc. To decrease the fat and sodium content in cheeses, different innovative techniques have been adopted successfully.

3.1 INTRODUCTION

Functional foods provide health benefits besides their energy content. Functional foods may be regarded as conventional food products with health-promoting ingredients or components that go beyond their traditional nutritive value [22]. The health-promoting ingredients are also known functional ingredients, and its incorporation is one of the methods to develop functional food. Some examples of these health-promoting ingredients are dietary fiber, conjugated linoleic acid (CLA), phytosterols, omega-3-fatty acids, bio-active peptides, prebiotics, etc.

Cheese is valued for its nutritional superiority, portability, long life, and high content of fat, protein, and minerals. A hundred types of cheeses are manufactured with wide-ranging flavors, textures, and forms. It contains several components with functional effects and makes cheese itself a functional food. Cheese contains high casein content, which is a source of several bioactive peptides that are released by a number of proteases of native and starter culture origin. These peptides have biological activities like anti-carcinogenic, anti-cariogenic, anti-hypertensive, anti-oxidative, opioid-activity etc. [4, 38, 41, 52, 69].

Apart from these bioactive peptides, cheese (particularly hard cheese and processed cheese) is a good source of Calcium having good bioavailability. Cheese also contains high amounts of CLA, which is responsible for modulation of immunity, a decrease of body fat, and inhibition of cancer. This content of CLA further rises during making the processed cheese from the natural cheese [9]. In addition to all these active components, the milk fat does not have negative effects on serum cholesterol level. In a study between intake of butter and cheese with the same quantity fat, it was reported that cheese intake was connected with lower blood cholesterol levels [39].

Different functional ingredients can be added to increase the functionality of cheese. Some of the suitable functional ingredients for incorporating into cheeses and other ways of enhancement of functionality of cheeses have been discussed in this chapter.

3.2 DIETARY FIBERS

Insufficient consumption of dietary fiber in our daily diet may raise the possibility of bowel abnormalities, constipation, and several related disorders. There is a positive relationship between the insufficient fiber and appearance of common illnesses like obesity, diabetes mellitus, hyperlipidemia, appendicitis, ischemic heart disease, diverticular disease, hemorrhoids, gall bladder disease, varicose veins, hiatus hernia, deep vein thrombosis, large bowel tumors and colorectal cancer [53, 67].

It has been shown that dietary fiber protects against cancers of mouth, throat, and esophagus. Now eating behaviors are altering, and the intake of ready-made foods has increased. This can lead to a reduction in dietary fiber consumption if these foods provide energy ingestion significantly. Simultaneously, the ingestion of traditional fiber sources is declining. In many developed countries, the consumption of dietary fiber is not sufficient, and an alternative way to increase consumption should be developed.

3.2.1 INULIN

Inulin, which is a soluble dietary fiber, is present in many cereals, fruits, and vegetables consumed regularly, such as wheat garlic, bananas, onions, leeks, chicory, artichokes, etc. Daily consumption of inulin is 3 to 11 g in Europe and is 1 to 4 g in the USA. It is recognized as "natural food ingredient" in most European countries, and it has GRAS (Generally Recognized as Safe) status in the USA. Inulin is manufactured from artichoke tubers and chicory roots and is applied as a functional food ingredient. This inulin is used for the nutritional benefit and some technological advantages in food. Inulin has sensory properties significantly for the formulation of food, allowing an improvement in mouthfeel and taste in an extensive range of uses.

3.2.1.1 ADVANTAGES OF INULIN

For digestion, insulin is not required. It is an excellent food for the person with diabetes because it inhibits hypoglycemia, and it delivers continuous blood sugar to the muscles for a long time.

- In the large intestine, inulin acts as prebiotic, nourishes the valuable bifidobacteria.
- In the colon, it increases the absorption of minerals.
- Increasing the lactic acid bacteria (LAB) by feeding inulin reduces the activities of enzymes involved in carcinogenesis.
- It eliminates constipation.
- It helps for a healthy circulatory (blood) system as it reduces total cholesterol (TC), low-density lipoprotein (LDL) and triglycerides, while increases high-density lipoproteins (HDL) in the blood.
- It slows down appetite because it is not digested in the stomach or small intestine, but provides a sense of fullness. Thus, obesity is reduced.

3.2.1.2 APPLICATIONS OF INULIN IN CHEESE

Incorporation of inulin in milk products is mostly because of two reasons: (1) It can be attributed towards the various physiological functions that it confers to the consumer (i.e., dietary fiber, prebiotic, etc.); and (2) Inulin has different technological properties and provides functionality in the food matrix (i.e., fat mimic, texture modifier etc.). Due to the water-holding property of inulin, it changes the rheology of the food to improve the texture, and it can be applied in food product especially in low and no sugar and low or non-fat systems.

Several studies show that the addition of inulin increases the viability of probiotic of several dairy products. Elewa et al. observed the prebiotic effect of inulin (2%) in UF-probiotic white soft cheese [16]. Araujo et al. noticed that the taste and texture of a symbiotic Cottage cheese remained unchanged up to 8% addition of inulin [6]. Koca and Metin observed the fat mimetic effect of inulin and found that adding 5% inulin to the low-fat Kashar cheese produced significantly less hardness than low-fat control cheese, but slightly higher as compared to full-fat control cheese [45].

El-Nagar and El-Aty reported that Karish cheese could be prepared from reconstituted skim milk by applying inulin as fat replacer (at the rate of 0.5, 1.0 and 1.5%). Inulin incorporation improved adhesiveness, cohesiveness, gumminess, and chewiness but decreased hardness and elasticity [17]. Addition of 10% of fat mimetic inulin in natural cheese did not affect the aroma of cheese [20].

Functional processed cheese spread (PCS) was prepared at National Dairy Research Institute (NDRI), Karnal [23, 26, 27, 29]. In the developed product, inulin was incorporated @ 6%. The hypocholesterolemic effect of the cheese spread has been validated through animal trials. The production cost of the product has also been estimated.

3.2.2   OAT FIBER

It is a water-soluble good dietary fiber rich in β-glucan. It can be incorporated due to its established health benefits. To balance blood glucose and insulin levels and to lower blood cholesterol levels, oat has an important role. However, it produces high viscosity in the food, and it may create a problem not only during food processing but also in sensory qualities [74].

3.2.3   SOY FIBER

Soy fiber raises stool quantity and reduces intestinal travel time. It contains mainly polysaccharides, and it is highly fermentable. It lowers blood cholesterol. Gahane [19] tried to develop a functional Quarg cheese, in which the addition of fibers varied from inulin 8–12%, and soy fiber and oat fiber 0–2% each. He found that best quality fiber-enriched Quarg cheese on the basis of sensory attributes, can be prepared using 10% inulin, 1.10% oat fiber and 0.95% soy fiber, on weight to weight basis of the curd [19].

3.3 PHYTOSTEROLS

Phytosterols are waxy substances with a structure similar to cholesterol, the only difference in the additional methyl or ethyl group in the side chain of cholesterol. More than 200 various types of phytosterols have been discovered, among them β-sitosterol, campesterol, and stigmasterol are principal. Saturated phytosterols are called phytostanols. There is no double bond in sterol ring of phytostanols. Phytosterols and phytostanols have similar beneficial effects.

Phytosterols are mainly found in vegetable oils. Till now, phytosterols, and stanols are manufactured commercially during vegetable oil refining and wood processing industries as a by-product. During refining (at deodorization step), phytosterols may be extracted from sunflower, corn, soybean, canola, peanut, cottonseed or palm oil. The phytosterols content of distillate of deodorization can be further increased by crystallization of solvents.

All phytosterols in the human body come from food because human beings are unable to produce it. Generally, people consume 100–500 mg phytosterol daily through a healthy diet [62]. Phytosterol and phytostanol decrease total blood cholesterol and LDL cholesterol up to 15 and 22%, respectively [43, 62].

3.3.1 *MECHANISM OF ACTION OF PHYTOSTEROLS*

Though several investigations have been conducted to find out the exact mechanism of cholesterol reduction property of phytosterols, yet exact action on a molecular level is not discovered till today. It is known that the main decrease of intestinal absorption of cholesterol is without reducing of triglycerides or HDL cholesterol [62]. The decrease in cholesterol absorption may be due to the structural similarity of cholesterol, phytosterol, and phytostanol [66].

The probable mechanism for interfering with the absorption of cholesterol may be due to the reduction or blockage of the hydrolysis of the cholesterol ester carried out by the lipases and the esterases; as a result, co-precipitation of phytosterols and/or stanols and cholesterol to form non-absorbable mixed compounds. [72].

3.3.2 APPLICATIONS OF PHYTOSTEROLS IN CHEESE

In 1995, a margarine incorporated with phytostanol was launched in Finland, and that was the first commercial application of phytostanol in food [60]. Phytosterol added margarine was launched to the US-market in 1999. After that, phytosterol and phytostanol incorporated foods are being marketed in several countries.

Both phytosterol and phytostanol esters do not offer energy to the body; therefore, these esters can be used to replace fat. In addition to this, the phytostanol ester can be applied in spreads to change the fatty acid composition of fat and exchange hard fat partly in spreads. These esters may give a crispy texture by preventing sogginess of the surface coating of cereal products [49]. In low-fat milk products, both esters provide an improved creamy texture. By covering the bitterness, they can enhance the taste of food products. Therefore, a lesser quantity of sugar or other sweetener is needed to get the pleasant taste [13].

Gahane did not observe any significant change in sensory quality up to 5% of phytosterol ester addition into fiber added Quarg cheese [19]. To find out the optimum phytosterols addition level in the PCS at NDRI, three levels of phytosterols were added @ 2, 3, and 4% during the manufacture of PCS [25, 28]. It was found that phytosterols added PCS samples were markedly higher in shearing action than the control. A sharp, steady, and significant decline in work of adhesion were observed with increase in addition of phytosterols in PCS form 0 to 4%. At 4% level of phytosterols addition, a significant reduction in flavor, color, appearance, and spreadability score was noticed than the control PCS. Therefore, PCS at a 2% level of phytosterols addition was optimum in this study.

3.4   OMEGA-3 FATTY ACIDS

Omega-3 fatty acids (ω-3 FAs) are long carbon chain poly-unsaturated fatty acids found in various plant and marine sources. They consist of α-linolenic acid (ALA), stearidonic acid, docosapentaenoic acid (DPA), eicosapentaenoic acid (EPA), and docosahexaenoic acid (DHA). The importance of ω-3 FAs was first recognized in research on the Eskimos. Authors assessed the eating behavior of Greenland Eskimos, who have a low death rate due to coronary heart disease (CHD). The importance of ω-3 FAs in growth and development of the brain and against atherosclerosis, cancer, Parkinson's disease, rheumatoid arthritis, and Alzheimer's diseases has been established [37, 40, 51, 70].

3.4.1   APPLICATIONS OF OMEGA-3 FATTY ACIDS IN CHEESE

Eating seafood or intake of 0.6–1g/day of ω-3 FAs is good for our health. However, most of the population is not following this recommendation. Therefore, the addition of ω-3 oil in different processed foods is an alternative idea. However, ω-3 FAs is very unstable for their autooxidation, which makes off-flavor. To avoid this, microencapsulation of ω-3 oils can be used.

The ω-3 FAs are used as liquids, capsules, or in other forms. If it is used as the liquid, then sometimes flavoring material is used to mask any off-flavor, which may be produced during prolonged storage. For the use of ω-3 FAs as therapeutic purposes, the ω-3 FAs in the oil may be concentrated.

Ye et al. [76] noticed that processed cheese samples with non-encapsulated fish oil emulsion resulted in higher oxidation than the encapsulated fish oil emulsion. Martini et al., reported that three months ripened 50% fat reduced Cheddar cheese might be used as a carrier for ω-3 FAs (DHA + EPA) without having off-flavors (oxidized, rancid, and fishy) [55]. Kolanowski and Weibbrodt reported that butter, processed, and spreadable fresh cheeses with ω-3 FAs can be stored up to 4 weeks when flavorings were present [46]. Giri and Kanawjia optimized addition of ω-3 FA (source: flaxseed oil) individually at a 2% level [24]. Giri developed a cheese spread, where inulin, phytosterol, and omega-3 oil were added in combination in cheese spread at 4%, 2%, and 4%, respectively [21].

3.5   WHEY PROTEIN CONCENTRATE (WPC)

Milk contains mainly two types of protein: casein and whey protein. Whey, once considered as a byproduct, is the aqueous product obtained during the production of cheese, *channa,* and *paneer.* Different constituents of whey are lactose, minerals, β-lactoglobulin (β-LG), α-lactalbumin, lactoferrin, bovine serum albumin (BSA), lactoperoxidase enzymes, glycomacropeptides, immunoglobulins (IGs), etc. Each component has its own physiological effect. For example, α-lactalbumin and β-LG are responsible for the repair and growth of tissue. While lactoferrin and lactoperoxidase confer anti-microbial benefits and IGs provide immune-modulating effects to the host. Before the advancement of ultra-filtration processing, the use of whey was limited despite the high nutritional property because of high lactose and low protein content. WPC is the product obtained by the ultra-filtration processing of whey. Typically WPC are categorized on the basis of protein content, which ranges from 28–89% with WPC-70 and WPC-80 being the most common.

Basically, WPC is the concentrated form of whey, which has all the components in the condensed form. Therefore, WPC represents a protein source of excellent nutritional and functional properties. When protein percentage in WPC increases, the lactose, mineral, and fat content decreases. Beyond nutritional benefit in different foods, WPC is widely applied as a functional ingredient for texture, binding, and aeration characteristics during food preparation [30].

3.5.1 APPLICATIONS OF WPC IN CHEESE

Addition of WPC as a fat replacer and texture modifier in some varieties of fresh cheese gives positive effects on the textural properties of cheese [50]. Low-fat PCS can be prepared with the addition of 20% WPC and 30% buttermilk; and the developed product is softer, smoother with pronounced flavor and more spreadable than the control [3]. Another study by Pinto et al., suggests that the addition of 4.5% WPC in PCS resulted in product enhanced meltability and spreadability [64].

Gupta and Reuter extensively studied the effect of the addition of WPC in processed cheese foods. Processed cheese food manufactured by 20% replacement of cheese solids resulted in much better scores in terms of flavor, consistency, and overall acceptability [34]. Kumar developed a functional cheese food, and it was found that the optimum levels of addition of phytosterol esters, WPC, and inulin were obtained at 4%, 2%, and 6%, respectively [48].

3.6 COCOA

The products of cocoa bean (*Theobroma cacao*) are cocoa powder, cocoa liquor, and dark chocolate contain polyphenolic compounds. In many studies, this polyphenolic compound proved itself as bioactive compounds [2]. Not only it is also rich in methylxanthines, but it has physiological effects on body systems, such as renal, cardiovascular, central nervous, respiratory, and gastrointestinal (GI) systems [7]. Cocoa powder may be incorporated in different food products to enhance its palatability, taste, and flavor, and to provide health benefits. Kumar developed a functional Quarg cheese in which cocoa, inulin, and fructooligosaccharides were successfully incorporated @ 6.67%, 4.16%, and 13%, respectively [47].

3.7 PROBIOTICS

Probiotics provide health beneficial effects. Food products with probiotic bacteria give several therapeutic benefits such as anti-carcinogenic property, relief from diarrhea and lactose intolerance, improves immunity, reduces blood cholesterol, etc. To get health benefit, the probiotic bacteria should be consumed daily, and the live probiotic bacterial count in the food matrix should be minimum 10^7 per g or per ml. Cheese with its slightly high pH, solid consistency, and high-fat content give protection during the passing of probiotics through the gastrointestinal tract (GIT).

Various experiments have been reported for production of probiotic-based cheeses such as with: Cheddar cheese [63], Cottage cheese [1], Gouda cheese [44], Canestrato Pugliese cheese [14], Crescenza cheese [12, 31], Fresco soft cheese [73], goat cheese [61], Minas fresh cheese [10, 11], and semi-hard cheese [8]. Shashikant noticed that probiotic *L. casei* (NCDC 298) added to Quarg cheese showed good survivability of probiotics during 30 day-storage. The survivability and performance of probiotic bacteria in cheese are satisfactory, but its capability differs from strain to strain.

3.8 POLYPHENOLIC COMPOUNDS

Polyphenols are functional ingredients in plant-derived foods and act as principal antioxidants in our diet. Besides this, they offer various therapeutic properties, such as anti-aging, anti-inflammation, apoptosis, anticancer, anti-atherosclerosis, cardiovascular disease, etc. [36]. Several polyphenols have been found in plant foods. In beverages, for example, coffee, tea, and red wine are the main sources of polyphenols; though, cereals, legumes, and vegetables contain a good amount of polyphenols [54].

Due to various health benefits, polyphenolic compounds have been tried to incorporate into different food products. Han et al., developed functional cheese product by adding different polyphenolic compounds (such as cate-chin, epigallocatechin, tannic acid, gallate, hesperetin, flavones acid and homovanillic) and natural crude composites (like cranberry powder, grape extract, and green tea extract). They observed retention of polyphenols in different levels in the cheese curds depending on hydrophobicity and gel-formation properties. It is reported that polyphenols @ 0.5 mg/mL in cheese resulted in satisfactory free radical-scavenging property [35].

3.9   LOW FAT CHEESE

Nowadays, consumers are trying to consume low-fat cheese because dietary fats are the cause of obesity, blood pressure, CHD, atherosclerosis, etc. If the fat content of cheese is reduced, then moisture percentage rises, and at that time the protein shows an important characteristic in texture modification. Cheesemakers are trying to prepare low-fat cheese, but generally, low-fat cheeses are poorer in taste as compared with full-fat cheese, due to a hard and rubbery texture and decreased flavor. To enhance the texture and flavor of low-fat cheese, several methods individually or in combination have been examined by using enzymes, specially designed starter cultures, and fat replacers, etc. [59].

The composition of milk used to preparation of low-fat cheese is different as compared to full-fat cheeses. As the fat content of milk is decreased, the protein percentage in milk is increased slightly. Therefore, the milk for cheese making has lesser milk solids. During low-fat cheese preparation, the casein to fat ratio in milk will be increased. The reduced fat in cheese is replaced with moisture. However, it is a fact that the fat removed is not the same as moisture added in the cheese, therefore the yield of cheese slower in case of low-fat cheese. Rudan et al., reported that Mozzarella cheese yield was decreased by 30% if the fat of cheese is reduced from 25% to 5%. [65].

Cheddar cheese with low fat has reduced and imbalance flavor due to a decrease of fatty acids such as methyl ketones, hexanoic acids, and butanoic acid. The fat in full fat cheese absorbs the bitter hydrophobic compounds; therefore, the low-fat cheese has a higher intensity of bitterness as compared to low-fat cheese.

In full-fat cheese, milk fat generally delivers smoothness because fats are scattered evenly in the casein matrix. In the absence of fat, insufficient breakdown of casein occurs as a result comparatively firm texture. Yates and Drake reported that the low-fat cheese was weak and rubbery. Due to fat reduction, there was an increase in Calcium retention that imparts firmness to cheese [75].

Some functional properties (Shredability, melting, oiling-off, and appearance) are changed in low-fat cheese. During heating, the whiteness of low-fat Mozzarella cheese was increased, and after cooling, it was decreased [58]. During heating, casein proteolysis products are produced.

In general, the meltability of low-fat Mozzarella cheese is poor, and free oil formation at the time of melting is decreased, causing unwanted browning during pizza baking [18]. Meilinger et al., developed both low fat and fat-free cream cheeses which had similar smoothness as traditional Cream cheese [57].

3.9.1 NOVEL MANUFACTURING TECHNOLOGY FOR LOW FAT CHEESE

For the preparation of low-fat cheese, the desired fat content of milk should be in the range of 0.05–1.8%. Milk may be incorporated as skim milk powder or ultrafiltered retentate. Starter culture plays an important role because it has an impact on proteolysis to control texture and flavor development. To control lactic acid production and proteolytic activity, thermophilic starters and Mesophilic *lactococci* in low-fat Mozzarella and Cheddar cheese, respectively, may be used.

Stabilizers have an important role to retain moisture and are used along with different fat replacers. Fat replacers (such as gums, carrageenan, cellulose gels, gelatin, etc.) are used in low-fat cheese. Good quality low-fat Cottage cheese may be prepared by cream dressing with little or no fat but attaining similar viscosity by using carrageenan, hydrocolloids, or locust bean gum.

3.10 LOW SODIUM CHEESE

Due to health concerns, the cheese industry is trying to manufacture cheeses without sodium chloride or with its alternatives. It is a real challenge for cheesemakers to reduce NaCl content in cheese. Salt is an essential component of cheese because it retains desirable flavor, body, and texture, and storage life by controlling the actions of microorganisms and enzymes. Also, some consumers prefer saltiness in cheese. Cheddar cheese prepared by adding sodium chloride and potassium chloride (1:1) was at par with the control [42]. Low-sodium Feta cheese can be prepared from UF retentate with 1% NaCl and 1% KCl [5]. El-Bakry et al., tried to decrease the addition of a percentage of NaCl in imitation cheese; and they found a decrease in cheese hardness. The product was optimized with 50% NaCl reduction [15].

3.11 SUMMARY

In this chapter, different innovative techniques have been discussed to increase the health benefit of cheeses. For this, different functional ingredients may be added in cheeses. The health benefit of these functional ingredients, their sources, the effect of applications in different cheeses, and their optimum addition level in cheeses have been discussed. There are several cheeses, which contain a high amount of fat and salt, and for this reason, consumers avoid to consume those cheeses. Different technologies have been discussed to make low fat and low sodium cheeses.

KEYWORDS

- Alzheimer's disease
- antiatherosclerosis
- anticarcinogenic
- antihypertensive
- anti-inflammation
- apoptosis
- *bifidobacteria*
- bovine serum albumin
- casein
- epigallocatechin
- fructooligosaccharide
- glycomacropeptides
- Hesperetin
- homovanillic acid
- lactoferrin
- lactoperoxidase enzymes
- methylxanthines
- phytostanol
- phytosterols
- prebiotics
- stearidonic acid
- stigmasterol
- *theobroma cacao*
- whey protein
- α-lactalbumin
- α-linolenic acid
- β-glucan
- β-lactoglobulin
- β-sitosterol

REFERENCES

1. Abadía-García, L., Cardador, A., Del Campo, S. T. M., Arvízu, S. M., Castaño-Tostado, E., Regalado-González, C., García-Almendarez, B., & Amaya-Llano, S. L., (2013). Influence of probiotic strains added to cottage cheese on generation of potentially antioxidant peptides, anti-listerial activity, and survival of probiotic microorganisms in simulated gastrointestinal conditions. *International Dairy Journal, 33*, 191–197.
2. Abbe, M. M. J., & Amin, I., (2008). Polyphenols in cocoa and cocoa products: Is there a link between antioxidant properties and health? *Molecules, 13*, 2190–2219.
3. Abd El-Aziz, M., El-Senaity, M. H., Shazly, A. B., Mahran, G. A., & Fatouh, A. E., (2011). Impact of using buttermilk curd and whey protein concentrates on the quality of low-fat processed cheese spread. *Egyptian Journal of Dairy Science, 39,* 153–164.

4. Sienkiewicz-Szłapka, E., Jarmołowska, B., Krawczuk, S., Kostyra, E., Kostyra, H., & Iwan. M., (2009). Contents of agonistic and antagonistic opioid peptides in different cheese varieties. *International Dairy Journal, 19*, 258–263.
5. Aly, M. E., (1995). An attempt for producing low-sodium Feta-type cheese. *Food Chemistry, 52*, 295–299.
6. Araujo, E. A., De Carvalho, A. F., Leandro, E. S., Furtado, M. M., & De Moraes, C. A., (2010). Development of a symbiotic cottage cheese added with Lactobacillus delbrueckii UFV H2b20 and inulin. *Journal of Functional Foods, 2*, 85–89.
7. Beaudoin, M. S., & Graham, T. E., (2011). Methylxanthines and human health: Epidemiological and experimental evidence. In: Fredholm, B. B., (ed.), *Methylxanthines* (pp. 509–548). Springer, Berlin- Heidelberg.
8. Bergamini, C., Hynes, E., Quiberoni, A., Suárez, V., & Zalazar, C., (2005). Probiotic bacteria as adjunct starters: Influence of the addition methodology on their survival in a semi-hard Argentinean cheese. *Food Research International, 38*, 597–604.
9. Bisig, W., Eberhard, P., Collomb, M., & Rehberger, B., (2007). Influence of processing on the fatty acid composition and the content of conjugated linoleic acid in organic and conventional dairy products-a review. *Le Lait, 87*, 1–19.
10. Buriti, F. C. A., Rocha, J. S., Assis, E. G., & Saad, S. M. I., (2005). Probiotic potential of minas fresh cheese prepared with the addition of *Lactobacillus paracasei. LWT - Food Science and Technology, 38*, 173–180.
11. Buriti, F. C. A., Rocha, J. S., & Saad, S. M. I., (2005). Incorporation of *lactobacillus acidophilus* in minas fresh cheese and its implications for textural and sensorial properties during storage. *International Dairy Journal, 15*, 1279–1288.
12. Burns, P., Patrignani, F., Serrazanetti, D., Vinderola, G. C., Reinheimer, J. A., Lanciotti, R., & Guerzoni, M. E., (2008). Probiotic crescenza cheese containing *Lactobacillus casei* and *Lactobacillus acidophilus* manufactured with high-pressure homogenized milk. *Journal of Dairy Science, 91*, 500–512.
13. Cantrill, R., (2018). Phytosterols, phytostanols and their esters. *Chemical and Technical Assessment for 69th JECFA*, pp. 1–13, www.fao.org/fileadmin/templates/agns/pdf/jecfa/./Phytosterols.pdf (Assessed on July 20 2019).
14. Corbo, M. R., Albenzio, M., De Angelis, M., Sevi, A., & Gobbetti, M., (2001). Microbiological and biochemical properties of *Canestrato Pugliese* hard cheese supplemented with bifidobacteria. *Journal of Dairy Science, 84*, 551–561.
15. El-Bakry, M., Beninati, F., Duggan, E., O'Riordan, E. D., & O'Sullivan, M., (2011). Reducing salt in imitation cheese: Effects on manufacture and functional properties. *Food Research International, 44*, 589–596.
16. Elewa, A. H. N., Degheidi, M. A., Zehan, M. A., & Malin, M. A., (2009). Synergistic effects of inulin and cellulose in UF- prebiotic white cheese. *Egyptian Journal of Dairy Science, 37*, 85–100.
17. El-Nagar, G. F., & El-Aty, A. M. A., (2004). Improving the quality of Karish cheese by using fat replacers (inulin and pea fibers). *Annals of Agricultural Sciences. 42*, 103–115.
18. Fife, R. L., McMahon, D. J., & Oberg, C. J., (1996). Functionality of low fat Mozzarella cheese. *Journal of Dairy Science, 79*, 1903–1910.
19. Gahane, H. B., (2008). *Development of Quarg Type Cheese with Enhanced Functional Attributes from Buffalo Milk* (pp. 162–164). M.Tech. Thesis, NDRI, Karnal, Haryana, India.

20. Gijs, L., Piraprez, G., Perete, P., Spinnler, S., & Collin, S., (2000). Retention of sulphur flavor by food matrix and determination of sensorial data dependent of the medium composition. *Food Chemistry, 69*, 319–330.

21. Giri, A., & Kanawjia, S. K., (2018). Cost estimation of newly developed functional processed cheese spread–An engineering approach. *Indian Journal of Dairy Science, 71*, 196–203.

22. Giri, A., (2015). Functional dairy foods against life-style disorders. In: Irani, R., (ed.), *Nutritional Deprivation in the Midst of Plenty* (pp. 26–30). J. B. Books & Learnings, Berhampore, West Bengal.

23. Giri, A., & Kanawjia, S. K., (2014). Cost estimation of inulin incorporated functional processed cheese spread. *Indian Journal of Dairy Science, 67*, 179–186.

24. Giri, A., & Kanawjia, S. K., (2013). Estimation of production cost for omega-3 fatty acid incorporated processed cheese spread. *International Journal of Science and Research, 2*, 278–282.

25. Giri, A., & Kanawjia, S. K., (2013). Evaluation of cost for phytosterols added cheese spread. *Indian Journal of Dairy Science, 66*, 527–534.

26. Giri, A., Kanawjia, S. K., & Khetra, Y., (2014). Textural and melting properties of processed cheese spread as affected by incorporation of different inulin levels. *Food Bioprocess Technology, 7*, 1533–1540.

27. Giri, A., Kanawjia, S. K., Pothuraju, R., & Kapila, S., (2015). Effect of inulin incorporated processed cheese spread on lipid profile of blood serum and liver in rats. *Dairy Science and Technology, 95*, 135–149.

28. Giri, A., Kanawjia, S. K., & Rajoria, A., (2014). Effect of phytosterols on textural and melting characteristics of cheese spread. *Food Chemistry, 157*, 240–245.

29. Giri, A., Kanawjia, S. K., & Singh, M. P., (2017). Effect of inulin on physicochemical, sensory, fatty acid profile and microstructure of processed cheese spread. *Journal of Food Science and Technology, 54*, 2443–2451.

30. Giri, A., Rao, H. G. R., & Ramesh, V., (2013). Effect of incorporating whey protein concentrate into stevia-sweetened Kulfi on physicochemical and sensory properties. *International Journal of Dairy Technology, 66*, 286–290.

31. Gobbetti, M., Corsetti, A., Smacchi, E., Zocchetti, A., & De Angelis, M., (1998). Production of Crescenza cheese by incorporation of bifidobacteria. *Journal of Dairy Science, 81*, 37–47.

32. Gomes, A. M. P., & Malcata, F. X., (1998). Development of probiotic cheese manufactured from goat milk: Response surface analysis via technological manipulation. *Journal of Dairy Science, 81*, 1492–1507.

33. Gomes, A. M. P., Malcata, F. X., Klaver, F. A. M., & Grande, H. J., (1995). Incorporation and survival of *Bifidobacterium spp.* strain Bo and *Lactobacillus spp.* strain Ki in a Cheese product. *Netherlands Milk and Dairy Journal, 49*, 71–75.

34. Gupta, V. K., & Reuter, H., (1992). Processed cheese foods with added whey protein concentrates. *Le Lait, 72*, 201–212.

35. Han, J., Britten, M., St-Gelais, D., Champagne, C. P., Fustier, P., Salmieri, S., & Lacroix, M., (2011). Polyphenolic compounds as functional ingredients in cheese. *Food Chemistry, 124*, 1589–1594.

36. Han, X., Shen, T., & Lou, H., (2007). Dietary polyphenols and their biological significance. *International Journal of Molecular Sciences, 8*, 950–988.

37. He, K., Song, Y., Daviglus, M. L., Liu, K., Horn, L. V., Dyer, A. R., & Greenland, P., (2004). Accumulated evidence on fish consumption and coronary heart disease mortality: A meta-analysis of cohort studies. *Circulation, 109*, 2705–2711.

38. Higurashi, S., Kunieda, Y., Matsuyama, H., & Kawakami, H., (2007). Effect of cheese consumption on the accumulation of abdominal adipose and decrease in serum adiponectin levels in rats fed a calorie dense diet. *International Dairy Journal, 17*, 1224–1231.

39. Hjerpsted, J., Leedo, E., & Tholstrup, T., (2011). Cheese intake in large amounts lowers LDL-cholesterol concentrations compared with butter intake of equal fat content. *The American Journal of Clinical Nutrition, 94*, 1479–1484.

40. Innis, S. M., (2007). Dietary (n-3) fatty acids and brain development. *Journal of Nutrition, 137*, 855–859.

41. Jarmołowska, B., Szłapka-Sienkiewicz, E., Kostyra, E., Kostyra, H., & Mierzejewska, D., (2007). Opioid activity of Humana formula for newborns. *Journal of the Science of Food and Agriculture, 87*, 2247–2250.

42. Johnson, M. E., Kapoor, R., McMahon, D. J., McCoy, D. R., & Narasimmon, R. G., (2009). Reduction of sodium and fat levels in natural and processed cheeses: Scientific and technological aspects. *Comprehensive Reviews in Food Science and Food Safety, 8*, 252–268.

43. Kanawjia, S. K., & Makhal, S., (2008). *Designing Novel Health Foods Using Phytosterols and Stanols* (pp. 57–64). Souvenir XXXVI Dairy Industry Conference, BHU (UP), Varanasi–UP.

44. Karimi, R., Mortazavian, A. M., & Da Cruz, A. G., (2011). Viability of probiotic microorganisms in cheese during production and storage: A review. *Dairy Science & Technology, 91*, 283–308.

45. Koca, N., & Metin, M., (2004). Textural, melting and sensory properties of low-fat fresh kashar cheeses produced by using fat replacers. *International Dairy Journal, 14*, 365–373.

46. Kolanowski, W., & Weibbrodt, J., (2007). Sensory quality of dairy products fortified with fish oil. *International Dairy Journal, 17*, 1248–1253.

47. Kumar, A., (2012). *Development of Sweetened Functional Soft Cheese* (p. 104). M.Tech. Thesis, NDRI, Karnal, Haryana, India.

48. Kumar, R., (2012). *Enrichment of Processed Cheese Food with Functional Ingredients* (p. 109). M.Tech. Thesis, NDRI, Karnal, Haryana, India.

49. Laakso, P., (2005). Analysis of sterols from various food matrices. *European Journal of Lipid Science and Technology, 107*, 402–410.

50. Lobato-Calleros, C., Reyes-Hernández, J., Beristain, C. I., Hornelas-Uribe. Y., Sánchez-García, J. E., & Vernon-Carter, E. J., (2007). Microstructure and texture of white fresh cheese made with canola oil and whey protein concentrate in partial or total replacement of milk fat. *Food Research International, 40*, 529–537.

51. Logan, A. C., (2003). Neurobehavioral aspects of omega-3 fatty acids: Possible mechanisms and therapeutic value in major depression. *Alternative Medicine Review, 8*, 410–425.

52. López-Expósito, I., Quirós, A., Amigo, L., & Recio, I., (2007). Casein hydrolysates as a source of antimicrobial, antioxidant and antihypertensive peptides. *Le Lait, 87*, 241–249.

53. Maconi, G., Barbara, G., Bosetti, C., Cuomo, R., & Annibale, B., (2011). Treatment of diverticular disease of the colon and prevention of acute diverticulitis: A systematic review. *Diseases of the Colon & Rectum, 54*, 1326–1338.

54. Manach, C., Scalbert, A., Morand, C., Remesy, C., & Jimenez, L., (2004). Polyphenols: Food sources and bioavailability. *American Journal of Clinical Nutrition, 79*, 727–747.

55. Martini, S., Thurgood, J. E., Brothersen, C., Ward, R., & McMahon, D. J., (2009). Fortification of reduced-fat Cheddar cheese with n-3 fatty acids: Effect on off-flavor generation. *Journal of Dairy Science, 92*, 1876–1884.

56. Mazumder, A., Prabuthas, P., Giri, A., & Mishra, H. N., (2014). Major food grade pigments from microalgae and its health benefits–A review. *Indian Food Industry, 33*, 19–30.

57. Meilinger, J. H., Brown, C. G., & Bohanan, M. A., (1995). Process for the manufacture of a fat-free Cream cheese product. *United States Patent, 5470593*.

58. Metzger, L. E., Barbano, D. M., Rudan, M. A., Kindstedt, P. S., & Guo, M. R., (2000). Whiteness change during heating and cooling of Mozzarella cheese. *Journal of Dairy Science, 83*, 1–10.

59. Mistry, V. V., (2001). Low fat cheese technology. *International Dairy Journal, 11*, 413–422.

60. Moreau, R., Whitaker, B. D., & Hicks, K. B., (2002). Phytosterols, phytostanols and their conjugates in foods: Structural diversity, quantitative analysis, and health-promoting uses. *Progress in Lipid Research, 41*, 457–500.

61. Oliveira, M. E. G. D., Garcia, E. F., Queiroga, R. D. C. R. D., & Souza, E. L. D., (2012). Technological, physicochemical and sensory characteristics of a Brazilian semi-hard goat cheese (coalho) with added probiotic lactic acid bacteria. *Scientia Agricola, 69*, 370–379.

62. Ostlund, Jr. R. E., (2002). Phytosterols in human nutrition. *Annual Review of Nutrition, 22*, 533–549.

63. Phillips, M., Kailasapathy, K., & Tran, L., (2006). Viability of commercial probiotic cultures (*L. acidophilus, Bifidobacterium*sp., *L. casei, L. paracasei* and *L. rhamnosus*) in cheddar cheese. *International Journal of Food Microbiology, 108*, 276–280.

64. Pinto, S., Rathour, A. K., Prajapati, J. P., Jana, A. H., & Solanky, M. J., (2007). Utilization of whey protein concentrate in processed cheese spread. *Natural Product Radiance, 6*, 398–401.

65. Rudan, M. A., Barbano, D. M., Yun, J. J., & Kindstedt, P. S., (1999). Effect of fat reduction on chemical composition, proteolysis, functionality, and yield of Mozzarella cheese. *Journal of Dairy Science, 82*, 661–672.

66. Salo, P., Wester, I., & Hopia, A., (2002). Phytosterols. In: Gunstone, F. D., (ed.), *Lipids for Functional Foods and Nutraceuticals* (pp. 183–224). The Oily Press, Bridgwater.

67. Satija, A., & Hu, F. B., (2012). Cardiovascular benefits of dietary fiber. *Current Atherosclerosis Reports, 14*, 505–514.

68. Shashikant, K. K., (2009). *Development of Functional Quarg Cheese with Extended Shelf Life* (pp. 3–9). M. Tech. Thesis, NDRI, Karnal, Haryana, India.

69. Shetty, S., Hegde, M. N., & Bopanna, T. P., (2014). Enamel remineralization assessment after treatment with three different remineralizing agents using surface microhardness: An *in vitro* study. *Journal of Conservative Dentistry, 17*, 49–52.

70. Siddiqi, H., (2007). Alzheimer's disease and omega-3 fatty acid consumption: A review. *Harvard Brain, 14*, 20–21.

71. Surette, M. E., (2008). The science behind dietary omega-3 fatty acids. *Canadian Medical Association Journal, 178,* 177–180.

72. Trautwein, E. A., Duchateau, G. S., Lin, Y., Mel'nikov, S. M., Molhuizen, H. O. F., & Ntanios, F. Y., (2003). Proposed mechanisms of cholesterol-lowering action of plant sterols. *European Journal of Lipid Science and Technology, 105,* 171–185.

73. Vinderola, C. G., Prosello, W., Ghiberto, D., & Reinheimer, J. A., (2000). Viability of probiotic (*Bifidobacterium, Lactobacillus acidophilus* and *Lactobacillus casei*) and nonprobiotic microflora in Argentinian Fresco cheese. *Journal of Dairy Science, 83,* 1905–1911.

74. Wood, P. J., (2007). Cereal β-glucans in diet and health. *Journal of Cereal Science, 46,* 230–238.

75. Yates, M. D., & Drake, M. A., (2007). Texture properties of Gouda cheese. *Journal of Sensory Studies, 22,* 493–506.

76. Ye, A., Cui, J., Taneja, A., Zhu, X., & Singh, H., (2009). Evaluation of processed cheese fortified with fish oil emulsion. *Food Research International, 42,* 1093–1098.

PART II
Novel Technologies in
Processing of Milk Products

CHAPTER 4

EMERGING TRENDS IN CONCENTRATION AND DRYING OF MILK AND MILK PRODUCTS

SUNIL KUMAR KHATKAR, JASHANDEEP KAUR, ANJU B. KHATKAR, NARENDER KUMAR CHANDLA, ANIL KUMAR PANGHAL, and GURLAL SINGH GILL

ABSTRACT

Milk is a complete source of nutrition, so in order to make it fully utilized, it needs to be preserved. Therefore, concentration is a way of preserving milk and milk products along with providing convenience for storage and transportation. Membrane technologies have a vital role in the concentration of dairy products. Drying is an inevitable process in the milk processing industry.

4.1 INTRODUCTION

Milk and milk products are an integral part of the food industry and have a high-flying position in global consumer goods. The changes in the demand patterns of the consumers greatly affect the milk production and the technologies related to it. Nowadays, high quality and convenience of milk products are considered as a priority for the development of better processing methods of production. Condensing and drying are both indispensable parts of the processing of milk products.

Milk processors convert raw milk to different products, using developed processing technologies, by value addition, thus, making it convenient for consumers to use it and store it. Traditional milk products have been improved to introduce newer ones, thus expanding the market for dairy products. Milk concentration has been one of the methods of preserving milk for a longer period.

The milk concentration reduces the volume and bulk and to preserve the milk has been known since decades. Milk is concentrated up to its flow limits (40–50%) by reducing the water content. Generally, this is done by evaporation of water using different evaporators, during which there is a reduction in mass and volume, and there is an increase in density and viscosity. It is done in order to preserve milk for a longer period as well as for its utilization into other products. Also, concentrated milk is easier to store or pack than the raw milk, whereas these two are different as condensed milk, which contains added sugar. Some people do confuse with concentrated, evaporated, and condensed milk, but these three are differing on the basis of different aspects:

- **In General Concentration of Milk:** It is defined as the partial removal of water employing evaporation phenomenon from whole milk, partially skim milk or fat-free milk. It contains not less than 7.5% milkfat and 25.5% (m/m) total milk solids. The product could be pasteurized, but as such, no heat treatment is given to prevent the spoilage.
- **The Evaporated Milk:** (Also known as unsweetened condensed milk) is a viscous product retained after partial evaporation of water from milk. The final product contains not less than 8% milkfat (m/m), and 26% total milk solids. The optional additives (equal to or less than 0.3% of the final product) like sodium citrate, calcium chloride, citric acid and sodium salts of orthophosphoric acid and polyphosphoric acid may be added. Evaporated milk is homogenized, and is heat-treated before or after filling in the sealed containers to preserve the product. The ratio of the concentration of milk solids is about 1:2.5.
- **Sweetened Condensed Milk:** It is prepared by partial removal of water from a mixture of milk and adding nutritive carbohydrate sweeteners, such as sugar. The product shall contain fat not less than 9% (m/m) and total milk solids not less than 31% (m/m). In order to prevent spoilage, sugar should be added in sufficient quantity (40% minimum of the final product), and it should not be too much to cause sugar crystallization. Sweetened condensed milk is pasteurized and may be homogenized. Sodium citrate, citric acid, calcium chloride, and sodium salts of orthophosphoric acid and polyphosphoric acid may also be added but within permissible limits (not exceeding 0.3% by weight of the final product). The ratio of the concentration of milk solids is about 1:3.

This chapter discusses: why there is a need to concentrate milk and milk products; the method for the production of sweetened condensed milk;

the role of different membrane technologies in the concentration of milk and milk products; the new types of emerging drying methods and dryers. Therefore, this chapter is of fundamental importance to get a glimpse of the emerging trends in the concentration and drying of milk products.

4.2 PREPARATION OF CONDENSED MILK

Sweetened condensed milk (Figure 4.1) has a significant role in confectionary, ice cream, frozen desserts, the chocolate industry, as well as for other dairy products. The method of preparation of condensed milk is presented in this section

4.2.1 RAW MILK PREPARATION

Raw milk of good quality is filtered and clarified before being standardized to specified fat content and SNF levels, as required for the condensed milk. Milk or concentrate is standardized by three-stage processes:

- To have the desired fat: SNF ratio;
- Sugar: Milk solids (before or during vacuum concentration); and
- Adjustment of total solids content (end of concentration and sugar incorporation).

4.2.2 PREHEATING/WARMING

The aims of this step is threefold: (1) first is for killing of pathogens, yeasts, and molds and thus for reduction of microbial load of the milk; (2) second is inactivation of certain enzymes, which if left activated may spoil the product; and (3) third is for controlling the age thickening as well as viscosity of milk. There are basically two types of heat treatment: direct heating (steam injection) and indirect heating.

Direct heating involves the HTST treatment, for example, 115–118°C for a few seconds or no hold or UHT heating for up to 5seconds at 130–140°C. The time-temperature relationship is selected on the basis of different factors: the tendency of milk for age thickening, the ultimate use of the product (like packaged in bulk or individual packs). Normally, a heat treatment of 76–79°C for 10–15 minutes is provided as it warrants adequate resistance to thickening during the aging process, whereas preheating at 82–100°C for

5–15 minutes supposed to promote age thickening. Due to this, generally heating through indirect mode is preferred over the direct mode like a steam injection. These two heat treatments could also be applied in combination.

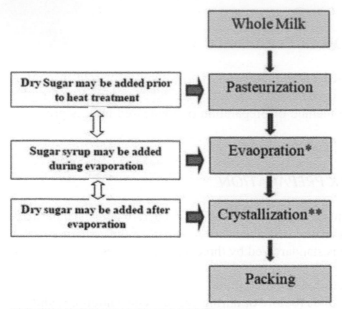

* Water is evaporated to give desired solid content
** To ensure excess sugar forms lactose crystals, fine seed crystals are added during cooling
** Crystallization at 30 degree C, Quantity of fine seeds: 375-500gms/ 1000 kg and Size of seed <200 microns

FIGURE 4.1   (See color insert.) Traditional process for the manufacture of sweetened condensed milk.

4.2.3   EVAPORATION AND ADDITION OF SUGAR

Sugar can be added to the milk before or after preheating, or it could be added near the end of the condensing process. It should be added after filtration and clarification. The stage, at which sugar is added as well as temperature during addition, plays a vital role in indicating the keeping quality of the product. The amount of sugar added should be sufficient to prevent the microbial spoilage, and it should not be too high to avoid sugar crystallization. Therefore, normally 43–45% (of finished product) sugar is preferred. The stage, at which sugar is added, is also important because if sugar is added before condensing, then it may result in a concentrate with higher viscosity, which will possibly hinder in further processing. It causes defects like age thickening, Maillard reaction and heat resistance in microbes, etc.

The pasteurized milk is fed into the vacuum pan or evaporator conveniently. Vacuum evaporators are used, when the product is needed to be prepared under higher sanitation and hygiene conditions, whereas an energy-efficient multiple affect evaporator is widely used when product is manufactured in bulk.

At the end of the evaporation process, the product is checked for desired consistency and should be analyzed for total solids.

4.2.4 HOMOGENIZATION

Due to high viscosity, fat separation or creaming is not prevalent, so homogenization is generally not required here. However, in which UHT treatment is given during preheating, the resulted product has a higher viscosity, which may lead to fat separation during storage. Therefore, to prevent this fat separation during storage, such product is homogenized at low pressure (2500 psi at 50 to 55°C).

4.2.5 COOLING AND CRYSTALLIZATION OF LACTOSE

Generally, the product's temperature is around 45°C, when there exits the vacuum pan or after the last effect of a multiple-effect evaporator, which corresponds to a vacuum operation at 63.5 cm Hg (16.6 kPa). Depending on how to vacuum cooling is carried out (either in the same pan or separate ones (provided with scraper agitator)); the temperature is lowered down to 32°C at the time of seeding. Seeding is done by blowing lactose dust through the sides of cooler. Further cooling of sweetened condensed milk is crucial processing parameter to control the texture and mouthfeel of the final product. The process is called "mass crystallization" of lactose. The excess amount of lactose always tends to crystallize out.

4.2.6 PACKAGING

After cooling, the sweetened condensed milk is filled in pre-sterilized tin cans (lacquered). In order to prevent mold growth during storage, a minimum amount of head-space is necessary. The can-filling environment should be aseptic (free from microorganisms), as aerial contamination may take place. Nowadays, there are different options available for the packaging of condensed milk, such as collapsible plastic tubes are used for retail packaging, and for bulk packaging plastic drums are generally utilized. Strict hygienic practices are required to protect the product against bacterial contamination. The packaged product should be stored at 10°C or less (RH should be below 50%).

4.3   APPLICATIONS OF MEMBRANE FILTRATION IN CONCENTRATION OF MILK

Membrane processes are used for the concentration or fractionation of liquid to yield two liquids of different compositions. The process of separation is entirely based on the permeation of selective components of the liquid constituents through a membrane. The mechanisms governing mass diffusion in different membrane processes vary as a function of the equipment configuration, membrane type, and as per processing conditions. The membranes, which play very important roles in the dairy industry, are classified into four major categories depending upon their pore size [28]:

- Ultrafiltration (UF: 0.002–0.1microns);
- Microfiltration (MF: 0.03–10microns);
- Nanofiltration (NF: 0.001 microns); and
- Reverse osmosis (RO: 5–30 nm).

Currently, some of the promising current applications of membrane technologies in dairy processing industry [28]: aim to increase the protein content of whey protein concentrates (WPCs); to enhance the manifestation of specific functional properties of whey proteins; and to fractionate whey proteins.

Membranes play a vital role in modern dairy processing, to increase the concentration of the milk or selected components of milk and clarification of the milk, or dairy by-products, etc. RO finds its application in the removal of potable drinking water from liquid milk, in order to reduce its volume thus, reduces the transportation costs. MF plants that are located close to large dairy farms can condense milk up to 45% of total solids. With RO, milk can be condensed/ concentrated by removal of about 70% of its water and sometimes with retaining all other components of milk without undergoing any thermal processing. Approximately, milk contains 87% water and

13% solids. RO reduces the volume; thus, the transportation costs. The technology of milk concentration works by removal of water from the cooled milk using very small pores of RO membranes.

4.3.1 REVERSE OSMOSIS (RO)

The RO membrane is permeable to solvents but not to the larger molecules in solution. The osmotic pressure of the solution is directly related to the temperature and solute concentration, whereas there is an inverse relationship with the molecular weight of the solute. Therefore, because of this low molecular weight, components make a greater contribution to osmotic pressure than large ones. That is why, the osmotic pressure in milk results from salts and lactose rather than proteins present [10].

The RO has low energy requirements as compared to other methods of water evaporation (Figure 4.2). It is an energy-efficient way of concentrating milk at ambient temperatures, thus avoiding the deterioration of milk due to heat treatment. Milk concentrates produced by this method have better nutritional and functional properties.

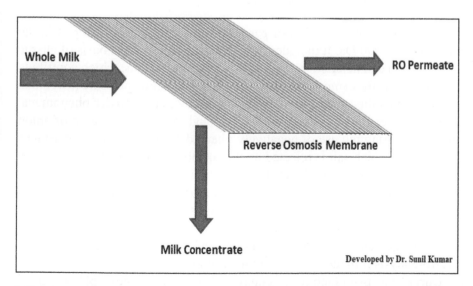

FIGURE 4.2 (See color insert.) Principle of reverse osmosis.

The main application of RO in the dairy industry is to concentrate whey or permeate to reduce transportation cost. In RO, the structure of the membrane is a key to effective fractionation and concentration. The active layer of these membranes is very thin (approx. 5–30 nm) with a pore size of 5–20°A. The most logical and probable mechanism for the separation process by RO is "preferential sorption–capillary flow." It is possible to concentrate milk to the total solid content of 24–30%. The concentration of milk by RO produces a concentrate with minimum cooked flavor. It finds its application in the manufacture of ice cream in place of milk powder, thus eliminating the powdery taste and giving a better texture. However, there are some limitations related to this RO process of concentration:

- First, there is a decrease in flux during the concentration process, which may be due to various reasons, such as concentration polarization, i.e., accumulation of materials at the surface of membrane, there may be a reduction in driving force due to the increase in concentration of the feed and membrane compaction at high pressures (HPs) of operation.
- Second is the damage to the milk fat at higher pressure.

4.3.2 NANOFILTRATION (NF)

NF undertakes the separation of particles having molecular weights in the range of 300-1000 Da. It includes the removal of ions on the basis of charge and diffusion characteristics. It is capable of rejecting ions, which contribute significantly to the osmotic pressure and allows operation pressure lower than RO. The separation through NF is controlled by mass transfer phenomena, which involves the diffusion and flow through pores. The degree of anion repulsion, during the use of negatively charged NF membranes, determines the extent to which salt is rejected by the membrane. The extent of rejection is directly proportional to the charge of the anion [28].

NF finds its application in concentration and partial demineralization of the whey. During the filtration process, the majority of monovalent ions, organic acids, and a portion of lactose will recover as permeate due to membrane selectivity. NF is a method of choice for ion exchange and electro-dialysis if moderate demineralization is required [20].

During the manufacture of cottage cheese, acid whey (pH 4.5–5) is produced in significant amounts, whereas during the production of cheddar cheese at stages after the application of salt to the curd salt-containing whey is generated. Both of these whey streams represent difficulties in relation to the functionality of their whey proteins as well as to their adverse effects on the environment. NF is used for the partial removal of acid from acid whey, removal of salt from salty whey, and the partial demineralization of sweet whey for production of lactose or demineralized whey [12]. About 35% reduction in ash content is the maximum amount of demineralization by NF. Demineralization can be increased up to 45% by applying a diafiltration (DF) step. As compared to other processes of membrane filtration, this is a very simple process, and concentration and partial demineralization of whey UF permeates takes place prior to the production of lactose and lactose derivatives with conversion of salt whey to normal whey while solving a disposal problem, thus treating cheese brine solutions for reuse [20].

Whey comprises of wastewater that shows a very high COD value (ranged from 40,000 to 60,000 mg O_2/l), which is due to the presence of compounds such as proteins, lactose, fat, vitamins, and mineral salts. Thus without a treatment, it cannot be drained. Bioactive substances that are present in whey (lactose and its derivatives) have many applications in pharmaceutical and food processing industries (diet foods: cheeses, drinks, soups, and infant formulations) [8]. Therefore, NF is utilized for partial demineralization and concentration of whey, whereas UF permeates in the recovery of lactose.

4.3.3 ULTRAFILTRATION (UF)

Over the last two decades, UF, a pressure-driven membrane process, has been used in various industrial applications. The dairy industry has been the major adopter of this technology. In UF, the pressure gradient is the driving force, and the mass transfer is dominated by the convective flux through pores [32]. UF consists of membranes with a molecular weight cutoff in the range of I-200 kDa and a pore size of 0.002–0.01 microns, and is performed at a pressure of 4000 kPa.

UF concentrates and separates molecules in solution based on differences in their molecular weight in conjunction with membrane molecular weight cut-off and other operation parameters. With the use of UF, the concentration and separation of whey proteins are possible in their un-denatured form; therefore, the recovered whey exhibits good functional property as compared to traditional sweet cheese whey [22]. Native WPCs and isolates are produced by UF by the concentration of native whey [23], showing excellent reconstituability, foaming, and gelling properties on drying [25]. With the application of membrane technology, it is possible to concentrate and fractionate individual whey protein, and also it facilitates the production of WPC with the enrichment of specific proteins or single proteins (Figure 4.3).

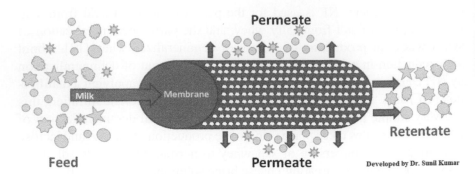

Principle of Membrane Filtration

FIGURE 4.3   **(See color insert.)** Principle of microfiltration.

4.3.4   MICROFILTRATION (MF)

MF processes are generally designed for the separation of particles in the range of 0.1–10 micrometers. It retains microorganisms, fat globules, and somatic cells, whereas it allows the passage of protein molecules along with minerals and lactose. It finds applications in clarification and purification of caseins derived from skim milk, in combination with other membranes. Therefore, MF is generally used for clarification before concentration (Figure 4.3).

4.4   ADVANCES IN DRYING OF DAIRY PRODUCTS

Drying is a fundamental process in food industries. The need to produce better quality products is a driving force in the development of efficient and improved drying technologies. Increase in the quality will ultimately increase their market value. Development of new drying technologies for improvement in the product quality is in line with the newer trends of quality enhancement, with a lesser impact on the environment [7].

Conventional drying methods such as hot air oven could lead to the deterioration of the product [24]. It may lead to changes in the flavor, color, and nutritional content of the product. The method of drying to be employed for a product depends on many other factors apart from the food characteristics. The newer drying methods are discussed in the following subsections.

4.4.1 HIGH-POWER ULTRASOUND DRYING (HP-USD)

HP-USD is very effective at low temperatures; thus, the deterioration of the product is probably reduced [21]. Thus, it represents a better option for heat sensitive foods. This method is based on the principle of acoustic cavitation. The potential of this technology relies on its ability to accelerate the mass transfer process in solid–liquid and solid–gas systems. HP-USD is utilized as a means of food dehydration without altering the quality characteristics of the product. It can be applied alone or in combination with other types of energy systems, such as hot-air. In combination, it reduces the treatment time and process temperature. When a high-intensity ultrasound wave (above 20 kHz) is coupled with the material to be dried, it produces a rapid series of alternative compressions and expansions.

In the drying process using ultrasound waves, moisture is removed due to the "vibration effect," "heating effect," and "synergistic effect" [15]. The forces involved in this process are generally higher than the surface tension, which therefore maintain the moisture inside the capillaries and thus creating microscopic channels for easy moisture removal. The cavitation process can also be utilized for the elimination of most attached water molecules; thus, it is possible to generate a product with very low moisture. It has been utilized in dairy processing [3, 30, 38].

4.4.2 MICROWAVE-RELATED DRYING (MW-D)

Reduced processing time, volumetric heating, and low energy consumption make this method attractive thermal energy source. The dielectric heating generates rapid energy coupling in the moisture of the food and leads to fast heating, and thus drying of the product [10]. It has been recommended that microwaves could be used in combination with conventional methods such as hot air, as microwaves alone cannot complete the drying [7]. As the microwave energy is absorbed by the material, a temperature gradient is created when the center of the food is at a higher temperature; thus forces the moisture out of the food [9]. Therefore, drying in the microwave that takes place at a falling rate principle [31]. This type of drying could be microwave convective or microwave vacuum drying.

It is suggested that MW-D of yogurt at low temperature is a useful alternative to freeze-drying and spray drying in terms of survival of starters and cost [13]. It has already been put into use in the dairy industry for yogurt drying [18].

4.4.2.1   MICROWAVE DRYING WITH CONVECTION

Heat is generated within the product due to molecular excitation, when the material couples with ultrasound energy. Water removal can take place effectively if the air is passed over the surface of the material to be dried. Therefore, this type of microwave drying is known as microwave convective drying due to air circulation [6].

4.4.2.2   MICROWAVE DRYING IN VACUUM

Integration of microwave with vacuum enables a faster moisture removal, and the combination is called as microwave vacuum drying. It has been proved very effective in drying of concentrated yogurt powder [14].

4.4.3   MULTI-STAGE SPRAY DRYER (MSS-D)

The complete dehydration process for the preparation of dairy powders is generally carried out using MSS-D [36]. Removal of internal moisture content is always tedious than removing the surface moisture; thus, it needs a larger dwell time and a larger dryer. Dryers such as fluid bed, spray dryer, flash, etc. are well-suited for the removal of surface moisture, whereas a longer residence time can be employed through air circulation in these dryers. Fluid bed or vibrated bed dryers have much longer dwell-times as compared to the dryers such as flash or spray dryers with shorter residence time. Thus in order to reduce the drying costs, spray drying could be followed with a fluid bed or vibratory bed dryer. This type of layout provides sufficient time for drying and without the use of high temperature. Therefore, it is an efficient process for the drying of coffee, skim milk, etc. [19]. The application of MSS-D in the dairy industry has been documented by many researchers [36, 37].

4.4.4   FOAM-MAT DRYING (FM-D)

Drying time can be reduced if liquid and semi-liquid foods are converted to foam, and it has been pointed out that the drying of foamed materials is faster than that of non-foamed ones [1]. Currently, researches are familiar with hot air drying of foamed materials but also for conventional drying or microwave drying [17]. Increased area of interface of foamed materials has been indicated as a factor responsible for the reduction in drying time, besides the increased transport of liquid water to evaporation front. A load of this type of dryer is also low as the density of the foam is lesser than the non-foamed ones (0.3–0.6 g/cm³). Therefore, there is a reduced dryer load as well as shorter drying time, and there is an increase in dryer throughput [27]. Yogurt powder has been successfully produced using FM-D [16, 33].

4.4.5 HYBRID DRYER (HY-D)

In this type of dryer, generally spray dryer is combined with a fluidized bed dryer, and the unit is combined with vibrated fluid bed dryer or cooler [19]. HY-D techniques are used commonly, as the combination of techniques have benefits of individual processes, e.g., the addition of a microwave system to sprouted bean system offers benefits of both the processes. Here the drying time is reduced by the microwaves whereas the drying uniformity is improved by sprouting system fluidization. There are unlimited combinations available that are being continuously developed with improved technology. The combination of a conventional drying process followed by microwave heating or microwave vacuum process has been proved very effective [31]. Use of HY-Din drying of dairy products has been documented by many scientists [29].

4.4.6 SPIN FLASH DRYER (SF-D)

This type of dryer has its application for the drying of high viscosity fluids and cohesive pastes to produce powders on a continuous basis. This dryer can be described as a fluid bed dryer with agitation. It can be effectively used for heat-sensitive foods due to: the dry powder is not re-introduced into the hot air zone, as soon as it becomes lighter, it is carried away; and the fluid bed mainly contains moist powder, which sweeps the bottom, thus keeping it at low temperature than the air temperature at the outlet [4]. SF-D has been successfully used to produce lactose powder [6].

4.4.7 FILTERMAT SPRAY DRYING (FMS-D)

It is a recommended system for the drying of high fat, fermented, sugar-based, as well as hydrolyzed products [26]. FMS-D is the combination of two drying technologies: spray drying and continuous flow belt drying [2]. Firstly the higher moisture feed is atomized into spray dryer, so that the water gets evaporated due to higher inlet temperatures. There is an advantage of shorter time and low material temperature; thus the product produced is of better quality; Secondly, the belt dryers provide continuous long time exposure to lower temperatures needed to complete the drying [17]. Use of FMS-D in the dairy industry has been documented by Woo [36].

4.4.8   VACUUM BELT DRYER (VB-D)

In this dryer, food slurry is distributed over the steel belt, which is then passed over two hollow drums in a vacuum chamber (1–70 Torr). Firstly the food is dried by steam-heated drums and then by steam-heated coils over the bands. The dried food is cooled by the cooling drum and removed using a doctor blade. These dryers are generally used to produce puffed dried products. There is partial drying of the foods, and then foods are sealed into the pressure chambers. The temperature and pressure are increased in the chamber and then are released instantly. The sudden release in the pressure expands the food, and there is the development of the porous structure. The nutritional and sensory characteristics of the food products are better.

4.5   SUMMARY

This chapter discusses: why there is a need to concentrate milk and milk products; the method for the production of sweetened condensed milk; the role of different membrane technologies in the concentration of milk and milk products; the new types of emerging drying methods and dryers. Therefore, this chapter is of fundamental importance to get a glimpse of the emerging trends in the concentration and drying of milk products.

KEYWORDS

- **crystallization**
- **filter-mat dryer**
- **foam-mat dryer**

- **microfiltration**
- **nanofiltration**
- **spin flash dryer**
- **ultrafiltration**
- **vacuum belt dryer**
- **value addition**

REFERENCES

1. Akintoye, O. A., & Oguntunde, A. O., (1991). Preliminary investigation on the effect of foam stabilizers on the physical characteristics and reconstitution properties of foam-mat dried soymilk. *Drying Technology, 9*, 245–262.
2. Anonymous, (2016). *Filtermat Spray Drying System* (p. 10). Technical information flyer. Fond Du Lac., Madison, WI–USA.
3. Ashokkumar, M., Bhaskaracharya, R., Kentish, S., Lee, J., Palmer, M., & Zisu, B., (2010). The ultrasonic processing of dairy products—an overview. *Dairy Science & Technology, 90*, 147–155.
4. AVP, (2000). Spin flash driers. In: *AVP Dryer Handbook, E-book.* https://userpages.umbc.edu/~dfrey1/ench445/apv_dryer.pdf (Accessed on July 20, 2019).
5. Carpin, M., Bertelsen, H., Dalberg, A., Roiland, C., Risbo, J., Schuck, P., & Jeantet, R., (2017). Impurities enhance caking in lactose powder. *Journal of Food Engineering, 198*, 91–97.
6. Changrue, V., Sunjka, P. S., Gariepy, Y., Raghavan, G. S. V., & Wang, N., (2004). Drying. *Proceedings of the 14th International Drying Symposium (IDS 2004)* (pp. 941–948). São Paulo, Brazil.
7. Chou, S. K., & Chua, K. J., (2001). New hybrid food technologies for heat sensitive foodstuffs. *Trends in Food Science & Technology, 12*, 359–369.
8. Cuartas-Uribe, B., Vincent-Vela, M. C., Álvarez-Blanco, S., Alcaina-Miranda, M. I., & Soriano-Costa, E., (2010). Application of nanofiltration models for the prediction of lactose retention using three modes of operation. *Journal of Food Engineering, 99*, 373–376.
9. Erle, U., (2005). *Drying Using Microwave Processing* (pp. 142–152). Woodhead Publ., Cambridge, England.
10. Feng, H., Yin, Y., & Tang, J., (2012). Microwave drying of food and agricultural materials: Basics and heat and mass transfer modeling. *Food Engineering Review, 4*, 89–106.
11. Gazzar, F. E., & Marth, E. H., (1991). Ultrafiltration and reverse osmosis in dairy technology: A review. *Journal of Food Protection, 54*, 801–809.
12. Kelly, P. M., Horton, B. S., & Burling, H., (1992). *New Applications of Membrane Processes* (pp. 130–140). International Dairy Federation, Brussels, Belgium.
13. Kim, S. S., Shin, S. G., Chang, K. S., & Bhowmik, S. R., (1997). Survival of lactic acid bacteria during microwave vacuum-drying of plain yogurt. *LWT–Food Science and Technology, 30*, 573–577.

14. Kim, S. S., & Bhowmik, R. S., (1994). Moisture sorption isotherms of concentrated yogurt and microwave vacuum dried yogurt powder. *Journal of Food Engineering, 21,* 157–175.
15. Kowalski, S. J., & Mierzwa, D., (2015). Ultrasonic assisted convective drying of biological materials. *Drying Technology, 33,* 1601–1613.
16. Krasaekoopt, W., & Bhatia, S., (2012). Production of yogurt powder using foam-mat drying. *AU J. T., 15*(3), 166–171.
17. Kudra, T., & Ratti, C., (2008). Process and energy optimization in drying of foamed materials. *Вестник ТГТУ, 2,* 14–18.
18. Kumar, Y., (2015). Application of microwave in food drying. *International Journal of Engineering Studies and Technical Approach, 1,* 9–24.
19. Kundra, T., & Mujamdar, A. S., (2002). Filtermatdrying. In: *Advanced Drying Technologies* (p. 364). Marcel Dekker Inc., New York.
20. Lipnizki, F., (2010). *Membrane Technology* (p. 318). Wiley-VchVerlag GmbH & Co., KGaA, Weinheim.
21. Mason, T. J., Riera, E., Vercet, A., & Lopez-Buesa, P., (2005). *Application of Ultrasound* (pp. 323–351). Elsevier, USA.
22. Maubois, J. L., (2002). Membrane microfiltration: A tool for a new approach in dairy technology. *Australian Journal of Dairy Technology, 57,* 92–96.
23. Maubois, J. L., Fauquant, J., Famelart, M. H., & Caussin, F., (2001). Milk microfiltrate a convenient starting material for fractionation of whey proteins and derivatives. In: *Proc. 3rd Int. Whey Conf.* (pp. 59–72). München. Germany.
24. Mujamdar, S. A., (1995). *Handbook of Industrial Drying* (2nd edn., p. 624). Marcel Dekker, New York.
25. Ostergaard, B., (2003). Adding value to whey by Pro-Frac. *Eur. Dairy Mag., 8,* 20–22.
26. Patel, R. P., Patel, M. P., & Suthar, A. M., (2009). Spray drying technology: An overview. *Indian Journal of Science and Technology, 2,* 110–118.
27. Rajkumar, P., Kailappan, R., Viswanathan, R., Raghavan, G. S. V., & Ratti, C., (2007). Foam-mat drying of alphonso mango pulp. *Drying Technology, 25,* 357–365.
28. Rosenberg, M., (1995). Current and future applications for membrane processes in the dairy industry. *Trends in Food Science and Technology, 61,* 12–19.
29. Sivakumar, R., Saravanan, R., Perumal, A. E., & Iniyan, S., (2016). Fluidized bed drying of some agro products–A review. *Renewable and Sustainable Energy Reviews, 61,* 280–301.
30. Soria, A. C., & Villamiel, M., (2010). Effect of ultrasound on the technological properties and bioactivity of food: A review. *Trends in Food Science & Technology, 21,* 323–331.
31. Soysal, A., Oztekin, S., & Eren, O., (2006). Microwave drying of parsley: modeling, kinetics, and energy aspects. *Biosystems Engineering, 93,* 403–413.
32. Strathmann, H., Giorno, L., & Drioli, E., (2006). *An Introduction to Membrane Science and Technology* (p. 315). Consiglio Nazionaledelle Ricerche., Rome-Italy.
33. Sulaksono, A. C., Kumalaningsih, S., & Wignyanto, S. I., (2013). Production and processing of yogurt powder using foam-mat drying. *Food and Public Health, 3,* 235–239.
34. Tarleton, E. S., & Wakeman, R. J., (1998). *Ultrasonic Assisted Separation Process* (p. 287). Springer, New York.
35. Uduwerella, G., Chandrapala, J., & Vasiljevic, T., (2017). Pre-concentration by ultrafiltration for reduction in acid whey generation during Greek yogurt manufacturing. *International Journal of Dairy Technology, 61,* 71–80.

36. Woo, M. W., (2017) *Recent Advances in the Drying of Dairy Products* (pp. 249–267). John Wiley & Sons, Chichester, UK, 2017.
37. Yazdanpanah, N., & Langrish, T. A. G., (2011). Crystallization and drying of milk powder in a multiple-stage fluidized bed dryer. *Drying Technology, 29,* 1046–1057.
38. Zisu, B., & Chandrapala, J., (2015). *High Power Ultrasound Processing in Milk and Dairy Products* (pp. 149–180). Chapter 6, John Wiley & Sons, Ltd., Chichester, UK.

CHAPTER 5

TECHNOLOGICAL AND BIOCHEMICAL ASPECTS OF GHEE (BUTTER OIL)

PINAKI RANJAN RAY

ABSTRACT

Ghee (butter oil) is one of widely consumed milk product in India and neighboring countries. It maintains a unique position among all edible fats and oils available in the market. Milk fat in ghee is available in its native state without causing much physicochemical changes, whereas refining is necessary in other vegetable oils. This makes ghee rich in naturally occurring nutritive and antioxidative properties with the specific fatty acid profile. A significant amount of short-chain fatty acids present in ghee make it inadequate for applications in baking and confectionary industry. This particular area of application of ghee requires physicochemical and biofunctional advancements. Different scientific findings suggest that the characteristic flavor of ghee and a mixture of biofunctional compounds make it suitable for consumption by persons of all ages for many health benefits. The biofunctional properties of ghee distinguish it from other milk products. The elixir properties of ghee have been regarded in *Ayurveda* since *Vedic* age for its medicinal role against several ailments.

5.1 INTRODUCTION

Ghee (butter oil) is one of the most popular traditional dairy products in India. Since Vedic era, it has been used for religious rites, cooking, cosmetic, and medicinal purposes. The importance of ghee in Indian diets has been recognized from prehistoric days because of its high nutritive value, pleasant aroma, and textural properties. Preparation of ghee is the best way to preserve the excess milk fat. Ghee is considered as an energy-rich food and rich in essential fatty acids, fat-soluble vitamins, and growth-promoting factors. Consumption of ghee in any form adds to the satiety feeling to meals. The pleasant aroma, good texture and to a lesser extent color of ghee are used as the criteria for judging the quality.

Approximately one-third of the total milk produced in India is used for making ghee and ghee prepared by rural milk producers constitutes about 80% of the total sales in the market. The organized dairies contribute about 20% of the consumer's needs of ghee [7]. Ghee is prepared from cream, derived from cow's, buffalo's or mixed milks or from cooking butter through a process of heat clarification. Several factors which promote the production of ghee in India are:

- Best utilization of unused milk fat;
- Easy storage without refrigeration;
- Longer shelf-life;
- Low cost and simple technology.

This chapter discusses different technological and biochemical aspects of ghee and its applications in the food industry.

5.2 GHEE PROCESSING TECHNOLOGY: HISTORY AND ADVANCEMENTS

Ghee making technology principally involves three distinct operations: Concentration of milk fat, clarification of fat-rich milk by heating and removal of the residue from fat after heat clarification.

The fat concentration in cream and creamery butter helps in reducing the load of ghee residue, fat loss in ghee residue and amount of water evaporation during heat clarification. Moisture removal by heat helps ghee acquiring its characteristic flavor, and solid-not-fat (SNF) contents are converted to the brown residue, which facilitates removal of maximum fat. In the third step, ghee residues are generally separated by decantation, cloth or pressure filtration or the centrifugal clarification techniques. Ghee manufacture techniques differ from each other essentially in the first step of fat concentration [9, 18] (Figure 5.1).

Ghee manufacturing process depends on the nature of raw material (milk, cream, butter) used, raw material treatments, and management of the ghee during different stages of processing. Batch Processing of ghee involves four different methods [23], such as:

- Creamery butter method;
- Direct cream method;
- Indigenous method;
- Pre-stratification method.

Ghee processing by the continuous method was developed to control the problems in the batch processing. Ghee making and filling units by the continuous method were developed later, but the extent of ghee production using this method is very limited [2, 3, 21]. The batch processes are used under different conditions for the different scales of ghee production. Manufacturing of ghee by batch processes is shown in Figure 5.1.

5.2.1 INDIGENOUS METHOD

The following steps are generally followed in the indigenous method of ghee processing:

- Raw milk is churned directly.
- Heat-treated milk is fermented by Lactic Acid Bacteria (LAB), which is followed by the churning of curd or
- Thick cream layers are removed from milk continuously by heating at around 80°C followed by grinding, dispersal in water and churning.

Lactic acid fermentation method for ghee preparation is mainly followed in rural India. Butter is separated by hand-driven wooden beaters. Butter accumulated is then melted in clay- pot or metal pan until all the moisture has been removed [23]. The contents in the vessel are kept undisturbed after heating. Decantation of the clear fat into ghee storage vessels is then done after the curd particles have settled at the bottom of a pan [24]. This method results in very low fat recovery (75–85%) and produces a considerable quantity of buttermilk.

5.2.2 DIRECT CREAM METHOD

This method has been technologically improved than the indigenous method and is generally practiced in small dairy industries. In this process, the cream is separated from the milk and is directly converted into ghee.

Cream is heated at about 115°C in a stainless steel jacketed ghee-kettle fitted with an agitator, temperature, and pressure gauges, and steam control

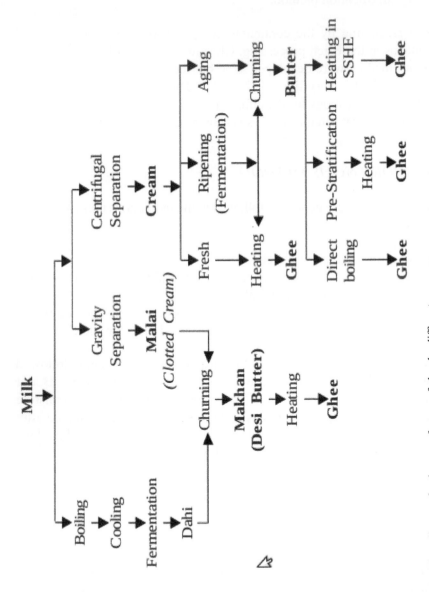

FIGURE 5.1   Flow diagram for the manufacture of ghee by different processes.

valve. Contents are emptied out by using a centrally bored movable, hollow, and stainless steel tube. Ghee residue turns to light brown or golden yellow, when the heating process is completed. Long heating time is required in this process for the removal of moisture, which limits the use of this process. A higher amount of serum solid results in strong caramelized flavor in ghee, which is a characteristic of this process. The extent of butter-fat loss during handling is limited to 4–6%. Fat loss and steam consumption may be minimized by the use of 75–80% plastic cream or washed cream. Low SNF (solids not fat) cream produces a less intense cooked flavor.

5.2.3   CREAMERY BUTTER METHOD

Organized dairies use this method as the standard method for ghee processing. In this method, ghee is prepared from white butter, unsalted creamery, or cooking butter. Nine units are generally present in creamery butter process namely: (1) cream separator; (2) butter churner; (3) butter melting accessories; (4) steam jacketed stainless steel made ghee kettle fitted with process controls and agitators; (5) oil clarifiers or ghee clarifiers; (6) storing tanks; (7) pipelines and pumps which interconnect the facilities; (8) tanks for crystallization; and (9) product filling and packaging lines. Clarification is done at about 110°C. Higher temperature results formation of 'cooked' flavor. The advantages of this method are:

- Ghee with flat or cooked flavor is produced by the creamery butter method.
- Less fat loss due to the low quantity of ghee.
- Less granulation is observed when compared to other methods.
- Relatively longer keeping quality.
- Require lower energy as compare to other methods.
- Storage of cooking butter requires less space.
- The quality of ghee is highly uniform.

5.2.4   PRE-STRATIFICATION METHOD

A characteristic aroma is generally developed in ghee by the pre-stratification method [25]. This method is cost-effective due to 35–50% savings in fuel consumption and approximately 45% saving in time and labor. This method also produces lower free fatty acid (FFA) in ghee and has lower acidity. This results in longer keeping quality of ghee. Residue formation in ghee is also reduced. Better quality ghee is produced due to the use of safety valves, temperature regulators, pressure gauge, and condensate outlet. Ghee with milder flavor is generally produced by this method.

5.3   DEVELOPMENT OF CONTINUOUS GHEE MAKING TECHNOLOGY

Ghee processing techniques are generally traditional, and very little changes have taken place over the ages. Batch methods of ghee processing are generally suitable for small and medium-scale manufacturers. Traditional method is used in processing almost 90% of the ghee in India. Development of continuous energy-efficient methods for ghee manufacture took center stage of attention by researchers in light of increased awareness in energy management and savings. The first attempt at mechanized/continuous ghee manufacture was made by Punjrath [21]. Other scientists later developed continuous ghee making process by following different techniques that included oil separator [4] for separating fat and serum phase or use of scraped surface heat exchangers (SSHE) [2]. Both the processes are energy efficient and produce ghee of comparable quality. Batch methods of ghee making have the following problems:

- Bulk spoilage of the product takes place as a function of large residence time and product inventory.
- Bulky equipment which results in low heat transfer.
- Chances of the accident are high due to the higher amount of product spillage.
- Equipment and processes are not suitable for large scale production.
- Lack of proper sanitation. Chances of contamination are high due to increased exposure to the environment.
- Strenuous cleaning and sanitation process; and poor performance of the equipment due to the formation of the obstinate scale of ghee and milk residues on the heating surface.
    Continuous ghee making techniques obviates all the limitations of the batch process of ghee making. The continuous method works on the following two basic principles:
- Evaporation of moisture from butter or cream using thin-film scraped surface heat exchanger (TSSHE); and

- De-emulsification of cream by high-speed clarifixator and oil concentrator followed by moisture evaporation (Figure 5.2).

Salient features of the ghee making by the continuous method are delineated as follows:

a. **Continuous Ghee Making by TSSHE:** This process was developed by using a thin film SSHE [1]. Butter is pumped to the TSSHE from a balance tank. A rotameter with a valve on the inlet line is used to control the flow rate of molten butter. The molten butter forms a film and spread uniformly on the heating surface of the SSHE by the centrifugal action of the rotor blade. Steam is admitted into the jacket of SSHE at a regulated rate. Rotating blades of the heat exchanger create turbulence, which causes the water evaporation from the butter film faster. A motor fitted with it controls the speed of the rotor blade. Vapors are released through an outlet located at the top of the SSHE. The released vapors are used for heating the butter in the balance tank for economizing the steam consumption. Molten butter and ghee temperature are controlled by a thermometer, which is adjusted by regulated steam supply with valves. A ghee tank is used to collect the ghee. An oil clarifier is used to separate the ghee. Residue-free ghee is finally transferred to the packaging line or stored in a tank.

b. **Cream De-Emulsification Method:** Cream de-emulsification method is a phase reversal process based on the de-emulsification of cream fat from oil-in-water phase to water-in-oil phase. A centrifugal separator is used to separate cream 40% fat from milk. A clarifixator converts the cream into plastic cream of 80% fat, which is further concentrated in a centrifugal concentrator. Clarifixator de-emulsify the fat. Flavor generation in ghee and removal of moisture from the fat concentrate is facilitated by an SSHE. Traces of moisture present in the ghee are removed by a vapor separator. An oil clarifier is used to remove the ghee residues. The process is described in Figure 5.2.

5.3.1 DEVELOPMENTS IN CONTINUOUS GHEE MAKING PROCESS

Research work has also been carried out to develop continuous ghee making with higher efficiency. Some notable research contributions to improve the continuous ghee making processes are:

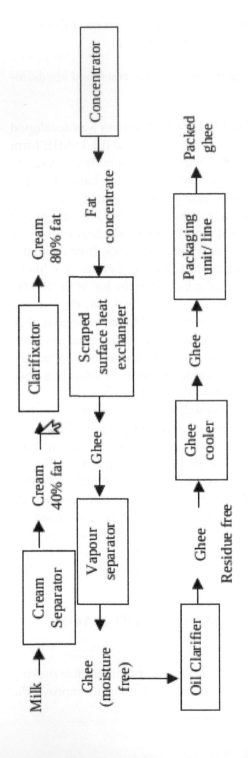

FIGURE 5.2 Cream de-emulsification method.

- Continuous ghee making with Flash evaporation method is based on the principle of flash evaporation [1]. It was conceived that the continuous ghee making process based on the principal of flash evaporation could be made compact and economically viable. The new plant was designed on falling film principle with a capacity of 100 kg/h.
- Continuous ghee making with energy conservation process is based on thin-film surface heat exchanger [1]. Advantages of this method are:
  - Fouling and foaming problems are minimum;
  - Heat transfer coefficient is high;
  - High capacity reduction;
  - Hygienic operation and better product quality;
  - Short residence time in a heated zone.

- An improved cost-effective method of ghee preparation at Sarvottam Dairy–Bhavnagar (Gujarat)–India was developed by using a serum separator [13]. The process reduced the fat, SNF losses during the production of ghee. The innovated method causes less scraping in ghee boilers due to less ghee residues. Sensory attributes of ghee samples were also at par with the other ghee samples prepared by conventional methods.

5.4   GHEE MAKING FROM SOUR CREAM: TECHNOLOGY AND ADVANCEMENTS

In India, most of the milk is produced in rural areas, where no-effective means of refrigeration are available. Therefore, a large amount of milk is generally soured, particularly during summer and monsoon months. A good amount of this milk may often be curdled during transportation to the dairy plants, and a large amount of this sour milk is rejected by the dairy to avoid possible difficulties in processing such milk.

The most common practice to utilize curdled milk, whenever received, is to churn it directly into butter for heat clarifying into ghee. However, only the plants equipped with butter churn can use the direct churning method for preparation of ghee from sour milk. Dairies without a butter churn sometimes neutralize curdled milk for separation into the cream. The practice of mixing small amounts of curd with a large portion of milk/cream for ghee making is also not uncommon. Curdled milk is generally converted either to cream or to butter, both of which are heat clarified to ghee. The cream can be obtained by the separation of neutralized curdled milk, while butter can be made from either curdled milk by direct churning or cream obtained from it. Normally two different methods are followed to prepare ghee from sour milk:

- Direct churning of curdled milk; and
- Neutralizing Sour/curdled milk.

5.4.1 DIRECT CHURNING OF CURDLED MILK

For *desi* ghee preparation, a smooth curd is set to about 0.8% lactic acidity. Milk that is received in the reception dock represents a varying state of coagulation ranging from a completely curdled mass, distinctly wheying-off, and to sour milk with definite signs of curd flakes. No systematic study is available regarding the effect of the variations of the curd on the churning efficiency of the curdled milk. Efficiency of churning may be lost during handling curdled milk in power churn designed specifically for 35–40% fat cream. It is not known if the process yields itself favorably to butter grain development and washing steps. Thus, dairies churning curdled milk directly to butter might not have yet optimized the process.

5.4.2 NEUTRALIZATION OF SOUR/CURDLED MILK

Neutralization of sour milk imparts enough physical stability for centrifugal cream separation. The neutralized sour or curdled milk should also have curd particles dissolved, enabling its warming up to a temperature of about 40°C for separation. The neutralization of cream for butter making has been studied with reference to the level of treatment, the use of sodium and calcium neutralizers, and the means of incorporating neutralizing solutions. The neutralization of curdled milk has not received similar attention.

Food grade alkali neutralizers can be used for reducing the acidity in sour milk. Soda neutralizers, when used, have the distinct advantage that they dissolve readily and are completely soluble in water and their action on the acid in the sour milk is faster than that of lime neutralizers. However, soda neutralizers develop brownish color more rapidly during heat processing of such milk. Sodium hydroxide is preferred for products requiring maximum solubility. Over neutralization should be avoided during neutralization.

The milk temperature should be less than 35°C, and at least 15 minutes should be allowed for the reaction before heating. Sodium bicarbonate has a balancing effect on Calcium, and it changes the reaction [10]. The temperature of the milk is lowered to 4°C or below, to avoid lapses at different stages of neutralization; otherwise, additional acidity is developed. Neutralized milk is heated to 80°C (Flash) for stopping further development of acidity. Carbon dioxide evolution during the process causes frothing, which needs careful handling. Completion of neutralization requires minimum boiling. Expulsion of carbon dioxide is continued even during the storage of boiled neutralized milk. After the sour milk is neutralized to the normal acidity, it behaves mostly like normal milk as far as heat processing and other operations are concerned.

Conversion of cream from neutralized curdled milk to butter should not be much different from the well-known procedure of making butter from the neutralized cream. Direct churning of curdled milk being the simplest process may be used in dairies having enough curdled milk for their power butter churn. Direct cream method via neutralization is the most suitable method, where butter churn is not available.

5.5 PACKAGING OF GHEE: TECHNOLOGICAL ADVANCEMENTS

Ghee contains 99.5% milk fat and rest 0.5% of material present in ghee is unsaponifiable matter, which is a complex mixture of substances like sterols, vitamins, etc. Moisture is also present in ghee in very small quantity, and it is not possible to eliminate this moisture from ghee during its preparation. Therefore due to low moisture content, chances of microbial infestation are very low. Utmost priority for packaging of ghee includes: protection from chemical spoilage, which is activated by oxygen, light, metals (Cu, Fe), humidity, and temperature, besides spoilage from the surroundings by absorption of foreign odors and also from the physical hazards.

5.5.1 HISTORY AND PRESENT STATUS

Lacquered or non-lacquered tin cans of various capacities are generally used for ghee packaging. Loose ghee is also sold to local consumers through their sales depots or stores, where the possibility of adulteration cannot be completely ruled out. Use of tin can is most advantageous. Tin can protect the product from tampering and helps in the transportation of the product to far-off places without much wastage. Apart from it, tin cans are attractive with colorful designs. But tin cans are expensive. Of late, some dairies have started packaging of ghee in simple containers, e.g., PE bags, multi-layer films, glass bottles, and cartons, etc.

5.5.2 PACKAGING MATERIALS FOR GHEE

Composition and characteristic of the product, shelf-life during storage, types, and causes of deterioration under different conditions, and the availability and functional properties of the packaging material are key factors, which determine the suitable flexible packages for ghee. Different packaging materials that have been used for packaging ghee more effectively are discussed in this section.

5.5.2.1 LIGHT TRANSMISSION BARRIER FILMS

Light can effectively catalyze many other promoters of rancidity, both hydrolytic and oxidative. Ghee exposed to oxygen under ordinary temperatures and illuminated places starts autooxidation without any induction period. The rate of autooxidation is directly proportional to the intensity of illumination. Therefore, selecting the packaging materials for ghee, which can effectively prevent the entry of light into the product, is very essential. Packaging materials having reflecting pigments and denser films like aluminum foils can fulfill the criteria. Entry of light into the product can be prevented by heavy overprinting.

5.5.2.2 AROMA AND GREASE BARRIER FILMS

Ghee is highly accepted by the users mainly due to its characteristic pleasing flavor. Hence, it is of utmost importance that the original 'ghee flavor' be protected or maintained. Ghee is very susceptible for picking up the foreign odor from packaging materials and their components or atmosphere. Loss of aroma can be prevented by selecting proper packaging materials, which should be 'flavor proof' and completely devoid of any inherent odor.

5.5.2.3 WATER VAPOR BARRIER FILMS

The agency of moisture or enzyme lipase is essential to affect hydrolytic rancidity in ghee. During the manufacture, ghee is subjected to such high heat treatment (110–120°C) that enzyme lipase is eliminated. In the process, the product gets contaminated from the lipase producing organisms, and they, in turn, produce the enzyme, which becomes active in the presence of moisture to liberate compounds responsible for hydrolytic rancidity. Therefore, the proper packaging material with excellent water vapor barrier properties can play a vital role in delaying this defect. Suitably laminated HDPE, PP, Al foil, multi-layer films, etc. could result in packages, which would be comparable to tin cans and can practically reduce water vapor transmission.

5.5.2.4   OXYGEN TRANSMISSION BARRIER FILMS

Due to the high amount of milk fat, ghee is very much susceptible to oxidative rancidity. Presence of oxygen in the packaging material initiates the chemical reactions with ghee, which ultimately results in the production of compounds gaining very strong off-flavors such as tainty, nutty, melon-like, grassy, tallowy, oily, fishy, etc. The undesirable odors of aldehydes and ketones of several types can be felt even at very low concentrations. Therefore, care should be taken to fill ghee up to the top of the container.

A packaging material with very low or negligible oxygen transmission rate (OTR) can also delay the autoxidation of ghee, i.e., if the package has a very low or negligible OTR, the diffusion of air/oxygen from the atmosphere into ghee can be prevented, and this issue can be deferred for a long time. These days, many indigenously available flexible materials with very low values for OTR (e.g., Polyester, Nylon-6, PVC (polyvinyl chloride) foils, and numerous laminates of certain flexible films) are available. These packaging materials provide excellent resistance to oils and fats, namely lacquered cellophane, polymer-coated cellophane, cellulose acetate, polyester, Nylon-6, PVC, Saran, etc., besides numerous laminates.

5.6   UTILIZATION OF GHEE RESIDUES: INNOVATIVE IDEAS

During ghee manufacturing, a brownish solid residue is obtained as a by-product, which is known as ghee residue. A considerable amount of milk fat, protein, and minerals are present in ghee residue. The intensity of the heat treatment used during the ghee manufacture determines the color of ghee residue, which may vary from light to dark brown. Excessive fat in the ghee residue makes the product glossy with a granular texture. An average estimate reveals that the total production of ghee residue in the country is approximately 3 million tons per annum [7].

5.6.1   GHEE RECOVERY FROM GHEE RESIDUE

Different processes have been innovated for recovery of the maximum amount of ghee from ghee residue. Two most widely used methods for recovery of ghee from ghee residue are centrifugal process and pressure technique [31]. The centrifugal process of ghee recovery from ghee residue involves heating of ghee residue in the water at 65°C for transferring occluded ghee present in the residue to water. Centrifugation is done thereafter for recovery of ghee. This method shows 46% efficiency with a yield of 25%. The pressure technique involves heating ghee residue at 65–70°C, which is followed by the exertion of defined pressure by a hand screw or hydraulic press. It has 67% extraction efficiency with a yield of 45%. The pressure technique method is an efficient, simple, economical, and practical method, which does not require any electricity or sophisticated equipment.

5.6.2   USE OF GHEE RESIDUES

Ghee residue has wide applications in the processing of different food items, such as:

- Bakery products preparation;
- *Burfi*-type confectionary preparation;
- Candy preparation;
- Chocolate preparation;
- Confectionary preparation;
- Edible pastes preparation.

5.7   BIOCHEMICAL ADVANCEMENTS IN GHEE PROCESSING

5.7.1   QUALITY OF GHEE

In most cases, a small amount of ghee obtained from curdled milk is blended with the bulk of dairy ghee. The influence of ghee made from sub-standard material remains unnoticed due to its blending with a large bulk. The nature of ghee spoilage, particularly oxidation and rancidity, is such that even a small amount of catalytic agents, such as FFA or copper ions, hastens the deterioration process. Thus, it is necessary to ascertain the quality of ghee obtained from a high acid and neutralized material such as curdled milk with particular reference to analytical constants, sensory attributes, shelf-life, and public health.

1. **Analytical Constants:** The major analytical constants of ghee remain unaffected by the preparation method. In the case of processing ghee from curd, it is possible only when the butter has undergone little deterioration on storage. However, in practice, *desi* ghee made from collected butter of indifferent quality, subjected to neutralization, washing (pre-stratification) and refining treatment may cause a varying amount of losses in water-soluble and volatile fatty acids affecting several analytical constants. Deterioration to such an extent is not normally expected in ghee made by direct churning route, if butter is clarified without delay. Other routes of ghee making involving neutralizing of curdled milk should not affect analytical constants.

   Physicochemical qualities of buffalo sour and curdled milk ghee, prepared through direct churning and reprocessing methods with 0, 1, 2, and 3 washings of cream/butter, were evaluated at NDRI, Karnal [10]. All experimental ghee samples had R.M. value, Polenske value, Iodine value, Saponification value, B.R. reading, % FFAs and Peroxide value within reasonable limits that were comparable with those of fresh buffalo milk ghee and conformed to PFA and Agmark Standards.

2. **Sensory Quality:** The sensory quality of ghee centers on flavor, texture, and color, of which flavor is considered to be the most important factor. Ghee develops its characteristic flavor during heat clarification. Ghee associated with lactic fermentation has the most appealing flavor characteristics. Curdled milk ghee is richer in flavor as compared to dairy ghee prepared from fresh products. Treatment standardized for neutralized cream for table butter may be suitably modified for neutralized curdled milk. *Desi* ghee has been claimed to have better textural properties than the dairy ghee that is known to exhibit often an excessive layering. It is not certain if this phenomenon is related to a more complete extraction of fat in direct cream and creamery-butter process used in dairies as against a partial removal of butterfat, particularly the high melting glyceride fraction, in butter obtained under practical village condition for *desi* ghee.

Studying the nature of fat obtained from buttermilk after direct churning process may provide a better explanation. Souring has been demonstrated to influence the whitish color of buffalo ghee due to the conversion of Biliverdin to Bilirubin, imparting a yellowish-greenish tinge. Researchers [6, 10] have evaluated the sensory qualities of buffalo sour and curdled milk ghee prepared by direct churning and reprocessing methods with 0, 1, 2, and 3 washings of cream/butter. Though direct churning method ghee samples were judged to have highly significantly (P<0.01) better appealing sensory qualities than the reprocessing method ghee, yet all samples were graded between "good to excellent." Washing of cream/butter did not much improve the quality of ghee and was thus found unnecessary. Some of the fermented products, particularly the water-soluble ones, contributing to desirable flavor in ghee, might be getting flushed away during washing treatment. Total sensory score of curdled milk ghee prepared through both the standardized methods (without washing of cream/butter) was found comparable with that of fresh 0.5% T.A. buffalo milk and NDRI ghee.

Based on color, ghee prepared through direct churning method was highly significantly (P<0.01) more liked than the one prepared through the reprocessing method. Probably sodium bicarbonate added during the reprocessing method imparted a relatively higher degree of brownish tinge to ghee. Washing of cream/butter was observed to improve the color characteristics of ghee prepared through both the methods, which may also be due to the partial removal of the serum portion through the washing of cream/butter. A greater amount of serum portion is understood to impart brownish color to the ghee due to the interactions, particularly, between casein and lactose during heat clarification.

3. **Keeping Quality:** Several factors, associated with the preparation of ghee from the fermented products, affect its keeping quality. The most important is the FFA content of ghee that is known to accelerate the development of tallowiness. However, washing of cream, as well as pre-stratification of melted butter, significantly reduces FFA with a resultant improvement of shelf-life.

4. **Safety and Nutrition:** From the public health view of point, the quality of ghee prepared from naturally curdled raw milk as against by lactic fermentation after an adequate heat treatment, may be questionable. This concern is based on the possibility of bio-toxin production during the uncontrolled curdling process and their subsequent transfer

to ghee. However, on closer scrutiny, the chances of bio-toxin production in milk during the natural souring process may be narrowed down essentially to bacterial toxins from a coliform group of organism and mycotoxins due to yeast and mold growth. It may be further realized that only fat-soluble toxins would be of any consequence in the manufacture of ghee. Furthermore, the fat-soluble toxins must be able to withstand the heat clarification treatment. Under these conditions, the probability of the production of fat-soluble, heat-tolerant bio-toxins during the normal milk curdling process can be established only on the basis of detailed toxicological studies conducted on a wider industrial basis. To date, no toxic effect accruing to ghee has been reported.

Ghee from direct churning method contains comparatively lower amounts of phospholipids. High acidity in ghee may also affect vitamin A potency. Thus, a lower phospholipid content and a slight loss in vitamin A potency may lower the nutritive value of curdled milk ghee in comparison to the best available product. However, ghee need not be considered a major source of these nutrients under Indian conditions. Of course, ghee samples of questionable safety value or which are nutritionally substandard can be put to several profitable non-edible uses, such as ceremonial lamp burning or in havans, etc. About 2% of the ghee produced in India is used for such purposes.

5.7.2 PRESERVATION OF GHEE

Ghee is more resistant to elemental and microbial attack than other milk products. When produced, packaged, and stored under controlled hygienic conditions, it is expected to keep in good condition for about 9 months at 21°C. Prolonged storage at ambient temperature results in oxidative changes in ghee, which may cause the following defects:

- Decrease in nutritive value;
- Destruction vitamins and carotene;
- Formation of toxic products;
- Loss of attractive color;
- Loss of unsaturated fatty acids;
- Production of objectionable flavor.

Presence of unsaturated fat, storage temperature, initial quality of ghee (particularly acidity and moisture content), presence of oxygen and catalytic salts (copper and iron in particular) and packaging conditions regulate the shelf life of ghee. Product durability can be increased by the adoption of the following practices:

5.7.2.1   USE OF ANTIOXIDANT

Ghee residue contains antioxidative substances, which prevent peroxide development in ghee samples. Both lipid and non-lipid constituents are responsible for this property [30]. Maximum antioxidant activity is exhibited by phospholipid followed by ∞-tocopherol and vitamin A. Cephalin shows the highest antioxidant activity among various phospholipid fractions of ghee. Oxidative stability of ghee is increased with an increase in phospholipid content. Enhancement of phospholipid content can be achieved either by solvent extraction or by the heat treatment process.

Maximum transfer of phospholipids from ghee residue to ghee can be obtained by heating ghee residue with ghee in the ratio of 1:4 at 130°C. Amino acids–present in ghee mainly lysine, proline, cysteine hydrochloride, and tryptophan–exhibit the maximum antioxidant properties. Antioxidant activity of proline is maximum among the amino acids but less than 0.02% BHA. The addition of glucose, galactose, lactose in ghee, and the products formed due to their interaction with protein and phospholipids also enhance the oxidative stability of ghee.

Free sulphydryls in ghee also contribute to the antioxidant properties of ghee residue [30]. Increase in the temperature of ghee clarification reduces the antioxidant activity of ghee residue [29]. Addition of low temperature (110°C) ghee residue results in lesser peroxide development than the addition of high temperature (150°C) ghee residue. Ghee residue of creamery butter method shows higher antioxidant activity as compared to direct creamery and *desi* butter ghee residues [29].

It can be said that ghee residue is a potential source of naturally occurring antioxidants and the antioxidant activity of the ghee residue is largely determined by the constituents present in it. Oxidative stability exhibited by the ghee residue to flavored butter oil was comparable with the antioxidative effect imparted by BHA and BHT [33, 34]. Therefore, the antioxidative potential of ghee residue can be explored for the improvement of the shelf-life of different food products.

Different synthetic compounds, namely gallates (ethyl, propyl, octyl), butylated hydroxytoluene (BHT), tertiary butyl hydroquinone (TBHQ), and butylated hydroxyanisole (BHA) were used as antioxidants [28]. Ascorbic acid, tocopherol, and phospholipids also play an important role as antioxidants [26] and some natural leaves and fruits such as curry leaves, soybean powder, safflower, betel leaves, and 'amla' (*Phyllanthus emblica*) are also permitted as antioxidants and can be used in small amounts for controlling the fat oxidation in ghee during storage. All these synthetic antioxidants come in the GRAS list. The addition of some of the natural preservatives to ghee has shown a very beneficial effect. The use of soybean and sunflower @ 0.5% has been found effective in delaying oxidative rancidity without influencing the texture, aroma, and composition of ghee. These seeds contain phospholipids, which act as an antioxidant in ghee and pure butterfat.

Curry and betel leaves (1% by weight of ghee added during boiling stage) exhibited their antioxidant property due to their phenolic compounds. Amino acids present in these leaves act as antioxidants. Betel and curry leaves are used to boil with 'desi' butter during clarification to improve the flavor, color, and shelf life of the ghee [19].

Ghee prepared with mango (*Mangifera indica L*) seeds kernels has better oxidative stability, perhaps due to the ghee phospholipids and phenolic compounds present in mango seed kernel [22]. Antioxidative compounds extracted from two tulsi varieties (namely *Ocimum sanctum* and *Ocimum tenuiflorum*) have been reported to have an effect on oxidative stability of ghee [14]. *Ocimum tenuiflorum* leaves extraction added to the creamery butter ghee at a level of 0.6 g/100 ml exhibited similar effectiveness as that of BHA when used @ 0.02 g/100 g to prevent oxidative damage.

The total phenolic content of herbs and their antioxidant potential have a positive correlation [16]. It was found that ethanolic extracts of herbs showed better antioxidant activity than the aqueous extract of the herbs in most cases in terms of total phenolic content and antioxidant activity determined by various methods such as the β-carotene linoleic acid model assay and DPPH radical scavenging activity. Though ethanolic extract of ashwagandha and aqueous extract of bidarikand exhibited different results [16]. Aqueous and ethanolic extracts of different herbs reduced oxidative damage in ghee, but the effectiveness of the herbs as an antioxidant was less than that of BHA. In terms of the induction period, ethanolic extracts of herbs showed a better result as compared to their aqueous extracts in the Rancimat [16].

5.8 BIOCHEMICAL ADVANCEMENT OF GHEE FLAVOR

Flavor is the main criteria of the food that determines its acceptability. Flavor of ghee has been contributed by the complex mixtures of organic compounds occurring in very minute quantities. There are always regional preferences for the ghee flavor.

5.8.1   ORIGIN OF GHEE FLAVOR

Milk fat is the sole source of most of the flavoring compounds present in ghee. They are generated as a result of the interaction of lipids, protein degradation products, lactose, minerals, and metabolites of microbial fermentation during heating. Flavor generation mechanisms leading to the development of ghee flavor are given in Figure 5.3 [32].

5.8.2   SIMULATION OF GHEE FLAVOR

Ripening of milk, butter, and *malai* as followed in the results of the village in curdy/acidic flavor. In commercial dairies, it is not possible to ferment milk and cream on account of additional storage space and energy required to achieve fermentation, and due to the problem of utilizing sour buttermilk. Simulation studies were limited to the treatment of plastic cream, butter, and ghee. Alternatively, butter oil may form the base for ghee making as it has longer shelf-life than butter. Flavor can be simulated as and when needed for the marketing of ghee in a given region. Viability of each treatment has to be viewed on the basis of yield, handling losses, processing costs, ease of operation, scale-up feasibility, quality of the finished product, and resulting shelf-life.

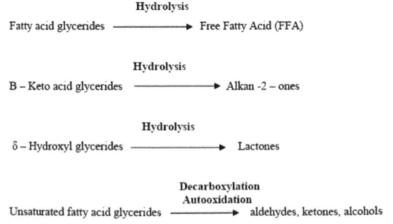

FIGURE 5.3   Flavor generation mechanisms leading to the development of ghee flavor.

The following approaches are suggested to manufacture ghee with curdy flavor:

- Differential blending of conventionally prepared ghee with ghee prepared from curdled and sour milk.
- Use of starter culture for fermenting dairy butter.
- Blending of butter with fermented skimmed milk/sour buttermilk/ curdled milk/lassi/sour skimmed milk powder/buttermilk powder/ lassi powder at one of the three stages, viz.: In churning working stage; In molten butter followed by storage overnight; In ghee boiler before clarification; and In ghee boiler at the final stage of clarification.
- Differential temperature and time clarification at 100, 105, 110 and 120°C for 5, 10, 15 and 20 minutes and blending in different proportions, especially when the cooked flavor was to be obtained.
- Clarification of ghee with vegetable leaves for improving flavor, shelf-life, and color.

Desi ghee, when mixed with dairy ghee in varying proportions, produces curdy of different intensity. Curdy flavor of ghee is increased by the addition of *dahi* to butter prior to heating or incorporation of *lassi* powder during the heating process.

5.8.3 SIMULATION OF COOKED FLAVOR

The cooked flavor will be more pronounced if the heat treatment is intense, and the solids-not-fat (SNF) is still present during heating. Ghee produced from cream is rich in cooked flavor because of the presence of higher amounts of SNF and a longer healing time for evaporation of moisture as compared with the butter process. Simulation of cooked flavor can be done by butter clarification at a temperature higher than 115°C for 10 min, 120°C for 5 min or 125°C with no holding time.

5.8.4 FLAVOR POTENTIAL OF GHEE AND GHEE RESIDUE

Ghee residue is produced during ghee processing as a by-product. Ghee-residue is rich in fat, proteins, and minerals and has antioxidative property. Recent studies have revealed that ghee residue also contains natural FFAs, carbonyls, and lactones, apart from its nutritional and antioxidant properties [8].

5.8.5 ROLE OF DAIRY STARTER ORGANISMS IN GHEE FLAVOR

The various biotechnological parameters have been optimized. In the optimized process, cream (40% fat) is steamed, cooled, and ripened with a culture DRC-1 at 3% level of inoculum at 30°C for 18 h, and clarified at 115°C/5 min. In a further modification, the ripening period of cream could be reduced by 5 h by using a starter concentrate (Viable count, 68 x 1010 cells/ml) at 1% level [34].

5.9 THERAPEUTIC POTENTIAL OF GHEE

Ghee is a rich source of energy and vitamin A, D, E, and K. Essential fatty acids present in ghee cannot be synthesized in our body. Classical texts of Ayurveda have classified medicinal properties of different ghee based species, manufacturing methods, and storage period. Cow milk ghee is mostly preferred for Ayurvedic applications. Cow milk ghee is considered good for eyes, light in digestion and strength-giving. It increases virility and appetite. The medicinal value of ghee also depends on the storage period of ghee. For external application, old ghee is preferred over new ghee.

Ayurvedic literature has reported about 50–60 types of medicated ghee. Herbal extracts were selectively fortified in the preparation of medicated ghee. All fat-soluble therapeutic components of the herbs are retained in ghee. Selected examples of the medicinal use of herbal ghee are as follows [17]:

- Amrutprash for anti-aging;
- Arjuna Ghrita for heart disease;
- Ashokaghrita for leucorrhoea;
- Ashwagandhaghrita for GI disorder;
- Kalyanghrita for madness.

Ayurveda suggests different treatment protocols for different ailments using medicated ghee manufactured with different herbal extracts. Different methods of preparation and suitability of different ghee with the specific process have been reviewed by investigators [20]. Medicated ghee has been used for various external and internal applications. Internal applications of medicated ghee include oral ingestion and an Ayurvedic treatment called *panchakarma*.

Scientific findings suggest the presence of conjugated linoleic acid (CLA), butyric acid, sphingomyelin, myristic acid, and vitamin A in ghee, which has functional properties and has potential to inhibit different degenerative disease. Ghee being richest source of CLA, has proven functional properties. Some recent research findings regarding the biofunctional properties of ghee are described in this section.

5.9.1   ANTICANCER ACTIVITY

Effect of feeding mixture of cow ghee and soybean oil on 7, 12-dimethylbenz (a)–anthracene (DMBA) induced carcinogenesis of mammary gland and expression of cox-2, and peroxisome proliferators activated receptors-γ (PPAR-γ) in rat mammary gland was studied [27]. The study revealed the anticancer potential of cow ghee. In the DMBA (a carcinogen) treated groups, the animals fed on soybean oil exhibited higher tumor incidence (65.4%), tumor weight (96.18g) and tumor volume (6285 mm^3) than the rats entirely fed on cow ghee (26.6%, 1.67 g, 1925 mm^3, respectively). Tumor latency period was increased to 27 weeks, when cow ghee was used as feed as compared to feeding with soybean oil, which exhibited a latency period of 23 weeks.

5.9.2   HEPATOPROTECTIVE ACTIVITY

Pachagavya, where cow ghee is a major ingredient, showed hepatoprotective effects in rats using carbon tetrachloride-induced toxicity [5]. Panchagavyaghrita @150–300 mg/kg/dpo reported prevention of CCl$_4$ induced elevation levels of SGPT, SGOT, AST, and ALP. The results were compared with standard drug Silymarin. Histopathological comparison of liver tissues exhibited almost normal architecture as compared to the control group.

5.9.3   ANTISTRESS ACTIVITY

Panchagavya ghrita, along with ethanolic extract of *Aloe babadensis*, exhibited antistress activity in mice using alprazolam as standard [12]. The combination was found significant ($p<0.01$) antistress potential as compared with the control and standards using a Tail suspension model. The synergistic action of panchagavya ghrita and aloe extract was attributed to the increase in the levels of GABA (gamma amino butyric acid) and decreased the level of dopamine and plasma corticosterone level.

5.9.4   EYE LUBRICANT ACTIVITY

Cow ghee is very useful for computer vision syndrome (CVS) [15]. The lubricating property of cow ghee may be useful for reduction of the symptoms of CVS. The product contains vitamin A, beta-carotene, and vitamin E. Vitamin A is responsible for the moistening of the outer lining of the eyeball and can prevent blindness. Beta-carotene and vitamin E are well known for their antioxidant activity. Hence, cow ghee can be used as eye lubricant and in CVS as an alternative treatment.

5.9.5   WOUND HEALING AND ANTIULCER ACTIVITY

A study of wound healing activity of a mixture consisting of *Aegle marmelos* leaves and cow ghee showed enhanced and rapid healing in buffalo [5, 11]. The effects produced by topical application of a combination of *Aegle marmelos* leaves extract and cow ghee in wound contraction, wound closure, surface area reduction of wound and tissue regeneration at the wound site were studied. The wound healing activity was found significant, and the wound was healed completely in eight days.

5.10   SUMMARY

The chapter includes discussions on: ghee and its production status; history and advancement of ghee processing technology; the continuous ghee making method; utilization of sour cream for ghee making; development of packaging technology for ghee; the technology of using ghee residue; the biochemical aspects of preservation of ghee; simulation of ghee flavor; the therapeutic potential of ghee. This chapter shall pave the way for further research on processing and biochemical advancements in ghee processing.

KEYWORDS

- **antioxidant**
- **Ayurveda**
- **butyric acid**

- **conjugated linoleic acid**
- **decantation**
- **hepatoprotective**
- **lactic acid**
- **pre-stratification**
- **scraped surface heat exchanger**
- **therapeutic potential**

REFERENCES

1. Abhichandani, H., Agarwala, S. P., Bector, B. S., & Verma, R. D., (1982). Design and development of falling film continuous ghee making machine. *Indian Journal of Dairy Science, 35*(4), 487–491.
2. Abhichandani, H., Sarma, S. C., & Bector, B. S., (1991). Continuous ghee manufacturing: An engineering solution. *Indian Food Industry, 10*(4), 35–37.
3. Agrawala, S. P., Prasand, S. A. D., & Nayyar, V. K., (1980). Development of ghee filling machine. *Indian Dairyman, 32*, 239–240.
4. Bhatia, T. C., (1978). *Ghee Making with Centrifugally Separated Butterfat and Serum* (p. 72). M.Sc. Thesis, Kurukshetra University, Haryana–India.
5. Biyani, D. M., Verma, P. R. P., Dorle, A. K., & Boxey, V., (2011). A case report on wound healing activity of cow ghee. *International Journal of Ayurvedic Medicine, 2*(3), 115–118.
6. Chandravandana, M. V., Daniel, E. V., & Dastur, N. N., (1974). Nature's watermark in ruminant milk. *Indian Dairyman, 26*, 233–236.
7. Dairy India, (2007). Milk utilization pattern. *Dairy India Yearbook* (p. 33). New Delhi–India.
8. Galhotra, K. K., & Wadhwa, B. K., (1993). Chemistry of ghee-residue, its significance and utilization–a review. *Indian Journal of Dairy Science, 46*, 142–146.
9. Ganguli, N. C., & Jain, M. K., (1973). Ghee: Its' chemistry, processing and technology. *Journal of Dairy Science, 56*, 19–25.
10. Gupta, V. K., Arora, K. L., & Chakraborty, B. K., (1986). Physicochemical and sensory qualities of ghee from curdled buffalo milk. *Asian Journal of Dairy Research, 5*(1), 49–55.
11. Gupta, A., & Gupta, S. K., (2014). Wound healing activity of topical application of *A. marmelos* and cow ghee. *International Journal of Drug Discovery and Herbal Research, 4*(2/3), 741–745.
12. Kumar, A., Kumar, R., Kumar, K., Gupta, V., Srivas, T., & Tripathi, K., (2013). Anti-stress activity of different compositions of Panchagavya and *Aloe barbadensis* Mill by using tail suspension method. *International Journal of Innovations in Biological and Chemical Sciences, 7*, 17–27.
13. Manikant, K., Pandya, H. B., Dodiya, K. K., Rajesh, B., & Mayur, M., (2017). Advancement in industrial method of ghee making process at Sarvottam Dairy, Bhavnagar, Gujarat (India). *International Journal of Science, Environment, 6*(3), 1727–1736.

14. Merai, M., Boghra, V. R., & Sharma, R. S., (2003). Extraction of antioxygenic principle from *Tulsi* leaves and their effects on oxidative stability of ghee. *Journal of Food Science and Technology, 40*, 52–57.

15. Mulik, S. S., & Bhusari, D. P., (2013). Conceptual study of Goghritaeye drops (*Aschyotana*) in computer vision syndrome. *Asia Journal of Multidisciplinary Studies, 1*(3), 1–6.

16. Nilakanth, P., Gandhi, K., Purohit, A., Arora, S., & Singh, R. R. B., (2014). Effect of added herb extracts on oxidative stability of ghee (butter oil) during accelerated oxidation condition. *Journal of Food Science and Technology, 51*(10), 2727–2733.

17. Pandya, N. C., & Kanawjia, S. K., (2002). Ghee: A traditional nutraceutical. *Indian Dairyman, 54*, 67–75.

18. Parekh, C., (1978). Ghee and its technology, *Dairy Technology, 9*, 32–35.

19. Patel, R. S., & Rajorhia, G. S., (1979). Antioxidative role of curry (*Murrayakoenigi*) and betel (*Piper betel*) leaves in ghee. *Journal of Food Science and Technology, 16*, 158–160.

20. Prasher, R., (1999). Standardization of Vasa ghrit and its form and their comparative pharmaco: Clinical study with special reference to SwasaRoga (Asthma). *M.D. Thesis* (p. 113). Gujarat Ayurved University, Jamnagar- India.

21. Punjrath, J. S., (1974). New developments in ghee making. *Indian Dairyman, 26*, 775–778.

22. Purvankara, D., Boghra, V., & Sharma, R. S., (2000). Effect of antioxidant principles isolated from mango (*Mangiferaindica L*) seed kernels on oxidative stability of buffalo ghee (butterfat). *Journal of the Science of Food and Agriculture, 80*, 522–526.

23. Rajorhia, G. S., (1993). *Encyclopedia of Food Science, Food Technology & Food Nutrition* (pp. 2186–2192). Academic Press, London.

24. Rangappa, K. S., & Achaya, K. T., (1974). *Indian Dairy Products* (pp. 221–226). Asia Publishing House, Mumbai.

25. Ray, S. C., & Srinivasan, M. R., (1975). *Pre-Stratification Method of Ghee Making* (p. 14). ICAR Res. Series No. 8, Krishi Bhawan, New Delhi.

26. Ramamurthy, M. K., Narayan, K. M., & Bhalerao, V. R., (1968). Effect of phospholipids on keeping quality of ghee. *Indian Journal Dairy Science, 21*, 62–63.

27. Rani, R., & Kansal, V., (2011). Study on cow ghee versus soybean oil on 7, 12–dimethylbenz(a)-anthracene induced mammary carcinogenesis and expression of cyclooxygenase-2 and peroxisome proliferators activated receptor in rats. *Indian Journal of Medical Research, 133*, 497–503.

28. Rao, C. N., Rao, B. V. R., Rao, T. J., & Rao, G. R. R. M., (1985). Shelf-life of buffalo ghee prepared by different methods by addition of permitted antioxidants. *Asian Journal of Dairy Research, 3*, 127–130.

29. Santha, M., & Narayanan, K. M., (1978). Antioxidant properties of ghee residue as affected by temperature of clarification and method of preparation of ghee. *Indian Journal of Animal Sciences, 48*, 266–271.

30. Santha, M., & Narayanan, K. M., (1979). Studies on the constituents–responsible for the antioxidant property of ghee residue. *Indian Journal of Animal Science, 49*, 37–41.

31. Viswanathan, K., Rao, S. D. T., & Reddy, B. R., (1973). Recovery of ghee from ghee residue. *Indian Journal of Dairy Science, 26*, 145–148.

32. Wadhawa, B. K., & Jain, M. K., (1990). Chemistry of ghee flavor–a review. *Indian Journal of Dairy Science, 43*, 601–607.

33. Wadhawa, B. K., Kaur, S., & Jain, M. K., (1991). Enhancement in shelf-life of flavored butter oil by natural antioxidants. *Indian Journal of Dairy Science, 44,* 119–121.
34. Wadhawa, B. K., Kaur, S., & Jain, M. K., (1991). Enhancement in shelf life of flavored butter oil by synthetic antioxidants. *Journal of Food Quality, 44,* 175–182.
35. Yadav, J. S., & Srinivasan, R. A., (1992). Advances in ghee flavor research. *Indian Journal of Dairy Science, 45*(7), 338–348.

PART III
Technological Advances in Fermented Milk Products

PART III

CHAPTER 6

FUNCTIONAL FERMENTED MILK-BASED BEVERAGES

NARENDRA KUMAR and VANDNA KUMARI

ABSTRACT

This chapter is an overview of dairy-based products in the functional food market [37], various traditional fermented beverages with their health benefits. Fermentation is a traditional method of food preservation, which further improves its nutritional and functional quality. Lactic acid and probiotic bacteria play a key role in the fermentation of dairy products, endowed with health beneficial and therapeutic properties like antidiabetic, cholesterol-lowering prevent tumor formation, etc. Bioactive components or functional ingredients are produced during milk fermentation, thus the term "dairyceuticals." Majority of the functional food market is contributed by probiotic (strains of *Lactobacillus* and *Bifidobacteria*) and prebiotic (GOS, FOS, lactulose, etc.) and dairy products serve as an ideal vehicle for delivery into the human gut for their maximum survivability. Fermented products in the form of beverages (yogurt-based drinks, kefir, Koumiss, and ymer) are most acceptable form, in which various bioactive ingredients can be easily fortified like vitamins, minerals, plant extract, antioxidants, and omega-3 fatty acids.

6.1 INTRODUCTION

In the last decade, consumer awareness regarding the direct role of food on physical and mental health has increased due to demand from high healthcare cost and an increase in life expectancy. Today, foods are not intended to fulfill appetite and source of essential nutrients, but also to prevent lifestyle-related diseases and to increase physical and mental health [58, 73, 82]. To meet the consumer demand for improving overall health and reducing the risk for specific diseases, the concept of functional foods has come into the limelight to provide health benefits beyond traditional nutritional values.

The Ministry of Health in Japan in 1991, initiated the approval of a specific health-related food category called FOSHU (Food for Specified Health Uses) [12, 44, 58]. Since then, this concept has spread worldwide especially in Europe and the United States with objectives of improvement of health, reduction in risk of some diseases and use for curing some diseases [52, 56, 85]. Functional foods include modified foods (fortified, enriched, or enhanced), conventional foods, foods for special dietary use, and medical foods [2].

There is no worldwide accepted definition of "functional food" so far as in most countries there is no any legal definition, which can differentiate between conventional and functional foods [1, 52, 62, 76]. A number of definitions for functional foods have been proposed by different national authorities, academic bodies and the industry ranging from simple one *"Foods that may provide health benefits beyond basic nutrition"* to *"fortified, enriched, or enhanced food with a component having a health benefits beyond basic nutrition can be considered functional foods"* [41]. The Functional Food Science in Europe (FuFoSE) defined functional food as *"a food product can only be considered functional if together with the basic nutritional impact it has beneficial effects on one or more functions of the human organism thus either improving the general and physical conditions or/and decreasing the risk of the evolution of diseases."*

The amount and form of intake of the functional food should be normal as expected for dietary purposes, not in the form of a pill or capsule [15]. However, in Japan since 2001, FOSHU products can also be taken in capsules and tablets form, although the majority of this food group were still in conventional forms [63]. The definitions of functional foods have the following three main concepts [10]:

- **Health Benefits:** Almost in all the definitions the health benefits of food that as has to be mentioned food has for the consumer in order to be labeled as a functional food, which is the most common concept in functional foods' literature;
- **Nutritional Properties:** All food to be functional must have some nutritional functions
- **Technological Process:** Some of the definitions emphasized fortification, enrichment or addition of ingredient the food, while others stress on the removal of allergens or of components that are

considered detrimental to the health if consumed in higher quantity (e.g., salt, sugar).

This chapter focuses on the functional aspect of various fermented milk beverages and ingredients using a wide range of lactic acid bacteria (LAB), including probiotic bacteria that have dominated the global market in the past decade.

6.2 MARKET TREND OF FUNCTIONAL FOODS

As there is no unitary accepted definition of functional food worldwide, the estimation of the market of functional foods is challenging. Based on the most accepted definition, as a food offering specific health, the estimated global market of functional food was $24.2 billion in 2011. In 2008 among the food industry, functional food was the fastest-growing sector, with an expected annual growth of 10% [94]. Currently, the United States represents 59% of the world's market for functional foods and supplements, followed by Europe and Japan [11, 87].

Being the birthplace of functional foods in Japan, more than 1700 functional food products have been launched between 1988 and 1998 with an estimated revenue of around 14 billion US$ in 1999 [56]. In 2003 and 2006, the functional foods market was estimated to be 5 billion and 5.73 billion US$, respectively, while more than 500 food products were labeled as FOSHU in 2005 [21, 85]. There is wide variation in the European market as there are many regional differences in the use and acceptance of functional foods. In Mediterranean countries, consumers' acceptance for functional foods is less as they value natural, fresh foods and consider them better for health than Central and Northern European countries [56, 95].

Functional foods have emerged in almost in all types of food categories, according to varying consumer preferences. Besides, food industries and the pharmaceutical industry have become the main suppliers of functional foods. Companies like Glaxo Smith Kine, Novartis Consumer Health, Abbott Laboratories, or Johnson & Johnson are investing in this market, due to shorter production time and lower product development costs as compared to pharmaceutical products. These companies also have immense experience in organizing clinical trials to validate health claims of a specific product. There are different types of functional foods available in the market (Figure 6.1). Kotilainen et al., [42], Sloan [87], and Spence [88] has classified these in four groups:

- **Altered Products:** Food from which a harmful component has been removed, reduced or replaced by another one with beneficial effects (labeled), e.g., fibers as fat releasers in meat or ice cream;
- **Enhanced Commodities:** Food in which one of the components has been naturally enhanced (labeled) for example, eggs with increased omega-3 content.
- **Enriched Products:** Food having additional nutrients or components which are not normally found in a particular food such as probiotics or prebiotics;
- **Fortified Products:** Food which is fortified with additional nutrients for example fruit juices fortified with vitamin E, vitamin C, calcium, zinc, and folic acid;

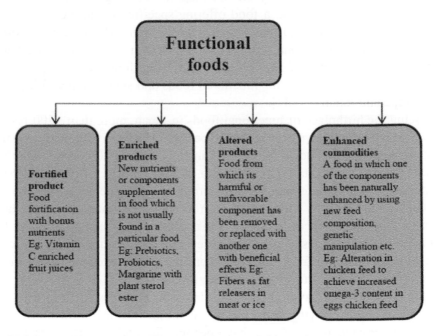

FIGURE 6.1   The main type of functional foods available on the market.

The market for functional food products includes products such as baked foods and cereals, baby foods, dairy foods, ready-to-eat foods, and confectionery. These are considered most dynamic functional foods category, because of several reasons such as [34, 80, 101]: (i) convenience to meet consumer demands for various container size, contents, shape, and appearance; (ii) ease of distribution and better storage and stability of products; (iii) more chance to incorporate desirable nutrients and bioactive compounds. Commercially available products can be grouped as (1) dairy-based beverages; (2) vegetable- and fruit-based beverages; and (3) sports and energy drinks.

6.3 DAIRY-BASED FUNCTIONAL FOODS

Milk in its natural form is considered as a functional food due to various health benefits contributed by bioactive peptides, saccharides, and lipids. It also contains various bioactive substances such as enzymes, hormones, immunoglobulins (IGs), antimicrobial peptides (AMPs), cytokines oligo-saccharides, and different growth factors [68]. The bioactive components in milk have various health benefits, which make milk as a complete and nutritional food (Figure 6.2).

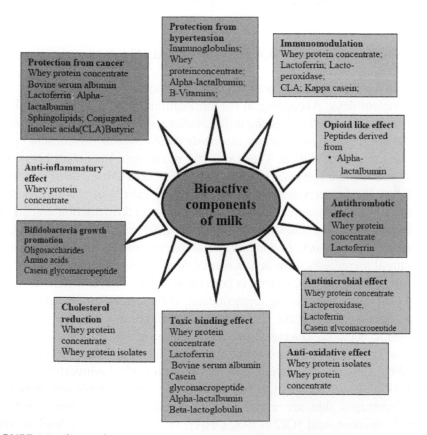

FIGURE 6.2   (See color insert.) Functions of bioactive components of milk.

Lactose is the predominant carbohydrate present in milk, which gives important nutritional benefits. It also helps in intestinal absorption of Calcium and Magnesium. Various antimicrobial agents are also found in milk that exhibits bacteriostatic and even bactericidal effects. Milk proteins, mainly caseins, may act as a precursor of a wide range of bioactive peptides having different physiological effects.

ACE inhibitory bioactive peptide converts angiotensin-I to angiotensin-II and degrades bradykinin by blocking its active site. Mammalian milk contains about more than 60 different types of enzymes, including digestive enzymes such as lipases, proteinases, phosphatases, amylases, and enzymes exhibiting antioxidant and antimicrobial properties (e.g., oxidoreductase, ribonuclease). These enzymes contribute to milk stability and protection against pathogenic bacteria [22]. Milk components (such as sphingomyelin, CLA, β-carotene, vitamins A and D, butyric acid) are found to have the anti-carcinogenic potential [38].

6.4   FERMENTATION PROCESS AND HEALTH

Fermented milks and milk products are an important part of our daily diet since ancient days. According to the International Dairy Federation (IDF), *fermented milk is defined as the milk product, which is fermented by using specific microorganisms that are desirably added to milk resulting in of pH reduction and coagulation.* These specific microorganisms shall be viable, active, and abundant (at least $10^7 CFU/g$) in the product to the date of minimum durability. The bacteria used for fermentation are mainly of two species such as LAB and Bifidobacteria. Some bacteria for use in fermentation are: *Lactobacillus acidophilus*, *Lactobacillus reuteri*, *Lactobacillus* GG (LGG), *L. caseisubsp. rhamnosus*, *B. longum*, *Bifidobacterium breve*. These bacteria use natural ingredients of milk as a source of nutrition during fermentation.

During fermentation, lactose is converted to lactic acid, which raises the acidity of milk. Also during fermentation, other bacteriostatic compounds are produced, which make the conditions unfavorable for the growth of pathogenic bacteria. During the fermentation process, proteolysis of milk proteins occurs to produce some peptides having bioactive properties providing immuno-stimulatory, opioid, or angiotensin-I converting enzyme (ACE-I) inhibitory activity [28, 97]. There are many available commercially dairy beverages that are enriched with bioactive components, such as docosahexaenoic acid (C22:6 n-3, DHA), ω-3 fatty acids, α-linoleic acid (C18:3 n-3, ALA), and eicosapentaenoic acid (C20:5 n-3, EPA) [64].

The fermentation process by specific LAB strains leads to removal of toxic or antinutritive factors present in milk, which make its suitable dairy product for lactose intolerance patients. Fermented milk products contain a higher amount of folic acid and other vitamins, including vitamin B, which is vital for developing a fetus in pregnant women. Fermented products are also rich in digestive enzymes, which help in the absorption of food. Fermentation can increase the flavor of food, which makes it more acceptable to consumers. LAB are natural inhabitants of gastrointestinal (GI) tract, having a number of traits, which make them to be used as "probiotic."

There are several health benefits of using probiotic-containing foods, such as: anticarcinogenic and antimutagenic activities, antimicrobial properties, antihypertension properties, aids in mineral metabolism such as Calcium, reduction in bowel and Crohn's disease symptoms, attenuation of food allergies symptoms and decreasing LDL-cholesterol levels [6, 45, 48, 55, 79, 86]. Some strains of *Lactobacillus* have also resulted in inhibition of pathogenic bacteria such as *Escherichia coli, Salmonella enteritidis, Serratiamarcescens*, and *Shigellasonnei* [17].

6.5   FUNCTIONAL ASPECTS OF FERMENTED DAIRY PRODUCTS

6.5.1   FERMENTED FORTIFIED-PRODUCTS

Fermented dairy foods act as a good matrix for fortification, which can be an efficient tool to combat micronutrient deficiencies. Compatibility of vehicle, fortificant, and process of fortification are important aspects to be taken into consideration for the successful application of food technology. Enrichment and fortification have been used interchangeably [32], but sometimes, it has been defined as the restoration of vitamins and minerals lost during processing [29].

Fermented dairy products, mainly yogurt and soft cheeses, have been used in iron fortification programs [48]. In the iron fortification of powdered nonfat dry milk, ferrous sulfate @ 10 ppm was found to be stable for a period of 12 months. Commonly added elements to dairy beverages are Calcium, Magnesium, Iron, and minerals to compensate for vitamin and mineral losses during processing, e.g., Dairyland Milk-2-Go® (Saputo, Canada), Zen R (Danone, Belgium) and Meiji Love R (Meiji Milk, Japan). Conjugated linoleic acid (CLA) is also used in milk products due to its anti-oxidative and anticancer effects, e.g., Natural Linea® (Corporacion Alimentaria Penanata S.A., Spain).

6.5.2   *APPLICATIONS OF PROBIOTICS*

Dairy foods are considered as excellent vehicles for probiotics having diverse health effects [24]. The technological properties of probiotic include stability and viability for long periods under storage conditions, ease of application, and stability in physicochemical damage during processing [69]. The probiotic bacteria should be acid and bile tolerant; adhesion to gut epithelial cells; able to colonize in the human intestine; possess antimicrobial property, e.g., bacteriocins production. These should be non-pathogenic, mutagenic or carcinogenic and non-toxic to the host organism, and without any plasmid transfer mechanism.

The minimum level of probiotics in the dairy product should be 10^6 CFU/g with a daily intake of about 10^8 CFU/g [82]. Some commercial strains of LAB are *L. fermentum, L. acidophilus, L. rhamnosus, L. casei, L. reuteri, L. plantarum, L. crispatus, B. longum, B. lactis* and *B. animalis*. To enhance the stability and viability of probiotics in food products and the intestinal tract, microencapsulation in hydrocolloid beads has been used [71]. Additionally, microencapsulation protects the cells inside the beads from bacteriophages, increased survival during freeze and spray drying [35, 83, 92]. The combined effect of probiotics and fermented milk has therapeutic significance (Table 6.1).

6.6   PREBIOTICS AND SYNBIOTIC

6.6.1   *FUNCTIONAL FERMENTED MILK-BASED BEVERAGES*

The market of fermented milk is estimated to be €46 billion, with Europe, North America, and Asia accounting for 77% of the market. Milk from various sources, including cow, goat, yak, camel, and sheep, can be used for the preparation of fermented beverages. Traditionally fermentation was done by back-slopping method, but at an industrial scale, starter cultures are used in the form of direct vat set (DVS). Fermented milk beverages are predominantly composed of LAB, mainly *Leuconostoc, lactobacilli, and lactococci* (Table 6.2).

The exact composition of starter culture varies depending on the source and treatment of the milk (e.g., pasteurization), the nature of the surrounding environmental bacteria, hygienic conditions, type, and treatment of containers used for fermentation, incubation temperatures and time. Under colder climates, fermentation is carried out by mesophilic bacteria,

TABLE 6.1 Probiotic Fermented Products and Their Therapeutic Significance

Clinical Condition/Symptoms	Health Benefits of Fermented Products	References
Hepatic encephalopathy: A neurological disorder in which rise in blood ammonia levels with advanced liver disease with impaired brain function	Alteration of the intestinal microflora A decrease in the amount of fecal urease, lowering of blood ammonia and clinical improvement when treated individually with *L. acidophilus*, *Lactobacillus GG*, and *E. faecium*. SF68	[46, 74, 84]
High serum cholesterol levels	Some strains of *L. acidophilus* and bifidobacteria species lower cholesterol	[26, 27, 30, 54, 93, 99]
Lactose intolerance: Manifested by an increase of glucose in the blood or escalation of hydrogen in the breath. It can causes bloating, abdominal pain, flatulence, and diarrhea	Presence of enzyme β-galactosidase in the yogurt culture, *L. acidophilus* aids lactose digestion.	[25, 39, 59]
Renal malfunction	Effect of fermented milk product on cholesterol metabolizing enzymes systems in liver Promotion of excretion of cholesterol through feces. Inhibition of cholesterol absorption by the binding of cholesterol to cells of LAB. The promotion of excretion by binding of bile salts to lactic acid bacterial cells. Reduce the level of toxic amines.	[25, 26, 99]
Side effects of radiotherapy: Radiotherapy alters the intestinal microflora, vascular permeability of the mucosa and intestinal motility.	Treatment with fermented milk containing NCFB 1748 significantly decrease pelvic radiotherapy associated diarrhea	[33, 78]
Tumor or defective immune systems	Lactobacilli and their metabolic products modify both the immune responses and antitumor activities. It is enhanced by macrophage activation and enhancing the activity of natural killer cells and T-cells, maintenance of normal intestinal flora Help the removal of dietary procarcinogens.	[3, 8, 33, 96]

TABLE 6.2 Selected Fermented Beverages and Their Corresponding Microflora; Modified from [53]

Fermented Beverage	Region	Microflora
Amasi	Africa (Zimbabwe)	*Lactobacillus, Lactococcus (L. lactis), Leuconostoc, Enterococcus.* Uncharacterized fungal component
Aryan	Turkey	*Lactobacillus bulgaricus, Streptococcus thermophilus*
Garris	Africa (Sudan)	*Lactobacillus (Lb. fermentum, Lb. plantarum and Lb. paracasei), Lactococcus, Leuconostoc, Enterococcus.* Uncharacterized fungal component
Kefir	Eastern Europe (Caucasian region)	*Lactobacillus, Acetobacter, Lactococcus, Leuconostoc;* Yeast: *Kluyveromyces, Kazachstania, Naumovozyma*
Kivuguto	Africa (Rwanda)	*L. lactis, Leuconostoc (Leu. pseudomesenteroides, Leu. mesenteroides) and* Uncharacterized fungal component
Koumiss	Russia /Asia	*Lactobacillus;* Yeast: *Saccharomyces, Kluyveromyces, and Kazachstania*
Nyarmie	Africa (Ghana)	*Lb. bulgaricus, Lb. helveticus, L. lactis, Leu. Mesenteroides;* Yeast: *Saccharomyces cerevisiae*
Rob	Africa (Sudan)	*Lb. acidophilus, Lb. fermentum, L. lactis, Streptococcus salivarius;* Yeast: *Candida kefir; Saccharomyces cerevisiae*
Shubat	China	*Lactobacillus (Lb. Helveticus, Lb. sakei, Lb. brevis) Leu. Lactis Weissellahellenica and Enterococcus (E. faecalis, E. faecium), Yeast: Candida, Kluyveromyces marxianus, and Kazachstaniaunisporus*
Suusac	Africa (Kenya)	*Lactobacillus (Lb. plantarum, Lb. salivarius, Lb. cruvatus, Lb. Raffinolactis), Leu. Mesenteroides;* Yeast:*, Geotrichum penicillatum, Candida krusei and Rhodotorula mucilaginosa*

e.g., *Lactococcus*, and *Leuconostoc*, whereas at higher temperatures thermophilic bacteria dominate such as *Lactobacillus* and *Streptococcus*.

6.6.1.1   KEFIR

Kefir is a refreshing carbonated fermented milk beverage originated from the Caucasus Mountains. It is produced by the symbiotic association in kefir grains consisting of a combination of bacteria and yeasts. The unique flavor of kefir is due to lactic acid, CO_2 and ethanol produced during fermentation, which provides acidity, viscosity, and a small amount of alcohol content to this beverage. Other than these end products, some components have also been found to contribute to flavors such as diacetyl, acetaldehyde, amino acids, and ethyl. Kefir is considered as "champagne" of fermented dairy products, which might be a better alternative of soft drinks. Starter culture of Kefir is known as "Kefir grains, "which is composed of various microorganisms immobilized on polysaccharide and protein matrix. Size of kefir grains varies from 0.3 to 3.0 cm in diameter, and irregular in shape with multilobular surface attached by a single central section, and having whitish to yellowish-white appearance. The grains have elastic, viscous, and firm texture [19, 50, 73].

LAB are predominant microorganism of Kefir, which converts lactose into lactic acid present in milk resulting in pH reduction, thus contributing to milk preservation. Other microorganisms of kefir constituent are: lactose and non-lactose fermenting yeasts, and acetic acid bacteria (AAB) [50, 72]. Major Kefir bacterial population includes [44, 47, 72]:

- Homofermentative LAB, mainly *Lactobacillus* species, such as *L. helveticus*, *L. delbrueckii* subsp. *bulgaricus*, *L. acidophilus; L. kefiranofaciens* subsp. *Kefirgranum* and *L. kefiranofaciens* subsp. *kefiranofaciens*, *Lactococcus* spp., e.g., *L. lactis* subsp. *cremoris* and *L. lactis* subsp. *lactis,* and *S. thermophilus.*
- Hetero-fermentative bacterial population includes *L. fermentum, L. brevis L. kefiri,* and *L. parakefiri*, citrate-positive strains of *L. lactis*, e.g., *L. lactis*subsp. *Lactis* biovar. *diacetylactis, Leuconostoc mesenteroides* subsp. *cremoris*, and *Leuconostoc mesenteroides* subsp. *mesenteroides.*

The typical flavor of kefir is contributed by metabolites produced during fermentation by citrate-positive strains of LAB [72]. Lactose consists of *Kluyveromyces lactis* var. *lactis, Kluyveromycesmarxianus, Debaryomyceshanseniie,* and *Dekkeraanomala*; while the non-lactose fermenting yeast includes: *Torulasporadelbrueckii, Saccharomyces cerevisiae, Saccharomyces turicensis, Kazachstaniaunispora, Pichia fermentans, Debaryomycesoccidentalis, and Issatchenkiaorientalis.*

Kefir is mainly produced by three methods [19, 72]: (1) the artisanal method; (2) Russian method of commercial production; and (3) the commercial method using pure cultures. Milk from other animal species, soybean milk, coconut milk, fruit juices, and molasses solutions may also be used in the preparation of Kefir [50]. Since ancient time, Kefir has been recommended for the treatment of several clinical conditions such as hypertension, GI problems, heart disease, and allergies [19, 72]. By using probiotic cultures in kefir grains, its functional properties have been enhanced while contributing to various types of health benefits such as antimicrobial activity, anti-inflammatory, anti-carcinogenic effects, hypocholesterolemic effect, and stimulation of the immune system.

6.6.1.2 KOUMISS

Koumiss is a traditional fermented milk beverage made from mare's milk, camel's milk, popular in Eastern Europe and Central Asiatic region. Koumiss is an acid-alcoholic fermented beverage, which contains approximately 0.5–1.5% lactic acid, 2% alcohol, and 2–4% lactose [14]. Traditionally, Koumiss is mainly prepared by back slopping method, i.e., inoculating fresh milk with a small quantity of previously fermented milk [14, 36]. Although the method of Koumiss production is not clearly understood, yet the fermentation of the product is a symbiotic action of two distinct types of microorganisms. The major microflora consists of lactobacilli and yeasts (*Kluyveromyces, Saccharomyces, Candida*), in which *Lactobacillus* is the major player in the fermentation process contributing to the aroma, texture, and acidity of the product. Lactose fermenting yeast converts lactose to alcohol, which gives a typical flavor and taste to Koumiss [58, 98].

Koumiss has been characterized into three types depending on the lactic acid content:

- **'Strong' Koumiss:** It is generated by LAB (*Lactobacillus rhamnosus* and *Lactobacillus bulgaricus*) having a pH of 3.3–3.6 and conversion of about 80–90% of lactose into lactic acid.
- **'Moderate' Koumiss:** It contains *L. fermentum, L. acidophilus, L. casei,* and *L plantarum*, which lower the pH to 3.9–4.5 after fermentation.

- **'Light' Koumiss:** It is prepared by *Streptococcus thermophilus* and *Lc. lactis* ssp. *cremoris* with a final pH of 4.5–5.0.

Yeast strains in Koumiss mainly have moderate to strong proteolytic and lipolytic activities, which contribute to the final sensorial property of the product [31, 89]. Due to various health benefits on the immune system, alimentary canal activity, circulatory, and nervous systems, kidney function and endocrine glands, the Koumiss have been recognized as a wholesome beverage [20, 90, 91].

6.6.1.3 YOGURT BEVERAGES

According to 1992 *Codex Alimentarius*, yogurt is defined as a coagulated milk product that results by the fermentation of lactic acid in milk by the symbiotic activity of *Lactobacillus bulgaricus* and *Streptococcus thermophilus*. Yogurt is also one of the best-known milk products containing probiotic bacteria. Yogurt based beverages are one of the fastest-growing fermented products in the functional food market, due to nutrient content, and health-promoting functions related to strengthening the immune response, ulcer prevention and increase in HDL-cholesterol [16, 18]. Commercial yogurt drinks are generally low fat (<2%) with 8–9% total solid-not-fat (SNF) and 8–12% sugar. Yogurt beverages are generally of two types:

- Regular with low total solid containing fruit juice or puree that are known as hybrid dairy products which provide health and flavor; and
- Another type of yogurt is fortified one containing physiologically active ingredients, such as omega-3-fatty acids, fibers, whey-based ingredients, phytosterols, vitamins, isoflavones, and antioxidants, which confer specific health benefits beyond nutrition.

To enhance the properties of yogurt, yogurt drinks containing prebiotics, probiotics, or synbiotics are being developed. FOS or inulin is used along with probiotic bacteria like *L. reuteri*, *L. casei*. Bifidobacteria, in addition to the starter culture, gives a synergistic effect. Fortification of yogurt drinks with Iron and Calcium are also done.

Due to the health effects of EPA and DHA and the wide consumption, yogurt has been used to develop a vehicle for consumption of fish oil, which is found to have positive effect on myocardial infarction, immune system, eye functions, and neurological disorders such as depression and Alzheimer's disease [51, 60, 81, 100].

6.6.1.4   CULTURED BUTTERMILK

Buttermilk is the aqueous phase of cream, which is released during its churning process. It contains all water-soluble components of milk (protein, lactose, and minerals), but more phospholipids content than milk alone due to the high content in MFGM [61]. Starter culture used for fermentation of buttermilk contains a combination of acid and flavor producing LAB in a ratio of 5:1. The acid-producing agents include *L. lactis* ssp. *cremoris* and *L. lactis* ssp. *lactis*, which are homolactic and produce acid that lowers pH. Flavor producing bacteria include homofermentative *L. lactis* ssp. *diacetylactis,* and heterofermentative *Leuconostoc mesenteroides* ssp. *cremoris* and *Leu. lactis* (which produce lactic acid along with acetic acid, ethanol, and CO_2).

Fermented buttermilk is a good source of Potassium, Phosphorus, vitamin $B_{12,}$ riboflavin, enzyme, Calcium, and protein [12]. Various trials have shown health benefits of fermented buttermilk, e.g., cholesterol reduction, anticancer, antiviral, and lowering of blood pressure. For lactose intolerance population, it serves as a better alternative to whole milk. Additionally, fermented buttermilk also helps in removal of undesirable acid in the stomach by creating a layer on the lining of the stomach. Studies have also proved to decrease the chances of bladder cancer by consumption of buttermilk and reduction of cholesterol level [13, 43]. It has been found that continuous consumption of cultured buttermilk reduces the arterial and systolic blood pressure.

The concentration of ACE-I is also found to be decreased, while there was no effect on the angiotensin-II converting enzyme (ACE-II). Another study has shown anti-rotavirus activity using milk fat globule membrane from buttermilk [23]. Nowadays, probiotic LAB and bifidobacteria are being used along with starter cultures adding more therapeutic effects to buttermilk.

6.6.1.5   ACIDOPHILUS MILK

Acidophilus milk is one of the first probiotic fermented milk produced by *L. acidophilus* isolated from the feces of breastfed infants. For the last two decades, the nutritional and therapeutic effect of fermented dairy products containing *Lactobacillus acidophilus* as a food or supplement has been studied [4, 5, 7, 77, 99]. Acidophilus milk is prepared by heating the milk to high temperature (95°C for 1 h), to decrease the microflora naturally present in milk. *Lactobacillus acidophilus* is inoculated to milk @ 2–5% and incubated at 37°C. The final acidity of acidophilus milk varies up to 1% lactic acid, but for therapeutic use 0.6–0.7% acidity is desirable. Alternatively, sweet acidophilus milk has been introduced, in which *Lactobacillus acidophilus* has been added without incubation, so that therapeutic effects of culture will be observed after reaching the gastrointestinal tract (GIT).

6.6.1.6 WHEY-BASED BEVERAGES

Whey is a subproduct of cheese and paneer industry, and it constitutes about 85–90% of the milk volume and about 55% of the milk nutrients. It is composed of lactose (5%), proteins (0.85%), minerals (0.53%) and fat (0.36%). Being a rich source of whey, it acts as a good medium for fermentation being rich in lactose. Lactose present in whey as a whole or deproteinized whey is hydrolyzed to glucose and galactose, which increase its sweetness. These hydrolyzed lactose syrups after condensation are used in ice cream and confectionery products. Due to the high content of essential amino acids (notably lysine, cysteine, and methionine) and cystin, whey proteins are one of the most nutritionally valuable proteins. The β-lactoglobulin (BLG) is one of the main proteins, which is considered as the major allergen of milk.

LAB is extensively used for fermentation of dairy products, as they can hydrolyze milk proteins and can degrade BLG during growth in whey and milk [9, 64]. Several strains of *Lactobacillus acidophilus, L. paracasei*, and *Bifidobacterium* have been reported to breakdown the BLG allergenic epitopes *in vitro* [65, 70].

Moreover, the use of probiotic strains of LAB and bifidobacteria has shown to induce oral tolerance to BLG and restoration of un-characteristic protein transport and its degradation in the intestinal mucosa [57, 67, 70]. It was also found that when infant formula is supplemented with viable LGG, then it could help in the prevention of allergy due to cow's milk [40].

Whey is considered as a suitable medium for preparation of lactic and alcoholic beverages due to the presence of valuable nutrients, which promote the growth of LAB and yeast during fermentation. A koumis-like beverage has also been prepared by combination of equal quantity of whey and buttermilk by a mixture of cultures containing *Lactobacillus acidophilus, Lactobacillus bulgaricus, and* koumis yeasts.

A variety of alcoholic and beer-like beverages have been prepared by using whey as a substrate. Beers like beverages preparing using whey are: a) A malted whey beer; b) A whey malt beer; c) An alcoholic whey beer;

and d) A whey nutrient beer. 'Acidowhey' is a fermented whey drink made by fermentation with an active culture of *L. acidophilus*. Antibacterial properties of 'Acidowhey' against Gram-positive and Gram-negative microorganisms such as *Escherichia coli, Micrococcus flavus, Bacillus subtilis, and Staphylococcus aureus* have also been demonstrated.

6.7 SUMMARY

The traditional fermented beverages should be inspired in the future of the fermented beverage for the development of novel products. With the advancements in probiotics, research, and characterization of the health claims and starter design will increase. In the dairy sector, probiotic has intrinsically been linked to the health claims of many beverages, such as probiotic soy beverages, and juice beverages. With the increasing knowledge and discovery of probiotics, alternative means of probiotic delivery is required. As research into the fermentation of waste and byproducts (e.g., whey) continues, there is a potential for a significant environmental impact. Due to increasing consumer health awareness, interest in functional foods, and fermented milk beverages is increasing.

KEYWORDS

- angiotensin-I converting enzyme (ACE-I)
- angiotensin-II converting enzyme (ACE-II)
- antimicrobial peptides
- casein
- conjugated linoleic acid
- cytokines
- cytosol
- dairyceuticals
- food beyond nutrition
- immunoglobulins
- intestinal microbes
- oligosaccharides

REFERENCES

1. Alzamora, S. M., Salvatori, D., Tapia, S., Lopez-Malo, M. A., Welti-Chanes, J., & Fito, P., (2005). Novel functional foods from vegetable matrices impregnated with biologically active compounds. *Journal of Food Engineering, 67*, 205–214.

2. American Dietetic Association (ADA), (2009). Position of the American Dietetic Association: Functional foods. *Journal of American Dietetic Association, 109*, 735–746.

3. Anand, S. K., Srinivasan, R. A., & Rao, L. K., (1985). Antibacterial activity with *bifidobacterium bifidum. Cultured Dairy Products Journal, 20*, 21–23.

4. Anderson, J. W., & Gilliland, S. E., (1991). Effect of fermented milk (yogurt) containing *Lactobacillus acidophilus* L1 on serum cholesterol in hypercholesterolemic persons. *The Journal of the American College of Nutrition, 18*(1), 43–50.

5. Andrade, S., & Borges, N., (2009). Effect of fermented milk containing *Lactobacillus acidophilus* and Bifidobacterium longum on plasma lipids of women with normal or moderately elevated cholesterol. *Journal of Dairy Research, 76*(4), 469–474.

6. Arunachalam, K. D., (1999). Role of *Bifidobacteria* in nutrition, medicine and technology. *Nutrition Research, 19*, 1559–1597.

7. Azlin, M., Jiang, T., & Savaiano, D. A., (1997). Improvement of lactose digestion by humans following ingestion of unfermented acidophilus milk: Influence of bile sensitivity, lactose transport, and acid tolerance of *Lactobacillus acidophilus. Journal of Dairy Science, 80*, 1535–1547.

8. Benno, Y., Sawada, K., & Mitsuoka, T., (1984). The intestinal microflora of infants: Composition of fecal flora in breast-fed and bottle fed infants. *Microbiology and Immunology, 28*, 975–986.

9. Bertrand-Harb, C., Ivanova, I. V., Dalgalarrondo, M., & Haertle, T., (2003). Evolution of β-lactoglobulin and α-lactalbumin content during yogurt fermentation. *International Dairy Journal, 13*, 39–45.

10. Bigliardia, B., & Galati, F., (2013). Innovation trends in the food industry: The case of functional foods. *Trends in Food Science and Technology, 31*, 118–129.

11. Blandon, J., Cranfield, J., & Henson, S., (2007). *International Food Economy Research Group: Department of Food, Agricultural and Resource Economics.* http://www4.agr. gc.ca/resources/prod/doc/misb/fbba/nutra/pdf/u_of_guelph_functional_foods_review_ final_25jan2008_en.pdf (Assessed on 20 July 2019).

12. Conway, V., Gauthier, S. F., Pouliot, Y., Cobo-Angel, C., Wichtel, J., & Ceballos-Márquez, A., (2014). Buttermilk: Much more than a source of milk phospholipids. *Animal Frontiers, 4*(2), 44–51.

13. Conway, V., Couture, P., Richard, C., Gauthier, S. F., Pouliot, Y., & Lamarche, B., (2013). Impact of buttermilk consumption on plasma lipids and surrogate markers of cholesterol homeostasis in men and women. *Nutrition, Metabolism & Cardiovascular Diseases, 23*(12), 1255–1262.

14. Danova, S., Petrov, K., Pavlov, P., & Petrova, P., (2005). Isolation and characterization of Lactobacillus strains involved in koumiss fermentation. *International Journal of Dairy Technology, 58*(2), 100–105.

15. Diplock, A. T., Aggett, P. J., Ashwell, M., Bornet, F., Fern, E. B., & Robertfroid, M. B., (1999). Scientific concepts of functional foods in Europe: Consensus document. *British Journal of Nutrition, 81*(1), S1–S27.

16. Douaud, C., (2007). Yogurt drinks are leading food and beverage product. Nielsen, A. C., (ed.), *Nutra Ingredients-USA.com Consumer Trends*. https://www.foodnavigator-usa.com/Article/2007/01/25/Yogurt-drinks-are-leading-food-and-beverage-product-ACNielsen# (Accessed on 20 July 2019).

17. Drago, L., Gismondo, M. R., Lombardi, A., Haen, C., & Gozzoni, L., (1997). Inhibition of enteropathogens by new *Lactobacillus* isolates of human intestinal origin. *FEMS Microbiology Letters, 153*, 455–463.

18. Fabian, E., & Elmadfa, I., (2006). Influence of daily consumption of probiotic and conventional yogurt on the plasma lipid profile in young healthy women. *Annals of Nutrition and Metabolism, 50*(4), 387–393.

19. Farnworth, E. R., & Mainville, I., (2008). Kefir-a fermented milk product. In: Farnworth, E. R., (eds.), *Handbook of Fermented Functional Foods* (2nd edn., pp. 89–127). Taylor & Francis Group, LLC, New York.

20. Fedechko, I. M., Hrytsko, R., & Herasun, B. A., (1995). The anti-immunodepressive action of *koumiss* made from cow's milk. *Likars' ka Sprava, 9,* 104–106.

21. Fern, E., (2007). Marketing of functional foods: A point of view of the industry. In: *International Developments in Science & Health Claims* (pp. 122–130). ILSI international symposium on functional foods in Europe, Paris–France.

22. Fox, P. F., (2003). Indigenous enzymes in milk. In: Fox, P. F., & Sweeney, P. L. H., (eds.), *Advanced Dairy Chemistry: Proteins* (Vol. 1, pp. 447–467). Kluwer Academic/Plenum Publishers, New York.

23. Fuller, K. L., Kuhlenschmidt, T. B., Kuhlenschmidt, M. S., Jiménez-Flores, R., & Donovan, S. M., (2013). Milk fat globule membrane isolated from buttermilk or whey cream and their lipid components inhibit infectivity of rotavirus *in vitro*. *Journal of Dairy Science, 96*(6), 3488–3497.

24. Gürakan, G. C., Cebeci, A., & Ozer, B., (2009). Probiotic dairy beverages: Microbiology and technology. In: Yildiz, F., (ed.), *Development and Manufacture of Yogurt and Other Functional Dairy Products* (pp. 165–195). Taylor & Francis Group, Florence, S. C.

25. Gilliland, S. E., (1989). Acidophilus milk products. A review of potential benefits to consumers. *Journal of Dairy Science, 72*, 2483–2494.

26. Gilliland, S. E., (1985). *Bacterial Starter Cultures for Foods* (pp. 233–245). CRC Press. Inc., Boca Raton, Florida.

27. Gupta, P. K., & Prabhu, T. R., (2004). Hypocholesterolaemic activity of *Lactobacillus acidophilus*. *Journal of Food Science and Technology, 41*, 695–603.

28. Haque, E., & Chand, R., (2006). Milk protein derived bioactive peptides. (On-line) UK, http://www.dairyscience.info/bio-peptides.htm (Assessed on 20 July 2019).

29. Hoffpauer, D. W., & Wright, S. L., (1994). Enrichment of rice. In: Marshall, W. E., & Wadsworth, J. I., (eds.), *Rice Science and Technology* (pp. 287–314). Marcel Rekker, New York.

30. Hosona, A., & Tono-oka, T., (1995). Binding of cholesterol with lactic acid bacterial cells. *Milchwissenchaft, 51*, 619–623.

31. Jakobsen, M., & Narvhus, J., (1996). Yeasts and their possible beneficial and negative effects on the quality of dairy products. *International Dairy Journal, 6*, 755–768.

32. Joint FAO/WHO Codex Alimentarius Commission, (1994). *Codex Alimentarius: Foods for Special Dietary Uses Including Food for Infants and Children* (p. 10). Food & Agriculture Organization, Rome–Italy.

33. Kansal, V. K., (2001). Probiotic application of culture and culture containing milk products. *Indian Dairy Man, 53*, 49–55.

34. Kausar, H., Saeed, S., Ahmad, M. M., & Salam, A., (2012). Studies on the development and storage stability of cucumber-melon functional drink. *Journal of Agricultural Research, 50*, 239–248.

35. Kearney, L., Upton, M., & Loughlin, A., (1990). Enhancing the viability of *Lactobacillus plantarum* inoculum by immobilizing the cells in Calcium- alginate beads. *Applied Environmental Microbiology, 56*, 3112–3116.

36. Kerr, T. J., & McHale, B. B., (2001). *Applications in General Microbiology: A Laboratory Manual* (p. 231). Hunter Textbooks, Winston-Salem.

37. Khan, R. S., Grigor, J., Winger, R., & Win, A., (2013). Functional food product development-opportunities and challenges for food manufacturers. *Trends Food Science Technology, 30*, 27–37.

38. Khanal, R. C., & Olson, K. C., (2004). Factors affecting conjugated linoleic acid content in milk, meat and egg: A review. *Pakistan Journal of Nutrition, 3*, 82–98.

39. Kim, H. S., & Gilliland, S., (1983). *Lactobacillus acidophilus* as a dietary adjunct for milk to aid lactose digestion in humans. *Journal of Dairy Science, 66*, 956–966.

40. Kirjavainen, P. V., Salminen, S. J., & Isolauri, E., (2003). Probiotic bacteria in the management of atopic disease: Underscoring the importance of viability. *Journal of Pediatric Gastroenterology and Nutrition, 36*, 223–227.

41. Kleinschmidt, A., (2003). When food is not just food. *Research Leaflet*, p. 2.

42. Kotilainen, L., Rajalahti, R., Ragasa, C., & Pehu, E., (2003). Health enhancing foods: Opportunities for strengthening the sector in developing countries. *Agriculture and Rural Development Discussion Paper* (p. 30). FAO, Rome.

43. Larsson, S. C., Mannisto, S., Virtanen, M. J., Kontto, J., Albanes, D., & Virtamo, J., (2009). Dairy foods and risk of stroke. *Epidemiology, 20*, 355–360.

44. Leite, A. M. O., Mayo, B., Rachid, C. T. C. C., Peixoto, R. S., Silva, J. T., Paschoalin, V. M. F., & Delgado, S., (2012). Assessment of the microbial diversity of Brazilian kefir grains by PCR-DGGE and pyrosequencing analysis. *Food Microbiology, 31*(2), 215–221.

45. Liong, M. T., Fung, W. Y., Ewe, J. A., & Kuan, L., (2009). The improvement of hypertension by probiotics: Effects on cholesterol, diabetes, renin, and phytoestrogens. *International Journal of Molecular Sciences, 10*(9), 3755–3775.

46. Loguerolo, C., Vechhio, B., & Collorti, M., (1987). *Enterococcus* lactic acid bacteria strain SF68 and lactulose in hepatic encephalopathy: A controlled study. *Journal of International Medical Research, 15*, 335–343.

47. Lopitz-Otsoa, F., Rementeria, A., Elguezabal, N. & Garaizar, J. (**2016**). Kefir: A symbiotic yeasts-bacteria community with alleged healthy capabilities. *Revista Iberoamericana de Micología* (Iberoamerican Magazine of Mycology), 23(2), 67–74.

48. Lourens-Hattingh, A., & Viljoen, B. C., (2011). Yogurt as probiotic carrier food. *International Dairy Journal, 1*(1/2), 1–17.

49. Lysionek, A. E., Zubillaga, M. B., Salguiero, M. J., Pinerio, A., Caro, A. R., Weill, R., & Boccio, R. J., (2002). Bioavailability of microencapsulated ferrous sulphate in powdered milk produced from fortified fluid milk: A prophylactic study in rats. *Nutrition, 18*(3), 279–281.

50. Magalhães, K. T., Dias, D. R., De Melo, Pereira, G. V., Oliveira, J. M., Domingues, L., Teixeira, J. A., De Almeida, S. J. B., & Schwan, R. F., (2011). Chemical composition and sensory analysis of cheese whey-based beverages using kefir grains as starter culture. *International Journal Food Science and Technology, 46*(4), 871–878.

51. Marchioli, R., Bomba, E., Chieffo, C., Maggioni, A. P., Schweiger, C., & Tognoni, G., (1999). Dietary supplementation with n-3 polyunsaturated fatty acids and vitamin E after myocardial infarction: Results of the GISSI-Prevenzione trial. *Lancet*, *354*(9177), 447–455.

52. Mark-Herbert, C., (2004). Innovation of a new product category–functional foods. *Technovation*, *24*, 713–719.

53. Marsh, A. J., Hill, C., Ross, R. P., & Cotter, P. D., (2014). Fermented beverages with health-promoting potential: Past and future perspectives. *Trends in Food Science Technology*, *38*, 113–124.

54. Marshall, V. M., (1999). Bioyogurt, How healthy? *Dairy Ind. Int.*, *61*, 28–29.

55. Marteau, P. R., Vrese, M., Cellier, C. J., & Schrezenmeir, J., (2000). Protection from gastrointestinal diseases with the use of probiotics. *American Journal of Clinical Nutrition*, *73*, 430–436.

56. Menrad, K., (2003). Market and marketing of functional food in Europe. *Journal of Food Engineering*, *56*, 181–188.

57. Mizumachi, K., & Kurisaki, J., (2003). Induction of oral tolerance in mice by continuous feeding with β-lactoglobulin and milk. *Bioscience, Biotechnology, and Biochemistry*, *66*, 1287–1294.

58. Montanari, G., Zambonelli, C., Grazia, L., Kamesheva, G. K., & Shigaeva, M. K., (1996). *Saccharomyces unisporus* as the principal alcoholic fermentation microorganism of traditional *koumiss*. *Journal of Dairy Science*, *63*, 327–331.

59. Morley, R. G., (1979). Potential of liquid yogurt. *Cultured Dairy Product Journal*, *14*, 30–33.

60. Morris, M. C., Evans, D. A., Bienias, J. L., & Tangney, C. C., (2003). Consumption of fish and omega-3 fatty acids and risk of incident Alzheimer disease. *Archives Neurology*, *60*(7), 940–946.

61. Mulder, H., & Walstra, P., (1974). The milk fat globule. In: *Emulsion Science as Applied to Milk Products and Comparable Foods: Report No. 4* (pp. 136–192). Centre for Agricultural Publishing and Documentation, Wageningen, Netherlands.

62. Niva, M., (2007). All foods affect health: Understandings of functional foods and healthy eating among health-oriented Finns. *Appetite*, *48*, 384–393.

63. Ohama, H., Ikeda, H., & Moriyama, H., (2006). Health foods and foods with health claims in Japan. *Toxicology*, *22*, 95–111.

64. Özer, B., & Kirmaci, H. Y., (2010). Functional milks and dairy beverages. *International Journal of Dairy Technology*, *63*, 1–15.

65. Pescuma, M., Hébert, E. M., Mozzi, F., & Font De Valdez, G., (2007). Hydrolysis of whey proteins by *Lactobacillus acidophilus, Streptococcus thermophilus,* and *Lactobacillus delbrueckii subsp. bulgaricus* grown in a chemically defined medium. *Journal of Applied Microbiology*, *103*, 1738–1743.

66. Pescuma, M., Hébert, E. M., Mozzi, F., & Font de Valdez, G., (2008). Whey fermentation by thermophilic lactic acid bacteria (LAB): Evolution of carbohydrates and protein content. *Food Microbiology*, *25*, 442–451.

67. Pessi, T., Sutas, Y., Marttinen, A., & Isolauri, E., (1998). Probiotics reinforce mucosal degradation of antigens in rats: Implications for therapeutic use of probiotics. *Journal of Nutrition*, *128*, 2313–2318.

68. Pouliot, Y., & Gauthier, S. F., (2006). Milk growth factors as health products: Some technological aspects. *International Dairy Journal*, *16*, 1415–1420.

69. Prado, F. C., Parada, J. L., Pandey, A., & Soccol, C. R., (2008). Trends in non-dairy probiotic beverages. *Food Research International, 41*, 111–123.

70. Prioult, G., Fliss, I., & Pecquet, S., (2003). Effect of probiotic bacteria on induction and maintenance of oral tolerance to beta-lactoglobulin in gnotobiotic mice. *Clinical and Diagnostic Laboratory Immunology, 10*, 787–792.

71. Rao, A. V., Shiwnarain, N., & Maharaj, I., (1989). Survival of microencapsulated *Bifidobacterium pseudolongum* in simulated gastric and intestinal juices. *Canadian Institute of Food Science and Technology Journal, 22*, 345–349.

72. Rattray, F. P., & O'Connell, M. J., (2011). Fermented milks–Kefir. *Encyclopedia of Dairy Sciences, 2*, 518–524.

73. Rea, M. C., Lennartsson, T., Dillon, P., Drinan, F. D., Reville, W. J., Heapes, M., & Cogan, T. M., (1996). Irish kefir-like grains: their structure, microbial composition and fermentation kinetics. *Journal of Applied Bacteriology, 81*(1), 83–94.

74. Read, A. E., Mc Carthy, C. F., & Heaton, K. W., (1968). *Lactobacillis acidophilus* (ENPAC) in treatment of hepatic encephalopathy. *British Medical Journal, 1*, 1267–1269.

75. Robertfroid, M. B., (2000). A European consensus of scientific concepts of functional foods. *Nutrition, 16*, 689–691.

76. Robertfroid, M. B., (2002). Global view on functional foods: European perspectives. *British Journal of Nutrition, 88*, S133–S138.

77. Salminen, S., Isolauri, E., & Salminen, E., (1996). Clinical uses of probiotics for stabilizing the gut mucosal barrier: Successful strains and future challenges. *Anton Leeuwenhoek, 70*(2–4), 347–358.

78. Salminen, S., Ouwehand, A. C., & Isolauri, E., (1998). Clinical applications of probiotic bacteria. *International Dairy Journal, 8*, 563–572.

79. Salminen, S., Deighton, M., Benno, Y., & Gorbach, S. L., (1998). Lactic acid bacteria in health and disease. In: Salminen, S., & Von Wright, A., (eds.), *Lactic Acid Bacteria: Microbiology and Functional Effect* (pp. 211–253). Marcel Dekker Inc., New York.

80. Sanguansri, L., & Augustin, M. A., (2009). Microencapsulation in functional food product development. In: Smith, J., & Charter, E., (eds.), *Functional Food Product Development* (pp. 3–23). John Wiley & Sons, New York.

81. Schmidt, E. B., Varming, K., Pedersen, J. O., & Lervang, H. H., (1992). Long-term supplementation with omega-3 fatty acids, II: Effect on neutrophil and monocyte chemotaxis. *Scand J. Clin. Lab. Invest., 52*(3), 229–236.

82. Shah, N. P., Ding, W. K., Fallourd, M. J., & Leyer, G., (2000). Improving the stability of probiotic bacteria in model fruit juices using vitamins and antioxidants. *Journal of Food Science, 75*, 278–282.

83. Sheu, T. Y., & Marshall, R. T., (1993). Microencapsulation of lactobacilli in calcium alginate gels. *Journal of Food Science, 54*, 557–561.

84. Shiby, V. K., & Mishra, H. N., (2013). Fermented milks and milk products as functional foods—A review. *Critical Reviews in Food Science and Nutrition, 53*(5), 482–496.

85. Side, C., (2006). Overview on marketing functional foods in Europe. *Proceedings Functional Food Network General Meeting* (pp. 8–10). New York.

86. Sindhu, S. C., & Khetarpaul, N., (2003). Effect of feeding probiotic fermented indigenous food mixture on serum cholesterol levels in mice. *Nutrition Research, 23*, 1071–1080.

87. Sloan, E., (2012). Beverage trends in 2012 and beyond. *Agro Food Industries, 23*, 8–12.

88. Spence, J. T., (2006). Challenges related to the composition of functional foods. *Journal of Food Composition and Analysis, 19*, S4–S6.

89. Spinnler, H. E., Berger, C., Lapadatescu, C., & Bonnarme, P., (2001). Production of sulphur compounds by several yeasts of technological interest for cheese ripening. *International Dairy Journal, 11*, 245–252.

90. Stoianova, L. G., Abramova, L. A., & Ladodo, K. S., (1998). Sublimation-dried mare's milk and the possibility of its use in creating infant and dietary food products. *Voprosi Pitania, 3*, 64–67.

91. Sukhov, S. V., Kalamkarova, L. I., Ilchenko, L. A., & Zhangabylov, A. K., (1986). Microfloral changes in the small and large intestines of chronic enteritis patients on diet therapy including sour milk products. *Voprosi Pitania, 7*, 14–17.

92. Sung, H. H., (1997). Enhancing survival of lactic acid bacteria in ice cream by natural encapsulation. *Dissertation Abstracts International, 13*, 5407–5413.

93. Usman., & Hosono, A., (2000). Effect of administration of *L. gasseri* on serum lipids and fecal steroids in hypercholesterolaemic rats. *Journal of Dairy Science, 83*, 1705–1711.

94. Valls, J., Pasamontes, N., Pantaleon, A., Vinaixa, S., Vaque, M., Soler, A., Millan, S., & Gomez, X., (2013). Prospects of functional foods/nutraceuticals and markets. In: Ramawat, K. G., & Merillon, J. M., (eds.), *Natural Products* (pp. 478–491). Springer-Verlag, Berlin.

95. Van Trijp, H., (2007). Consumer understanding and nutritional communication. In: *International Developments in Science & Health Claims* (pp. 133–139). ILSI international symposium on functional foods in Europe. Paris.

96. Varnam, A., & Sutherland, J., (1994). *Milk and Milk Products Technology, Chemistry and Microbiology* (p. 451). Chapman and Hall, London.

97. Vasijevic, T., & Shah, N. P., (2008). Probiotics–from Metchnikoff to bioactives. *International Dairy Journal, 18*(7), 714–728.

98. Viljoen, B. C., (2001). The interaction between yeasts and bacteria in dairy environments. *International Journal of Food Microbiology, 69*, 37–44.

99. Walker, D. K., & Gilliland, S. E., (1993). Relationships among bile tolerance, bile salt deconjugation, and assimilation of cholesterol by *Lactobacillus acidophilus*. *Journal Dairy Science, 76*(4), 956–961.

100. Werkman, S. H., & Carlson, S. E., (1996). A randomized trial of visual attention of preterm infants fed docosahexaenoic acid until nine months. *Lipids, 31*(1), 91–97.

101. Wootton-Beard, P. C., & Ryan, L., (2011). Improving public health? The role of antioxidant-rich fruit and vegetable beverages. *Food Research International, 44*, 3135–3148.

CHAPTER 7

BIOFUNCTIONAL YOGURT AND ITS BIOACTIVE PEPTIDES

JAGRANI MINJ, SUBROTA HATI, BRIJ PAL SINGH, and SHILPA VIJ

ABSTRACT

Fermented milk and milk products like yogurt provide health benefits to the consumer. This also contains various components that positively affect many physiological functions in our body. Research studies have shown that fermented functional yogurt and their bioactive components are useful for the treatment of many life-style related diseases, e.g., obesity, cancer, osteoporosis, coronary heart disease (CHD) and hypertension. Yogurt containing probiotic bacteria generated bioactive peptides, and other metabolic end-products contribute to the functional aspects of biofunctional yogurt. These scientific findings need to explore proper validation in the near future by investigation of multifunctional aspects of yogurt and yogurt derived bioactive peptides and their mechanisms of action.

7.1 INTRODUCTION

Yogurt is a popular fermented dairy product since ancient times and is a mostly consumable product in western countries. Yogurt is also popular as dahi or curd in India. Most households prepare dahi by the traditional method by using previous day culture. Daily consumption of yogurt leads to an improvement in human health and also cures various diseases. Many researchers have studied yogurt for its maximum benefits using additionally lactic acid bacteria (LAB) and probiotic bacteria. It is well known that yogurt is made up of two bacteria *Streptococcus thermophilus* and *Lactobacillus bulgaricus* ssp. *delbrueckii* (Figure 7.1). Both these bacteria have mutual benefits from each other by the production of specific compounds. Yogurt bacteria show synergistic interaction by the production of lactic acid, acetaldehyde, and exopolysaccharides (EPS). *L. bulgaricus* exhibits proteo-lytic activity by breaking down into smaller peptides and amino acids, which are easily utilized by *S. thermophilus*. These yogurt bacteria also produce a typical yogurt flavor acetaldehyde. Although other flavoring compound diacetyl, acetoin, etc. are also present in yogurt, but their concentration is low as compared with acetaldehyde.

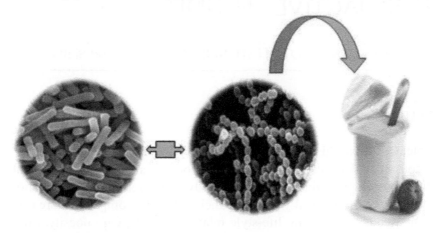

FIGURE 7.1   (See color insert.) Yogurt cultures *Lactobacillus delbrueckii* ssp. *bulgaricus* and *Streptococcus thermophilus*.

Nowadays, the consumer demands dairy products with maximum health benefits. And yogurt is one of them that fulfill the criteria to the consumers. Therefore, manufacturers are developing milk products with more than one biofunctional property. The biofunctional properties of yogurt can be increased by using probiotic yogurt culture itself or by addition of *lactoba-cillus* probiotic cultures. Some other ingredients such as prebiotics (which are also called dietary fibers) are beneficial for the growth and promotion of probiotic bacteria by acting as a nutrient supplement. The dietary fibers, which are frequently used as prebiotics, are inulin, galactooligosaccharides, fructooligosaccharides, maltose, starch, etc. Some other important bioactive components in yogurts are listed in Table 7.1.

This chapter mainly focuses on the importance of fermented dairy product yogurt in the promotion of a healthy lifestyle, and to provide more information on the nutritional and therapeutic values of the yogurt and its bioactive peptides. As fermented milk product yogurt has been studied widely for its biofunctional properties, therefore past research studies on yogurt are included in this chapter to elaborate the collective information.

TABLE 7.1 Examples of Bioactive Components in Yogurt

Category	Examples
Bioactive lipid	May become useful: Conjugated linoleic acid
Bioactive peptides	Bioactive peptides of milk proteins are listed in Table 7.6 to 7.9.
Major milk proteins	αs1-casein, αs2-casein, β-casein, κ-casein, α-lactalbumin, β-lactoglobulin, proteose-peptone, glycomacropeptide
Minerals	Calcium, phosphorus, magnesium, and zinc
Minor proteins; Naturally occurring bioactive peptides	Adrenocorticotropic hormone, calcitonin, plasmin enzymes (plasmin, catalase, glutathione peroxidize, lactoperoxidase), growth factors, immunoglobulins, insulin, lactoferrin, lactoperoxidase, luteinizing hormone-releasing hormone, lysozyme, prolactin, relaxin, somatostatin, thyroid-stimulating hormone, thyrotropin-releasing hormone, and transferrin.
Prebiotics	Those being tested are inulin, beta-glucans, pectin, gums, and resistant starch and fiber.
Probiotics	Main cultures: *Lactobacillus bulgaricus* and *Streptococcus thermophilus* Supplementary cultures: *L. acidophilus*, *L. casei*, *Bifidobacterium bifidum*, *B. longum*, *B. infantis*, and *B. breve*.
Vitamins	Vitamin D, vitamin B_{12}, thiamine, riboflavin, niacin, and folate

Source: Reprinted with permission from [123]. © 2009 John Wiley and Sons.

7.2 NUTRITIONAL VALUE OF YOGURT

When it comes to the nutritional value of yogurt, it is one of the best product having maximum nutritional components are available. Fermented milk yogurt is an exceptional source of minerals such as calcium, protein, phosphorus [75], and B-vitamins (B-12, B-6), folic acid, Niacin, and riboflavin.

Milk itself is a good precursor of many nutritional components and further LAB, and probiotic bacteria enhance more valuable components by synthesizing vitamin B12, folate, etc. [79]. Folate production can be increased six-folds by using a combination of *Bifidobacterium animalis* and *S. thermophilus* during fermentation [30]. As it has been known that folate is synthesized in the mammalian cells, hence the requirements of this vitamin can be fulfilled by the exogenous supplementation. Folate has been reported for many important metabolic pathways, nucleic acids, and some other amino acids biosynthesis, and vitamins, DNA replication and also acts as a cofactor [55, 85]. Many *Lactobacillus* species are reported for their folate production such as *L. brevis* [65], *L. bulgaricus* [86, 93, 158], *L. fermentum* [65], *L. plantarum* [65, 66], *S. thermophilus* [63, 67, 93, 158], etc.

Similarly, vitamin B12 synthesis has also seen by LAB. Vitamin B12 is also called riboflavin, a water-soluble vitamin. Riboflavin acts as coenzymes, especially in the physiological role of flavin mononucleotide (FMN) and flavin adenine dinucleotide (FAD), which has an important role in the oxidation-reduction reactions as electron carriers. Reported LAB for riboflavin synthesis are: *L. reuteri* [53, 87], *L. plantarum* [16, 89], *L. rossiae* [33], and *L. fermentum* [11].

Nutritional value of yogurt is affected by milk-based sources, processing factors, yogurt culture, and type of LAB and probiotic bacteria, which is used in the fermentation. Mazza [106] reported that probiotic bacteria directly affect the physiological and nutritional value of the final yogurt. Bronner and Pansu [21] mentioned that low pH of yogurt enhances the ionization of Calcium and enables its absorption in the intestine. Therefore, yogurt is one of the best sources of Calcium bioavailability than any other dairy product. Yogurt contains a higher amount of proteins than that of milk, due to the addition of extra solid-not-fat (SNF) for the standardization purpose. These proteins easily get digested by proteolytic LAB and their active enzymes such as peptidases, which contribute to the rise of concentration of free amino acid proline and glycine in the final product [17]. These proteolytic bacteria also generate the bioactive peptides from the available proteins in the yogurt, such as: casein, lactoferrin, lactoglobulin, and serum albumin. These peptides promote and enhance gut microflora [130].

Yogurt also contains a higher concentration of conjugated linoleic acid (CLA) than the milk. Production of short-chain fatty acids such as butyrate, acetate, and propionate occurs in the colon by colonic bacteria that may reduce the circulatory serum cholesterol concentration [71] and alleviates the level of total cholesterol (LDL) in the blood [170].

Production of nutritional components is also influenced by a total number of yogurt bacteria present in the product. Actually, the total number of bacteria in the yogurt ranges from 10^8 to 10^9 colony forming units (CFU) per g during the production. However, CFU decreases with storage time and temperature. Nutritional components of different types of yogurt vary according to the fat content and added ingredients. The details of the nutritional components of different types of yogurts (per 100 g) are mentioned in Tables 7.2 and 7.3. The energy obtained from the yogurt is mainly influenced by the available fat, protein, and carbohydrate contents in that particular yogurt.

TABLE 7.2 Nutritional Composition of Different Types of Yogurt (per 100 g)

Component	Whole Milk Yogurt	Low Fat Yogurt	Non-Fat Yogurt	Greek Style Yogurt	Drinking Yogurt
Calcium (mg)	200	162	160	126	100
Carbohydrate (g)	7.8	7.4	8.2	4.8	13.1
Carotene (µg)	21	Trace	Trace	Trace	Trace
Energy (Kcal)	79	56	54	133	62
Fat (g)	3.0	1.0	0.2	10.2	Trace
Folate (µg)	18	18	8	6	12
Niacin (mg)	0.2	0.1	0.1	0.1	0.1
Phosphorus (mg)	170	143	151	138	81
Potassium (mg)	280	228	247	184	130
Protein (g)	5.7	4.8	5.4	5.7	3.1
Riboflavin (mg)	0.27	0.22	0.29	0.13	0.16
Thiamin (mg)	0.06	0.12	0.04	0.12	0.03
Vitamin B12 (mg)	0.2	0.3	0.2	0.2	0.2
Vitamin B6 (mg)	0.10	0.01	0.07	0.01	0.05
Vitamin D (mcg)	0	0.1	Trace	0.1	Trace

Source: The Dairy Council [161]

TABLE 7.3 Average Composition and Energetic Percentage Value of Cow's Milk and Selected Types of Yogurt

Parameter	Milk		Yogurt	
	Full Cream	Skimmed	Skimmed White	Full Cream White
Dry residual, %	12.5	9.5	14	22
Energetic value, kcal/ 100g	62	35	47	102
Fat, %	3.4	0.2	0.2	4.5
Lactic acid, %	0.003	0.003	0.5	0.5
Protein, %	3.3	3.4	6.7	4.3
Total sugar, %	4.8	5.0	4.9	15.7

Source: Agroscope Composition [3]; Agricultural Research Service [2].

Furthermore, to increase the nutritional value and biofunctional properties of normal yogurt, fortification is done by incorporating either fruit pulps or raw fruits. Because fruits are a rich source of antioxidants (such as polyphenols, anthocyanins, flavonoids), dietary fibers and minerals, these components along with probiotic yogurt bacteria double enhance its nutritional and therapeutic values. Recently Harvard University researchers collected the data and analyzed from the nurses' health study and concluded that high consumption of dairy fat is linked with a greater risk of ischemic heart disease, especially in women. To overcome this issue, they suggested that low-fat content and fruit fortified yogurt may be the best choice to overcome these problems. Data pertaining to low-fat yogurts, major components, and their calorific values are described in Table 7.4.

7.3   BIOFUNCTIONAL YOGURT

Biofunctional yogurt or any other fermented milk product has a specific biological effect. The biofunctional properties of yogurt can be increased by supplementing with biofunctional ingredients like probiotics, prebiotics, fruits, etc. Probiotic itself is well established for its various health benefits to the host. On the other hand, prebiotics is also well known for its health-promoting effects. Prebiotics are dietary fibers, which are not digested directly by the commensal and probiotic bacteria; therefore it directly enters into the colon and stimulates the growth of probiotic bacteria by playing a role as a nutrient component to the bacteria in the colon.

Yogurt products have positive health effects, including anti-cancerous effect, cholesterol assimilation, prevention from gastrointestinal (GI) infection, and blood pressure-lowering effects [139]. Makino et al., [100] reported that consumption of *L. delbrueckii* ssp. *bulgaricus* OLL1073R-1 fermented yogurt reduced the risk of infection to the common cold in the elderly by augmenting the activity of natural killer (NK) cells. Li and Shah [90] reported the antiproliferative and anti-inflammatory activities of natural and sulfonated EPS from *S. thermophilus* ASCC1275. Meydani and Ha [112] reviewed the therapeutic and preventive properties of yogurt and LAB, which are mainly studied for their immune-stimulatory effects. Therefore, yogurt has many beneficial effects on immune-compromised people, infection, GI disorders, cancer, asthma, etc.

Yogurt is the first choice among consumers, because it is rich in major protein and other nutrients found in milk products such as vitamins (vitamin B-2, vitamin B-9, and vitamin B-12), Calcium, Magnesium, and Potassium.

TABLE 7.4 List of Low Fat Yogurt and Their Major Components Along with Calorific Value

Yogurt Type	Calories	Fat (g)	Saturated Cholesterol (g)	Calories from Sugar (%)	Calcium (% Daily Value)	Vitamin D (% Daily Value)
Low-fat Yogurt (for 6 Ounces)						
Dannon Activia, Blueberry flavor	165	3 1.5	7.5	62	23	–
Dannon creamy fruit Blends, Strawberry flavor	170	1.51	10	70	20	–
Stony field Farms Organic Low-Fat, Fruit flavored	130	1.51	5	68	25	–
Yolapit original 99% fat-free, Fruit flavored	170	1.51	10	63	20	20
Yoplait Yo Plus	165	2.2	15	58	23	15
Raspberry		1.5				
"Light" Yogurt (for 6 Ounces)						
Dannon Activia Light Fat-Free Raspberry	105	0	0	<546	22.5	–
Dannon Light &Fit 0% Fat Plus	75	0	0	<556	15	15
Dannon Light'n Fit, Fruit flavored	60	0	0	<547	20	20
Weight Watchers Nonfat Yogurt (with 3g of fiber)	100.5	0		548	30	30
Yoplait Fiber One Non-fat Yogurt (with 5g fiber)	120	0	0	<555	15	22.5
Yoplait Light, Fruit flavored	100	0	0	<556	20	20

Source: Harvard University Research Data.

Excluding all of these nutrients, sometimes yogurt contains live and healthy bacteria, which are regarded as probiotic bacteria. These probiotic bacteria are "friendly bacteria" that are naturally present in the human digestive system by birth itself. These bacteria are involved in several gut functions, such as digestion of lactose by producing lactose dehydrogenase (LDH) enzymes/lactase, immune stimulation by producing anti-inflammatory cytokines thus making a proper balance between the pro- and anti-inflammatory cytokines. These pro-inflammatory cytokines are dominant under disease conditions.

Biofunctional property of good bacteria can also be enhanced by exhibiting the pathogen/harmful bacteria exclusion by competitive exclusion mechanisms. Other health benefits of probiotic yogurt are also documented, such as antioxidant, anti-obesity or weight management, antidiabetic properties, etc. These are generally called as lifestyle-related diseases. A sedentary lifestyle and diet habits play a major role in these diseases and related complications. To overcome this, fermented milk like yogurt and other fermented products are suggested for the treatment of these diseases.

Mozaffarian et al., [116] reported that yogurt consumption prevents age-associated weight gain through the unknown mechanism. They also suggested that the mechanistic action of probiotic bacteria may be responsible for this. Probiotic yogurt or the purified *L. reuteri* bacteria fed mice did not alter their calorie intake; on the other side, it benefited mice in terms of general demeanor, skin, and hair coat and reproductive performance [88].

All beneficial aspects of probiotic bacteria are being explored more and more by addition in food and food products. Many manufacturers are claiming that these probiotic bacteria are live in the product so that consumer can get healthy bacteria for maximum health benefits. One of the topmost brands Dannon Co. is producing and marketing yogurt as DanActive and Activia, which contains probiotic bacteria. Daily consumption of Activia up to two weeks has proven for regulating the digestive system. On the other side, DanActive is a yogurt drink, which has been clinically proven in strengthening the body's defense systems.

7.4 BIOFUNCTIONAL YOGURT AND ITS HEALTH BENEFITS

The beneficial effect of yogurt has been recognized since the ancient era. It has been mentioned in the Bible about the use of yogurt, and Hippocrates also described that fermented milk was considered as medicine. He described that fermented milk is very useful for intestinal disorder and stomach care. Later in the 20th century, scientist Elli Metchnikoff proposed that lactobacilli present in fermented milk are mainly responsible for the healthy and long lifespan of Bulgarian people. That is why the typical name of lactobacilli was given as *Lactobacillus bulgaricus* [97].

Further, the health benefits of *L. bulgaricus* were studied, and 73% of *L. bulgaricus* was found in fecal samples from yogurt consumers than the non-consumers [7]. By consideration of the therapeutic properties of fermented milk, researchers are targeting each and every biological activity of fermented milk in details. For this *in vitro*, *in vivo* study in an animal model and clinical trials on human are basic key steps to understand the mechanistic action of biofunctional properties of the particular probiotic bacteria, which were used to make that products. Some of the biofunctional properties of yogurts and related fermented milks are mentioned in this section.

7.4.1   ANTIOXIDANT ACTIVITY

The oxidative stress caused by various reactive oxygen and nitrogen species, which include: hydroxyl radicals, hydrogen peroxide, and the peroxide radicals, superoxide radicals, nitric oxide, nitrogen dioxide, etc. These accumulated reactive oxygen species play an important pathological role in various diseases such as diabetes, cancer, atherosclerosis, hypertension, and arthritis, etc. [45]. Although our body has an inherent antioxidative system to protect itself from damage, yet continuous increasing attack of free radicals is responsible for slowing down and retarding the function of antioxidant system. The antioxidant activity in terms of α, α–diphenyl-β-picrylhydrazyl (DPPH) free radical and inhibition of linoleic acid peroxidation has been showed by *B. longum* ATCC 15708 and *L. acidophilus* ATCC 4356 [91, 94]. It is suggested that mechanisms for antioxidant effects are still not clear.

Lin and Yen [92] studied the ability of yogurt organisms for their radical scavenging activity by using the intracellular extracts of yogurt bacteria. They tested five strains of *S. thermophilus* and six strains of *L. delbrueckii* ssp. *bulgaricus*. These strains were very effective to inhibit the linoleic acid peroxidation. The intracellular extracts of yogurt bacteria (approximately 10^8 cells) exhibited the antioxidant effect, which was almost equivalent to 25–96 ppm butylated hydroxytoluene (BHT), thus implying good antioxidant activity in all 11 strains. The maximum hydroxyl radical scavenging ability was noted at 234 mM for *L. delbrueckii* ssp. *bulgaricus* Lb whereas *L. delbrueckii* ssp. *bulgaricus* 448 and 449 and *S. thermophilus* MC and 821 demonstrated the highest scavenging ability to hydrogen peroxide at 50 m

M. S. *thermophilus* CNRZ368 genes in yogurt bacteria have been proven for their powerful antioxidant activity against free radicals [162].

Gjorgievski et al., [48] found that the highest antioxidant activity was observed in milk, which was fermented with probiotic strain L. *acidophilus* by DPPH assay (63.99%), while the lowest value (39.43%) was observed in milk fermented with symbiotic cultures L. *delbrueckii* ssp. *bulgaricus,* and S. *thermophilus*. The fermentation process released many compounds such as peptides, free form amino acids, flavoring compounds, etc.

Shori and Baba [151] prepared the yogurt using the *Azadirachta indica* neem tree to determine the antioxidant potential and key enzyme inhibition, which has been linked to hypertension and type-2 diabetes. A. *indica* yogurt had the highest DPPH inhibitory action, i.e., 53.1 ± 5.0% on 14th day, the highest total phenolic content (74.9 ± 5.1 µg GAE/mL) on 28th day compared to plain yogurt (35.9 ± 5.2% and 29.6 ± 1.1 µg GAE/mL), respectively. A. *indica* yogurt also exhibited maximum inhibition of the ACE enzyme (48.4 ± 7.2%), α-amylase (47.4 ± 5.8%) and α-glucosidase activity (15.2 ± 2.5%). However, they suggested that A. *indica* yogurt could be used in the development of functional yogurt having antihypertensive, antidiabetic, and antioxidant activities.

7.4.2 ANGIOTENSIN-CONVERTINGENZYME(ACE)INHIBITORY ACTIVITY

In case of hypertension, ACE (angiotensin-converting enzyme) has an important role in the rennin-angiotensin system (RAS), which is known for the arterial blood pressure regulation and maintains the salt and water equilibrium in the body. It has been seen that when the enzyme catalyzes the hydrolysis of angiotensin I to angiotensin II, then it increases the blood pressure. Angiotensin II, which is known for vasoconstrictor agent, degrades bradykinin to a greater extent [29]. Yogurt and their peptides can be useful in lowering the blood pressure in hypertensive patients. The suggested mechanisms are to inhibit the production of angiotensin II, vasoconstrictor, and to degrade the bradykinin metabolism [98, 173]. This vasodilation helps to reduce the arterial pressure and promotes the water and sodium excretion.

Mostly *Lactobacillus helveticus*-based fermented milks having antihypertensive actions are commercially available, but fermented milks having this property with other lactobacilli cultures are still scarce. Thus, there is potential for utilizing proteolytic LAB for the development of new and innovative functional dairy products, which may be useful in the reduction of high blood pressure and their consequences. Therefore, the consumer may be ready to pay more for these functional foods. Fermented milk with wild strains of *Lactococcus lactis* in a single dose [143] and after long administration [144] reduced hypertension in spontaneously hypertensive rats. The peptide sequence IPP and VPP have been reported for their presence in the fermented milk for the ACE-inhibitory activity [138, 145].

Although most researchers attribute the hypotensive effect of fermented milk to ACE-I peptides, yet many other possible mechanistic actions may be involved in the antihypertensive activity [77, 99, 102, 165]. A comprehensive antihypertensive action of fermented functional foods has also been reviewed by researchers [13]. In another study, commercial yogurt cultures *Lactobacillus delbrueckii* ssp. *bulgaricus*, *Streptococcus thermophilus*, *Lactobacillus acidophilus* La5 or *Bifidobacterium animalis* BB-12 were used to prepare two types yogurts such as plain yogurt and probiotic yogurt. The ACE-inhibitory activities were seen favorable in both types of yogurts. However, the comparative study showed that yogurt prepared with probiotic bacteria had even higher inhibitory action than the non-probiotic yogurt [150].

In the case of hypertension, the blood pressure exceeding the normal range [27] affects more than 1 billion people worldwide [171], which includes not only adults but children also [13]. Moreover, it is considered a significant risk factor for the development of heart diseases, cerebrovascular accidents, renal failure, strokes, and related many complications [22]. It has been seen that control of high blood pressure and vascular tone are an important metabolic pathway of the RAS [31].

The role of ACE for the regulation of blood pressure is well proven. It controls the blood pressure in many ways such as: conversion of ACE-I to ACE-II and hydrolysis of vasodilators. The potent vasoconstrictor and vasodilators like kallidin and bradykinin are the main components for the elevation of high blood pressure. The inhibition of ACE will cause a vasodilator response, which lowers blood pressure. Although pharmacological therapies are most widely used to treat hypertension, yet they have long-term secondary side effects. Thus, fermented milk is recommended as a non-pharmacological treatment for hypertension, mainly because it lacks undesirable side effects [44]. The dairy-based bioactive components like peptides, fatty acids; calcium, etc. exhibit beneficial effects. Therefore, the development of biofunctional yogurt may lower the blood pressure and support a healthy heart.

7.4.3   IMMUNE-STIMULATORY ACTIVITY

Fermented product yogurt comprises of live microorganisms, such as *S. thermophilus* and *L. delbrueckii* ssp. *bulgaricus*. These live bacteria may be probiotic in nature and however produce immune-stimulating factors by suppressing pro-inflammatory cytokines and help to maintain a balance between pro- and anti-inflammatory cytokines. It boosts the immune response, which would further increase the resistance to diseases, which are immune-related. The immunostimulatory properties of yogurt are due to yogurt cultures. However, the mechanism for immune stimulation and other immune-related improvements have not been completely determined [34, 137].

When active LAB enters into the gastrointestinal tract (GIT), these activate specific and nonspecific immune responses of gut-associated lymphoid tissue. Another mechanism for immune-modulatory action is the protection of tissues against microbial infections. The degree of contact with the gut lymphoid tissue significantly affects the immune-stimulating properties of intestinal bacteria [4, 131].

Many immune-stimulatory actions of yogurt consumption have been seen, when cells were exposed to the LAB *in vitro* or in the human studies with changes in the immune-related mechanisms such as antibody production, cytokine production, phagocytic activity, NK cell activity, T-cell function, etc. Incidence of many diseases or disease like conditions such as GI disorders, cancers, and allergic symptoms have been reduced with the immune-stimulating properties of yogurts [112].

7.4.4   CHOLESTEROL LOWERING PROPERTY

Fermented milk yogurt containing probiotic bacteria have the ability to assimilate the cholesterol from the body. A lab experiment was conducted to evaluate the impact of health status on consumers' choices for functional foods. Consuming yogurts with added plant sterols revealed for a reduction in cholesterol [101]. Yogurts containing probiotic bacteria *Bifidobacterium longum* BB536 or *Bifidobacterium pseudocatenulatum* G4 have been studied for their cholesterol-reducing ability in terms of lipid peroxidation, plasma lipid content reduction and fecal excretion of bile acids in rats fed with high cholesterol-enriched diets. However, it was reported that significantly lower very-low-density lipoprotein (VLDL) cholesterol, low-density lipoprotein cholesterol (LDL-C), and plasma total cholesterol (TC). The bile acids excretion via fecal were significantly increased in the yogurt fed rats [6].

Yogurt cultures *S. thermophilus* 2 (Sc.t2), and *L. bulgaricus* 2 (Lb.b2) have been proven for the resistance of bile salts and assimilation of cholesterol [37]. Pigeon et al., [132] also reported that yogurt cultures *L. delbrueckii* ssp. *bulgaricus* Lb-18 and Lb-10442 have the ability to bound 15.3% cholic acid, which was significantly higher than the *S. thermophilus* strains. These strains have been unable to bind the conjugated bile acid and glycocholic acid. *L. delbrueckii* ssp. *bulgaricus* have the capability to uptake cholesterol from MRS broth and artificial GIT [174]. Consumption of yogurt regularly resulted in the reduction of blood serum cholesterol. It has been anticipated that yogurt containing LAB might be responsible for the cholesterol-lowering effects. Assimilation of cholesterol by lactic acid bacterial cells has been proposed for the mechanistic action of cholesterol-lowering properties [24].

Recently, Megalemou et al., [107] reported the anti-thrombotic and cardio-protective property of Greek yogurt. Probiotic yogurt consumption is also beneficial for body mass index reduction and body weight, and serum levels of fasting insulin [117]. Similarly, the interaction of yogurt and maternal overweight status was positive in the non-overweight women during the pre-term delivery [82]. The total yogurt consumption is inversely related with type-2 diabetes [36]. The potent PAF-inhibiting properties of ovine and caprine milk yogurts were also observed [107].

7.5 BIOACTIVE PEPTIDES

Bioactive peptides have some specific protein fragments, which directly show the beneficial impact on physiological functions [95]. Bioactive peptides are comprised of many groups of peptides, which may be: hormonal peptides, neuroactive peptides, physiologically active peptides, immunoactive peptides, antimicrobial peptides (AMPs), enzyme inhibitors and regulators [47].

Small length amino acid (2–20) having a molecular mass less than 6 kDa [157] have been reported for the most biologically active peptides [109]. In a protein molecule, theses peptides are latent, but when exposed to suitable conditions it gets released that is called hydrolysis of proteins. Thus, a dietary protein has been recognized as the best source of amino acids and provides many health benefits as protein hydrolysate or in the intact form itself. However, the interest has been in the fermented food products for a therapeutic purpose to treat and prevent many lifestyle-related diseases.

When it comes to fermented foods, a dairy-based fermented food comes on the topmost due to milk proteins exhibiting many biological effects. The fermentation with proteolytic LAB has the ability to break downs these proteins into smaller bioactive peptides. The food processing and GI digestion also release biologically active peptides (Figure 7.2). It was noted that milk protein hydrolysate or fermentate have mostly been studied for their antihypertensive peptides under *in vitro* and *in vivo* experiments.

The functional aspects of bioactive peptides are significantly present in fermented foods like yogurt, dahi, fermented milk, cheese, etc. Thus, fermentation of milk by proteolytic LAB can lead to the production of biofunctional compounds, which have specific health benefits beyond their basic nutritional components. The production of bioactive peptides from fermented milk varies from strain to strain. Thus, it affects the concentration of bioactive peptides and its function in the fermented milk products. Therefore, the increase of concentration of bioactive peptides in the fermented food can be achieved by incorporation of the highly proteolytic strain of LAB.

It is also very important to note that proteolytic activity of strain should not be too much higher, otherwise product will have a bitter taste, and consumer acceptance and demand will automatically be decreased. Microbial fermentation of milk protein by LAB generates many biofunctional bioactive peptides (Figure 7.2) that include antioxidative, immunomodulatory, antimutagenic, and antihypertensive or ACE-inhibitory peptides [80, 81, 49, 105].

7.5.1 YOGURT–DERIVED BIOACTIVE PEPTIDES

The primary structure of milk proteins has several regions that encrypt for several latent biological actions. The bioactive peptides released by microbial fermentation have been associated with beneficial health effects. Many peptide sequences for their hypotensive activity, mineral binding activity, cytomodulatory, and opioid activities have been identified in the fermented milk products like yogurts and cheese. Bioactive peptides directly alter the physiological functions by their drug-like or hormone-like activity. These peptides bind to the specific receptors on target cells and induce the physiological responses (Figure 7.2).

Many peptides and peptide fractions have been isolated from fermented dairy products having various physiological roles, such as immune-modulatory, anti-cancerous, hypocholesterolemic, antimicrobial (including bacteriocin), mineral binding, and opiate-like activity, peptidase inhibitory and bone formation activities. Therefore, the pursuit for ACE-inhibitory substances, such as peptides, in food has been conducted so that such peptides could be used in the prevention and treatment of hypertension [165]. Several ACE-inhibitory peptides have been identified in the fermented milk and cheese [61, 134], although their actual antihypertensive mechanism is still unclear [68].

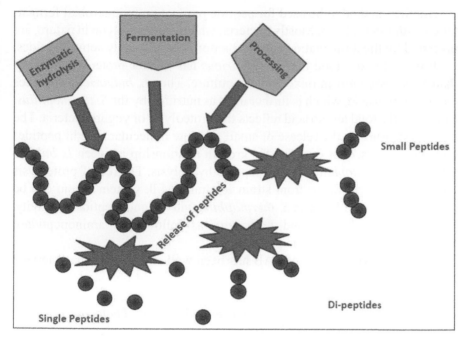

FIGURE 7.2   (See color insert.) Methods to release bioactive peptides from native protein.

It has been suggested that bioactive peptides of yogurt and other milk products have various physiological functions. These peptides are encrypted in their native form, and they may get released during milk fermentation. In yogurt, the release of amino acids and peptides from the milk proteins is due to the proteolytic activity of yogurt bacteria, i.e., *Streptococcus thermophilus* and *Lactobacillus delbrueckii* ssp. *bulgaricus*.

Bioactive peptides are biologically active, thus exert a physiological effect in our body. Milk as such contains various milk proteins like casein, immunoglobulin (IGs), α-lactalbumin, β- lactoglobulin, lactoferrin, etc. These proteins have their own function in our body. It has been found that milk proteins as such have less physiological activities than small and low molecular weight bioactive peptides. The generation of low molecular weight and smaller bioactive peptides need more proteolytic action by proteolytic LAB. Except for proteolytic action, the functionality of protein can be further increased by the breakdown of native proteins into their smaller peptides (Figure 7.2). These peptides can be generated by enzymatic hydrolysis using different enzymes such as trypsin, pepsin, chymotrypsin, protease, rennin, etc.

Another method to be used for peptide production by microbial fermentation with food is LAB. Mostly bacteria, which are proteolytic in nature, are preferred for the fermentation and production of biologically active peptides.

It is well understood that *L. bulgaricus* has greater proteolytic activity than *S. thermophilus* in mixed yogurt culture. Thus *L. bulgaricus* produces the free amino acid, which is further used as nutrition by the *S. thermophilus*. In yogurt, the total amino acid reflects the proteolysis of yogurt bacteria. The proteolysis leads to the release of small and low molecular weight peptides and amino acids. This explains the growth relationship between *L. bulgaricus* and *S. thermophilus* due to peptide hydrolysis. Degree of proteolysis by yogurt bacteria varies from strain to strain. Milk protein casein can be degraded by yogurt bacteria *S. thermophilus* via the endopeptidase activity. *Lactobacillus helveticus* and *L. bulgaricus* follows the aminopeptidase activity for casein catabolism pathway.

The proteolytic system of LAB has been well studied and characterized for the generation of bioactive peptides in starter and non-starter bacteria, e.g., *Lactobacillus delbrueckii* ssp. *bulgaricus*, *L. plantarum*, *L. helveticus*, *L. rhamnosus*, *L. acidophilus*, *S. thermophilus*, and *Lactococcus lactis*. Among all these LAB, *L. heleveticus* is well proven for highly proteolytic activity. The peptide released by *L. helveticus* proteolytic system is widely useful for the many biofunctional aspects. This proteolytic system mainly consists of a number of unique intracellular peptides such as dipeptidases, tripeptidases, aminopeptidases, and endopeptidases. This system has also been known for consisting of cell wall-bound proteinase.

The milk protein and casein get degraded into oligopeptides by extracellular proteinases. Oligopeptides occur in a long chain, but to some extent, it could be a source of bioactive peptides. Additionally, these long-chain oligopeptides are degraded by intracellular peptidases of lysed-LAB. The small chain peptides exhibit more biological activities than the long-chain peptide sequences.

7.6 BIOFUNCTIONAL PROPERTIES OF BIOACTIVE PEPTIDES FROM YOGURT

7.6.1 ANTIOXIDATIVE PEPTIDES

The main source of antioxidant peptides is plant resources and animal proteins. Food protein source of antioxidant peptides include: milk casein, egg-yolk or egg proteins, milk-kefir or milk proteins, soymilk-kefir or soymilk protein, casein, berry fruits, etc. Several other factors may affect the antioxidant potential of bioactive peptides such as enzymes and their degree of hydrolysis, microbial fermentation, and processing parameters. Excluding all these parameters, the antioxidant activity of generated peptides is also affected by their resultant structural properties that may be due to their molecular weight, size, hydrophobicity pattern, amino acid sequence and their composition [133]. Due to this, the uniqueness of fermented milk containing peptides exhibits various biological activities. Some fermented milk containing peptides exhibit only single biofunctional activity, but some others may exhibit multifunctional properties.

Research has shown several properties of yogurt, such as antioxidative, antithrombotic, antimicrobial, immunomodulatory, iron-binding, opioid, antagonistic activities, and ACE inhibitory activity [125].

Oxidative stress is one of the major causes of various diseases [133]. Many pathological conditions such as rheumatoid arthritis, atherosclerosis, cancer, diabetes, etc. are interlinked with free radicals, which are caused by oxidative stress [1, 54]. These serious diseases and disease-like conditions can be protected by the inhibition and reduction of the free radicals from the foods and living body as well. Milk proteins and peptides are generally known for their multifunctional properties. One example of the multifunctional properties of milk peptide αs1-casein fraction f(194–199) has antihypertensive and immunomodulatory activities. The caseinophospho peptides have been identified for their immune-modulatory and mineral carrier, whereas α- and β-lactorphin have shown opioid and antihypertensive activities, respectively [81].

Like other fermented dairy products, yogurt bacteria also release many functional peptides. Yogurt bacteria are known for their synergistic action and thus exhibit the symbiotic association. This leads to the release of many biologically active peptides and amino acids sequences, which show the stimulatory peptides among them and also act as growth promoters [167].

Aloglu and Oner [5] reported that the HPLC fractions F2 of commercial and traditional yogurt exhibited higher antioxidant activity. Perna et al. [128] also reported the antioxidant potential of yogurts made from milk of two different breeds-Italian Holstein and Italian Brown, and they found that the maximum antioxidant activity in yogurt produced with Italian Brown milk in comparison with Italian Holstein milk and that antioxidant activity was increased during storage up to 5 days.

Kunda et al., [83] used micro liquid chromatography coupled to time-of-flight mass spectrometry (microLC-TOF-MS) for the separation and identification of bioactive peptides from yogurt sample, which was marketed as an antihypertensive functional food. Results showed that only 50 bioactive peptides were identified and confirmed as functional peptides. Among them, a significant number of ACE-inhibitory peptides were reported, and nine of them were recognized as antihypertensive peptides. They also confirmed that these peptide sequences were also involved with other physiological activities such as immune-modulatory, antioxidant, antimicrobial, phagocytosis stimulation, and opioid-like activities.

Farvin et al. [41] studied the antioxidant activity of yogurt by isolation, the separation of yogurt derived bioactive peptide. The peptide sequence having 3–10 kDa was linked with the antioxidant activity. The free amino acids Tyr, Lys, His, and Thr have been reported for their antioxidant potentials, which came under <3 kDa peptide fractions. The antioxidant peptide fractions were identified by LC-MS/MS and showed that several proteins like β-casein, κ-casein, αs1- and αs2- casein having N-terminal fragments have been contained in these peptides.

It has been noted that at least one proline residue was must in all these peptides sequences. Apart from the proline residue, each sequences having antioxidant potentials contained hydrophobic amino acid residues Leu or Val at the N-terminus and His, Pro or Tyr in the amino acid sequence. Athira et al., [10] studied the antioxidant effects of whey protein hydrolysate (WPH) on paracetamol-induced oxidative stress mice. WPH has been seen for the increase in liver and erythrocytes antioxidant enzymes level, such as catalase (CAT), superoxide dismutase (SOD) and glutathione peroxidase (GPx). They found that WPH had been effective against paracetamol-induced oxidative stress and hepato-nephrotoxicity. Some of the peptide sequences and their bioactivities are given in Table 7.5.

In our body, the proteins and peptides get digested into the stomach and small intestine because many digestive enzymes and microbes stimulate the degradation process. The gastric enzymes break down the proteins into peptides, which are further broken down into smaller subunits that can be easily absorbed. A variety of specialized enzymes hydrolyzes the peptide bonds; thus, proteins get digested, and this is called peptidases. These peptidases are known for their most important barrier limiting in the absorption of biologically active peptides. Further, these bioactive peptides influence our health. Bioactive peptides can be taken orally to observe the substantial effect on human physiological systems.

Different peptides having amino acid sequences show different physiological effects. Many research reports suggested that these physiological effects might include antioxidative, antihypertensive, immune-modulatory,

TABLE 7.5 Peptides with Antioxidative Properties from Milk Proteins by Fermentation and Other Digestive Enzymes

Microorganisms/ Enzymes	Protein	Peptide Sequence	Bioactivity	References
B. longum	–	RELEELNVP-GEIVE	ABTS radical scavenging activity	[25]
Corolase PP	β-lg f(145–149) β-lg f(19–29) β-lg f(42–46)	MHIRL WYSLAMAASDI YVEEL	Radical scavenging activity (ORAC)	[62]
Fermented milk	β-CN f(199–208)	GPVRGPFPII	ABTS	[60]
Human milk simulated digestion	–	YGATGA ISELGW	Radical scavenging activity (ORAC)	[164]
L. delbrueckii, *S. thermophilus*	β-CN f(178–182) β-CN f(176–182) β-CN f(169–175) β-CN f(200–208) β-CN f(26–30)	VPYPQ KAVPYPQ KVLPVPE GVRGPFPII IPIQY	DPPH radical scavenging activity, Fe2+ ion-chelating capacity	[40]
Pepsin	αS2-CN f(174–181) αS2-CN f(203–208) k-CN f(28–30) k-CN f(30–32)	FALPQYLK PYVRYLIQY YVL	Radical scavenging activity (ORAC)	[95]
Thermolysin	α-la f(115–118) α-la f(101–104)	LDQW INYW	ABTS radical scavenging activity	[148]
Trypsin, Clostripain, and Subtilisin	β-CN f(162–177)	VLPVPQKKV-LPVPQK	Linoleate oxidation	[141]

anti-thrombotic, and anti-microbial activities [43]. Hartmann and Meisel [56] reported that peptide having 2–20 amino acid residues exert multifunctional properties. Fermented milk product such as yogurt is a good source of multifunctional bioactive peptides as it contains enough protein contents that can be hydrolyzed in the presence of various proteolytic bacteria and enzymes.

7.6.2 ANTIMICROBIAL PEPTIDES (AMPS)

AMPs are proteins and peptides molecules, which are found both in prokaryotes and eukaryotes and provide the protection against pathogens. These AMPs have a lot of applications in the food as well as in the dairy industry. AMPs provide the enhancement of shelf-life of products. ADP database has shown that approximately 2400 naturally isolated or synthesized AMPs have been listed in AMPs database records [64].

Antimicrobial activity of milk proteins and peptides are mainly due to protein IGs and non-immune proteins, namely lactoferrin, lysozyme, and lactoperoxidase. A potent antimicrobial peptide from a fragment of lacto-ferrin is lactoferricin. Apart from this lactoferrin, other potential precursors for bactericidal activity in whey proteins are β-lactoglobulin (β-LG) and α-lactalbumin. Similarly, αs1-, αs2-, β- and κ-caseins are also the good precursors of AMPs fragments. Almost all of these peptides have been reported for their antibacterial activity against a broad range of pathogenic organisms such as *L. monocytogenes, E. coli, B. cereus, S. typhi, S. aureus, Helicobacter*, etc.

Hayes et al., [59] isolated three potent antibacterial peptides namely casei-cins A, B, and C from the sodium caseinate fermentate, which was fermented by *L. acidophilus* DPC6026 and purified by RP-HPLC method. However, the fermentate not passed through 10 kDa did not show any antibacterial activity. The improved antimicrobial activity has been seen in case of RP-HPLC puri-fied samples in comparison to normal filtrate samples. The differences in anti-microbial activity of fermentate, membrane filtration, and RP-HPLC may be due to the electrostatic interactions between the charged peptides, which showed a negative effect on the antimicrobial activity of fermentate. Whereas, reducing interactions in terms of reduced size and charges of peptide content to ≥ 10 kDa exhibited the improved antimicrobial activity of the peptide fractions. Antimicrobial activity was also shown by casein-derived peptides (casecidins).

Minervini et al., [115] hydrolyzed human milk with a purified proteinase of *L. helveticus* PR4 and isolated an antimicrobial protein from human milk β-casein. A broad spectrum of antibacterial activity against Gram-positive and Gram-negative bacteria have been observed in the human milk casein β-CN f(184–210). The bacterial genera, which are of potential clinical interest, were used in the experiment such as *Salmonella* spp., *Escherichia coli, Listeria innocua, Staphylococcus aureus, Yersinia enterocolitica, Enterococcus faecium*, and *Bacillus megaterium*. One of the most important properties of bioactive peptides is their resistant to further degradation by any proteolytic enzymes and bacteria.

In a similar study, different antimicrobial fragment from α, β, and κ-casein of nine Italian cheese varieties were identified for their potential inhibitory activity against pathogenic bacteria [142]. AMPs have also been suggested for their ability to modulate immune response [49].

Plaisancie et al. [136] also identified a novel biologically active peptide TPP from yogurt sample that modulated the mucin gene expressions MUC2 and MUC4. This peptide also enhanced the Paneth and goblet cell populations in the small intestine and also protected them. They suggested that yogurt peptide β-casein (94–123) was responsible for the maintaining of intestinal homeostasis, protection against damaging agents in the lumen and restoring them.

Antimicrobial activity of fermented milk and peptides is an important characteristic because of lactoferrin, which is known as iron-binding or chelating protein. It has bactericidal and bacteriostatic properties. Bactericidal mechanism is not dependent on iron, because identified peptides are different from the iron-binding site of the molecules. The bactericidal nature of protein is due to disrupting the essential membrane function. It was found that pepsin digested hydrolysate has higher antimicrobial potency than undigested lactoferrin. Antimicrobial activity against pathogenic and food spoilage bacteria of N-terminus lactoferrin have also been reported [23].

It was reported that pathogenic strains were inhibited by as2-casein or lactoferricin [26]. The first defense peptide cathelicidin obtained from chymosin-mediated digestion of casein protein at neutral pH was exhibited *in vitro* antimicrobial activity against *Bacillus subtilis*, *Staphylococcus aureus*, *Streptococcus pyogenes*, *Diplococcus pneumonia*, *Sarcina* spp. [84]. Many AMPs have been known to be present in the milk itself such as αS1-casein, αS2-casein, α-lactalbumin, β-LG, κ-casein, lactoferrin, and lysozyme. These specific peptides have been reported for their antibacterial actions against pathogenic and spoilage causing bacteria. Incorporation of these bioactive peptides in the food products could enhance the antimicrobial activity [39]. Bovine lactoferrin has been reported to have multifunctional (antimicrobial, anticancer, immunomodulatory) properties [8, 32].

Many peptide sequences with antimicrobial action have been reported in the primary structure of casein (Table 7.6). Milk peptide sequences and their specific antibacterial action against Gram-negative and Gram-positive bacteria are listed in Table 7.7.

TABLE 7.6 Presence of Antimicrobial Peptide Sequences in Casein

Casein Type	Bioactive Peptide Sequence	Amino-Acid Segment	References
αs1 (bovine)	1–23	RPKHPIKHQGLPQEVLNENLLRF	[84]
αs2 (bovine)	164–179	LKKISQRYQKFALPQY	[140]
	165–203	KKISQRYQKFALPQYLKTVY-QHQKAMKPWIQPKTKVIPY	[175]
	183–207	VYQHQKAMKPWIQPKTKVIPYVRYL	[140]
β (bovine)	193–209	YQEPVLGPVRGPFPIIV	[149]

7.6.3 ACE-INHIBITORY PEPTIDES

Like other functional peptides, fermented milk-derived bioactive peptides also exhibit ACE-inhibitory activity. ACE enzymes have a significant role in the control and reduction of hypertension, due to the connection of ACE-inhibitory peptides with the kinin-nitric oxide system (KNOS) and RAS. In the case of homeostasis of electrolyte and regulation of peripheral blood pressure, ACE acts like a key enzyme. Mainly casein derived peptides withIC_{50} values 5 to 720 μm have been involved in the antihypertensive actions. In some cases, casein and whey fermentation have been employed with the yogurt or 'ropy' starter cultures. However, it becomes necessary to subsequently incubate with the trypsin and pepsin enzymes so that it can release ACE-inhibitory peptides. Many LAB have been studied for their hypotensive effect such as *Lactobacillus helveticus* LBK 16H [154, 153], *Lactobacillus helveticus* and *Saccharomyces cerevisiae* [118], *L. helveticus* R211 [155], *L. helveticus* CP790 [172], and *Lactococcus lactis* NRRLB-50571 or NRRLB-50572 [143].

Milk fermented with *B. bifidum* MF 20/5 strain showed the presence of a novel peptide sequence LVYPFP (IC_{50}=132 μM) with ACE- inhibition property. Another bioactive peptide LPLP (IC_{50}=703 μM) has also been identified for their ACE-inhibitory property [52]. Hydrolysis of casein produced two unique antihypertensive peptide sequences such as αs1-CN f(90–94) RYLGY and αs1-CN f(143–149) AYFYPEL. The oral administration of these peptides @ 200–800 mg/kg of body weight showed a significant reduction in the systolic blood pressure of spontaneously hypertensive rats. The stability of these peptides was also good during the processes of pasteurization, homogenization, and atomization. Further, the incorporation of these peptides into the liquid yogurt did not show any significant reduction of the peptide during the storage [35].

Ruiz-Gimenez et al. [147] studied the hypotensive effect of bovine lactoferrin hydrolysate. This hydrolysate was prepared by pepsin digestion, and <3 kDa molecular mass size peptides were orally fed to the spontaneously hypertensive rats to see the effect of reduction in the high blood pressure. HPLC identified 38 different peptides, which were further identified by the mass spectrometer. Based on peptides abundance, the ACE-inhibitory peptides were tested against 11 chemically synthesized peptides. The peptide sequences RPYL, LIWKL, and LNNSRAP showed the ACE-inhibitory activity, and IC_{50} values were 56.5, 0.47, and 105.3 μM, respectively. They suggested that the lactoferrin derived peptides and pepsin lactoferrin hydrolysate <3 kDa reduced high blood pressure.

Ashar and Chand [9] also identified an ACE-inhibitory peptide from fermented milk, and this fermented milk was prepared with the *L. delbrueckii*

TABLE 7.7 List of Antimicrobial Milk Peptides and Their Activity Against Pathogens

Milk Protein Type	Peptide Sequence	Release Protease	Action Against Gram-Positive Bacteria	Action Against Gram-Negative Bacteria	Yeast and Fungi
Casecidin	κ-CN (f 17–21) αs1-CN	Chymosin and Trypsin	*Staphylococcus aureus, Sarcina, Bacillus subtilis, Diplococcus pneumoniae, Streptococcus pyogenes*	–	–
Caseicin	αs1-CN A (f 21–29) B (f 30–38) C (f 195–208)	Synthetic peptide Synthetic peptide Synthetic peptide	*Listeria innocua*	*E. coli, E. sakazakii, E. coli, E. sakazakii*	–
Casocidin-I	αs2-CN (f 165–203)	Synthetic peptide	*Staphylococcus carnosus*	*E. coli*	–
Isracidin	αs1-CN (f 1–23)	Chymosin and Trypsin	*Staphylococcus aureus*	–	*Candida albicans*
Kappacin	κ-CN (f 106–169)	Chymosin	*Streptococcus mutans*	*E. coli*	–
Lactoferrampin Lactoferrin	(f 265–284)	Pepsin	*Streptococcus mutans*	*E. coli*	*Candida albicans*
Lactoferricin B Lactoferrin	(f 17–41)	Pepsin	Bacillus Listeria Streptococci Staphylococci	*E. coli* 0111 *E. coli* 0157:H7 Klebsiella, Proteus Pseudomonas Salmonella	*Candia albicans Dermatophytes, Cryptococcus unigulattulus Penicillium pinophilum Trichophyton mentagrophytes*

Source: Tidona et al., [163].

ssp. *bulgaricus* starter culture. ACE-inhibition was seen due to the peptide sequence Ser-Lys-Val-Tyr-Pro-Phe-Pro-Gly Pro-Ile from β-casein with an IC_{50} value of 1.7 mg/mL. Fermentation was with starter culture *Lactococcus lactis* biovar. *diacetylactis* and *S. thermophilus,* and it generated β-casein derived peptide sequence Ser-Lys-Val-Tyr-Pro with an IC_{50} value of 1.4 mg/mL. These two peptides were also stable in the presence of alkaline and acidic pH, digestive enzymes, and during storage at refrigeration temperature for 4 days.

Paul and Somkuti [126] added 11 mer antimicrobial, and 12 mer antihypertensive to protein-derived peptides, which were incubated with *L. bulgaricus* and *S. thermophilus* (yogurt cultures) mid-log cells to restrain the overall extent of proteolysis. They found that these peptides can act as a good supplement in the liquid and semi-solid dairy foods like yogurts. Therefore, the consumption of functional foods enriched with food proteins and derived peptides having ACE-inhibitors reduces the risk of developing cardiovascular diseases. Fermented milk prepared with *B. bifidum* MF 20/5 was identified for the presence of angiotensin-I converting enzyme (ACE-I) inhibitory peptide sequence VLPVPQK [52].

In another study, two types of sheep milk yogurt were prepared by fermentation with *L. delbrueckii* subsp. *bulgaricus* Υ 10.13, *S. thermophilus* Υ 10.7 and the second type of yogurt was prepared with additional culture *L. paracasei* subsp. *paracasei* DC412 along with yogurt cultures to check the proteolytic activity. It has been found that *L. delbrueckii* subsp. *bulgaricus* have the most proteolytic activity among all cultures. The peptide content and its action to inhibit ACE-enzymes were increased throughout the storage. Identified peptide β-CN f(114–121) exhibited ACE-inhibitory and opiate-like activity in the second type of yogurt [122].

Pihlanto et al. [134] studied the antihypertensive activity of fermented milk prepared with 25 different LAB and found that only 5 cultures showed low IC_{50} in whey fraction namely *L. acidophilus* ATCC 4356 (0.42 mg/mL), *Leuconostoc mesenteroides* 356 (0.44 mg/mL), *L. mesenteroides* 358 (0.48 mg/mL), *Lactococcus lactis* ssp. *lactis* ATCC 19435 (0.5 mg/mL) and *L. jensenii* ATCC 25258 (0.52 mg/mL).

Rodriguez-Figueroa et al. [146] reported that artisanal dairy products isolated 20 wild strains of *Lc. Lactis* exhibited highest ACE-I activity and the lowest IC_{50} (13–50 μg/mL) in whey fractions <3 kDa. Furthermore, two *Lc. Lactis* strains fermented milk fractions obtained through reversed-phase HPLC showed low IC_{50} for strains NRRL B-50571 (0.034 μg/mL) and NRRL B-50572 (0.041 μg/mL) [145]. Similarly, fermented milk was prepared for 24 h with different *Lactobacillus* strains; and *L. fermentum* ATCC 11976, *L. reuteri* 14171 and *L. johnsonii* ATCC 33200 exhibited 42.04 to 83.36% ACE-I activity, though *L. fermentum* exhibited lowest IC_{50} (21 mg/mL) in whey fraction <3 kDa [51].

In another study, yogurt milk was treated with trypsin before the fermentation released phosphopeptide-rich fractions. Mainly calcium phosphopeptides (CPP) sequences β-CN(1–25)-4P and αs1-CN(43–79)-7P were released during trypsin treatment. The yogurt proteolysis caused by peptidases did not show any significant results [96]. The fermented milk prepared with *L. helveticus* CNRZ32 has been reported the release of ACE-inhibitory peptides. The general peptidases aminopeptidase (PepN) and X-prolyl dipeptidyl aminopeptidase (PepX) have been released during the fermentation process and showed the effect on ACE-inhibitory activity [76]. Some of the antihypertensive peptide sequences present in the caseins are given in Table 7.8.

TABLE 7.8 Presence of Antihypertensive Peptide Sequences in Caseins

Casein Type	Peptide Sequence	Amino Acid Segment	References
αs1 (bovine)	23–24	FF	[103, 104, 111]
	23–27	FFVAP	[103, 104, 111]
	102–109	KKYKVPQ	[50]
	142–147	LAYFYP	[135]
	157–164	DAYPSGAW	[135]
	194–199	TTMPLW	[103, 104, 111, 135]
αs2 (bovine)	174–179	FALPQY	[160]
	174–181	FALPQYLK	[160]
	189–197	AMKPWIQPK	[98]
	189–193	AMKPW	[98]
	190–197	MKPWIQPK	[98]
	198–202	TKVIP	[98]
β (bovine)	74–76	IPP	[119]
	84–86	VPP	[119]
	108–113	EMPFPK	[135]
	177–183	AVPYPQR	[103, 104, 111]
	193–198	YQEPVL	[135]
	193–202	YQEPVLGPVRGPFPI	[103, 104, 111]
	199–204	GPVRGPFPIIV	[119]
κ (bovine)	108–110	IPP	[119]
ϒ-casein (bovine)	108–113	EMPFPK	[129]
	114–121	YPVEPFTE	[129]
Sheep milk yogurt peptides		DKIHPFAQ	[122]
		TQTPVVVP	[122]
		KAVPQ	[122]
		RPKHPIKH	[122]

7.6.4   IMMUNO-MODULATORY PEPTIDES

Milk derived peptides are reported for stimulation of the immune system. Immunomodulatory peptides have been found for their stimulation of anti-body synthesis, macrophage activation, and lymphocyte proliferation. As milk is rich in casein and whey protein, thus it contributes to immune-stimulating factors. Mostly, immunomodulatory activities of peptides are evaluated for their influence on lymphocyte proliferation and activity, production of antibodies, and cytokinin secretion [38, 46, 156].

The milk-derived peptides also protect from the harmful and pathogenic bacteria via proliferation and maturing of NK and T-cells [168]. Caseino-phosphopeptides β CN f(1–25)4P, αs1-CN f(59–79)5P and αs2-CN f(1–32)4P have been reported for their immune-enhancing properties in mouse spleen cells by enhanced IG IgG production. The IgA production was also higher in serum and intestine in the caseino-phosphopeptides fed mice in comparison to the control group [58]. Milk derived peptide sequences, and their immunomodulatory actions are listed in Table 7.9.

7.6.5   ANTI-CANCER PEPTIDES

Hartmann et al., [57] reported that the milk peptides seemed to have a certain effect on cancer cells. The lactoferrin and its derivative peptide lactoferricin and related fragments were found to be particularly highly effective. In another study, experimentally-induced colorectal carcinoma was reduced, and tumor growth was inhibited by yogurt consumption. They concluded that the evidence of anticancer activity is increasing by consumption of milk and milk-derived peptides [127]. Many immune peptides from fermented milk have also been documented for their antitumor effects [105].

7.6.6   OPIOID PEPTIDES

Fermented milk and milk products contain many opioid peptides. The peptides beta-casomorphin 5 (BCM5) and beta-casomorphin 7 (BCM7) have been investigated for their opioid-like activity in yogurts. BCM5 concentration was decreased from 1.3 ng/g to1.1 ng/gin milk and yogurt, respectively. And this yogurt was prepared with 0-day storage milk [120]. Similarly, BCM7 concentrations were decreased to 1.9 ng/g from milk

TABLE 7.9 Immunomodulatory Peptide Sequences and Their Precursor Proteins

Precursor Protein	Fragment	Peptide SEQUENCE	Name	Function	References
Bovine lactoferrin	17–41	FKCRRWQWRMKKLGAPSITCVRRAF	lactoferricin	Immunomodulation activity	[12, 14]
	17–26	FKCRRWQWRW			[166]
Human lactoferrin	268–284	KWNLLRQAQEKFGKDKS	Lactoferrampin		[159]
α-lactalbumin	18–20	YGG	α-lactorphin		[74]
αs1-casein	158–162	YVPFP	αs1-casomorphin		[72, 169]
	194–199	TTMPLW	α-casokinin-6	Immunostimulatory activity	[42, 103]
β-casein	1–25	RELEELNVPGEIVES (P)LS(P)S(P)S(P)EESITR	Casein phosphor peptide	activity	[73]
	54–59	VEPIPY			[113, 124]
	60–66	YPFPGPI	β-casomorphin-7	Immunomodulation activity	[19, 20, 74]
	63–68	PGPIPN	β-casomorphin-7		[113, 114]
	60–70	YPFPGPIPN	β-casomorphin-11	Immunostimulatory activity	[110]
	114–118	YPVEP	β-casochemotide-1	Innate immune response	[78]
	191–193	LLY	β-casokinin-7	Immunomodulation activity	[15, 113]
	193–202	YQQPVLGPVR	β-casokinin-10	Immunostimulatory activity	[74]

to yogurt. The processing conditions significantly affected these peptide concentrations. The processing and storage reduced the concentration of both opioid peptides in the yogurt. BCM7 was reported for presence in bovine milk [28] and commercial yogurts [69]. Peptide sequences YPFPGPI have also been reported for their opioid-like activity [70, 108].

The human and bovine milk β-casein were encrypted with opiate receptor ligands, which exhibit the opioid-like activity and get released by the enzymatic action [18]. The presence of opioid receptors has been seen in the immune and nervous systems, endocrine, and in the mammal's GIT. The opioid peptide sequence AQTQSLVYPFPGPIPK has also been reported in the buttermilk (fermented milk product *lassi*) [121].

7.7   COMMERCIALLY AVAILABLE DAIRY PRODUCTS AND HEALTH CLAIMS OF THEIR BIOACTIVE PEPTIDES (Table 7.10)

Table 7.10 shows the commercially avaiable dairy prodcuts and the health claims of their bioacrtive peptides.

7.8   SUMMARY

Based on this information on nutritional and therapeutic aspects of fermented milk and milk products, especially yogurt, is one of the best dairy products for several health benefits. Making awareness about the biofunctional attributes (like antioxidant, anti-cancerous, antimicrobial, antimicrobial, antihypertensive, ACE-inhibitory, etc.) to the consumer is the best good idea to promote fermented milk and milk products. Nowadays, delivering food itself as a medicine is the best way to treat and cure health-related problems and its complications.

The fermented dairy product, yogurt prepared with well-proven probiotic bacteria, prebiotics, and many other biofunctional ingredients all exhibit comprehensive favorable results in the physiological functions of our body. However, biofunctional yogurt and derived bioactive peptides can be considered as a therapeutic food product. However, more studies are needed on the molecular level to understand the exact mechanisms of these products.

TABLE 7.10 Bioactive Peptides Based Health Benefits of Selected Commercially Available Dairy Products and Ingredients; Modified from Korhonen and Pihlanto [80]

Commercial Name of Product	Name of the Product/Ingredients	Functional Peptides	Functions	Manufacturers
"BioPue-Alpha-lactalbumin"	Whey protein isolate	α-lactalbumin	Sleep and memory improvement	Davisco, USA
"BioPure-GMP"	Whey protein isolate	κ-casein f (106–169) (Glycomacropeptide)	Dental caries prevention, Effects on the clotting of blood, Protective action against harmful microbes	Davisco, USA
"Biozate"	Hydrolyzed whey protein isolate	β-lactoglobulin fragments	Blood pressure lowering	Davisco, USA
"C12"	Ingredient/hydrolysate	Casein derived peptide	Blood pressure lowering	DMV International, The Netherlands
"Calpis"	Sour milk	Val-Pro-Pro, Ile-Pro-Pro, derived from β-casein and κ-casein	Blood pressure lowering	Calpis Co., Japan
"Capolac"	Ingredient	Casein phosphor peptide	Helps mineral absorption	Arla Foods Ingredients, Sweden
"Cystein Peptide"	Ingredient/hydrolysate	Milk protein derived peptide	Aids to raise energy level and sleep	DMV International, The Netherland
"Evolus"	Calcium enriched fermented milk drink	Val-Pro-Pro, Ile-Pro-Pro, derived from β-casein and κ-casein	Blood pressure lowering	Valio Oy, Finland
"Festivo"	Fermented low-fat hard cheese	αs1-casein f (1–9) αs1-casein f (1–7) αs1-casein f (1–6)	No health claim as yet	MTT Agrifood Research Finland
"PeptoPro"	Ingredient/hydrolysate	Casein derived peptide	Improves athletic performance and muscle recovery	DSM Food Specialties, The Netherlands
"Prodilet F/200 Lactium"	Flavored milk drink, confectionary, capsules	αs1-casein f (91–100) Tyr-Leu-Gly Tyr-Leu-Glu-Gln-LEU-Leu-Arg)	Stress reducer	Ingredia, France
"Vivinal Alpha"	Ingredient/hydrolysate	Whey derived protein	Aids relaxation and sleep	Borculo Domo Ingredients (BDI), The Netherlands

KEYWORDS

- bioactive peptides
- biofunctional peptides
- fermented milk product
- lactic acid bacteria (LAB)
- yogurt

REFERENCES

1. Abuja, P., & Albertini, R., (2001). Methods for monitoring oxidative stress, lipid peroxidation and oxidation resistance of lipoproteins. *Clinica Chimica Acta*, *306*, 1–17.
2. Agricultural Research Service–USDA, (2014). *National Nutrient Database for Standard Reference, Release*, p. 26. http://www.ars.usda.gov/ba/bhnrc/ndl (Accessed on 20 July 2019).
3. Agroscope., (2014). Yogurt's composition, FOAG Swiss Federal Office for Agriculture, 2007, http://www.agroscope.admin.ch/jogurtsauermilchprodukte/01368/index.html? lang=en (Accessed on 20 July 2019).
4. Alm, L., & Patterson, L., (1980). Survival rate of lactobacilli during digestion: An *in vitro* study. *The American Journal of Clinical Nutrition*, *33*, S2543.
5. Aloglu, H. S., & Oner, Z., (2011). Determination of antioxidant activity of bioactive peptide fractions obtained from yogurt. *Journal of Dairy Science*, *94*, 5305–5314.
6. Al-Sheraji, S. H., Ismail, A., Manap, M. Y., Mustafa, S., Yusof, R. M., & Hassan, F. A., (2012). Hypocholesterolaemic effect of yogurt containing *Bifidobacterium pseudocatenulatum* G4 or *Bifidobacterium longum* BB536. *Food Chemistry*, *135*, 356–361.
7. Alvaro, E., Andrieux, C., Rochet, V., Rigottier-Gois, L., Lepercq, P., Sutren, M., Galan, P., Duval, Y., Juste, C., & Dore, J., (2007). Composition and metabolism of the intestinal microbiota in consumers and non-consumers of yogurt. *British Journal of Nutrition*, *97*(1), 126–133.
8. Artym J., (2012). *Lactoferrin–an unusual protein*. Borgis Ltd., (in Polish), *Warsaw, 80*, 109–20, 142.
9. Ashar, M. N., & Chand, R., (2004). Antihypertensive peptides purified from milks fermented with *Lactobacillus delbrueckii* ssp. *bulgaricus*. *Milchwissenschaft*, *59*, 14–17.
10. Athira, S., Mann, B., Sharma, R., & Kumar, R., (2013). Ameliorative potential of whey protein hydrolysate against paracetamol-induced oxidative stress. *Journal of Dairy Science*, *96*, 1431–1437.
11. Basavanna, G., & Prapulla, S. G., (2013). Evaluation of functional aspects of *Lactobacillus fermentum* CFR 2195 isolated from breast fed healthy infants' fecal matter. *Journal of Food Science and Technology*, *50*(2), 360–366.
12. Bellamy, W., Takase, M., Yamauchi, K., Wakabayashi, H., Kawase, K., & Tomita, M., (1992). Identification of the bactericidal domain of lactoferrin. *Biochimica et Biophysica Acta*, *1121*, 130–136.

13. Beltran-Barrientos, L. M., Hernandez-Mendoza, A., Torres-Llanez, M. J., Gonzalez-Cordova, A. F., & Vallejo-Cordoba, B., (2016). Invited review: Fermented milk as antihypertensive functional food. *Journal of Dairy Science, 99*, 4099–4110.

14. Berge, G., Eliassen, L. T., Camilio, K. A., Bartnes, K., Sveinbjornsson, B., & Rekdal, O., (2010). Therapeutic vaccination against a murine lymphoma by intratumoral injection of a cationic anticancer peptide. *Cancer Immunology, Immunotherapy, 59*, 1285–1294.

15. Berthou, J., Migliore-Samour, D., Lifchitz, A., Delettre, J., Floch, F., & Jolles, P., (1987). Immunostimulating properties and three-dimensional structure of two tripeptides from human and cow caseins. *FEBS Letters, 218*, 55–58.

16. Bhushan, B., Tomar, S. K., & Chauhan, A., (2017). Techno-functional differentiation of two vitamin B_{12} producing *Lactobacillus plantarum* strains: An elucidation for diverse future use. *Applied Microbiology and Biotechnology, 101*(2), 697–709.

17. Bos, C., Gaudichon, C., & Tome, D., (2000). Nutritional and physiological criteria in the assessment of milk protein quality for humans. *Journal of the American College of Nutrition, 19*, 191S–205S.

18. Brantl, V., (1984). Novel opioid peptides derived from human β-casein: Human β-casomorphins. *European Journal of Pharmacology, 106*, 213–214.

19. Brantl, V., Teschemacher, H., Blasig, J., Henschen, A., & Lottspeich, F., (1981). Opioid activities of beta-casomorphins. *Life Sciences, 28*, 1903–1909.

20. Brantl, V., Teschemacher, H., Henschen, A., & Lottspeich, F., (1979). Novel opioid peptides derived from casein (beta-casomorphins). I. Isolation from bovine casein peptone. *Hoppe-Seyler's Zeitschrift für Physiologische Chemie* (Hoppe-Seyler's Journal of Physiological Chemistry), *360*, 1211–1216.

21. Bronner, F., & Pansu, D., (1999). Nutritional aspects of calcium absorption. *Journal of Nutrition, 192*, 9–12.

22. Bruce, K. D., & Hanson, M., (2010). The development origins, mechanisms, and implications of metabolic syndrome. *Journal of Nutrition, 140*, 648–652.

23. Bruni, N., Capucchio, M. T., Biasibetti, E., Pessione, E., Cirrincione, S., Giraudo, L., Corona, A., & Dosio, F., (2016). Antimicrobial activity of lactoferrin-related peptides and applications in human and veterinary medicine. *Molecules, 21*(6), E752–758.

24. Buck, L. M., & Gilliland, S. E., (1994). Comparisons of freshly isolated strains of *Lactobacillus acidophilus* of human intestinal origin for ability to assimilate cholesterol during growth. *Journal of Dairy Science, 77*, 2925–2933.

25. Chang, O. K., Seol, K. H., Jeong, S. G., Oh, M. H., Park, B. Y., Perrin, C., & Ham, J. S., (2013). Casein hydrolysis by *Bifidobacterium longum* KACC91563 and antioxidant activities of peptides derived there from. *Journal of Dairy Science, 96*, 5544–5555.

26. Chierici, R., (2001). Antimicrobial actions of lactoferrin. *Advances in Nutritional Research, 10*, 247–269.

27. Chobanian, A. V., Bakris, G. L., Black, H. R., Cushman, W. C., Green, L. A., Izzo, J. L., et al., (2003). Joint National Committee on Prevention, Detection, Evaluation, and Treatment of High Blood Pressure, and National High Blood Pressure Education Program Coordinating Committee. The Seventh Report of the Joint National Committee on Prevention, Detection, Evaluation, and Treatment of High Blood Pressure (JNC 7). *Hypertension, 42*, 1206–1252.

28. Cieslinska, A., Kostyra, E., Kostyra, H., Olenski, K., Fiedorowicz, E., & Kaminski, S., (2012). Milk from cows of different b-casein genotypes as a source of b-casomorphin-7. *International Journal of Food Sciences and Nutrition, 63*(4), 426–430.

29. Coates, D., (2003). Molecules in focus: The Angiotensin converting enzyme (ACE). *International Journal of Biochemistry & Cell Biology*, *35*, 769–773.

30. Crittenden, R. G., Matinez, N. R., & Playne, M. J., (2003). Synthesis and utilization of folate by yogurt starter cultures and probiotic bacteria. *International Journal of Food Microbiology*, *80*, 217–222.

31. Daien, V., Duny, Y., Ribstein, J., Du Cailar, G., Mimran, A., Villain, M., Daures, J. P., & Fesler, P., (2012). Treatment of hypertension with renin-angiotensin system inhibitors and renal dysfunction: A systematic review and meta-analysis. *American Journal of Hypertension*, *25*, 126–132.

32. Darewicz, M., Dziuba, B., Minkiewicz, P., & Dziuba, J., (2011). The preventive potential of milk and colostrum proteins and protein fragments. *Food Reviews International*, *27*(4), 357–388.

33. De Angelis, M., Bottacini, F., Fosso, B., Kelleher, P., Calasso, M., Di Cagno, R., Ventura, M., Picardi, E., Van Sinderen, D., & Gobbetti, M., (2014). *Lactobacillus rossiae*, a vitamin B12 producer, represents a metabolically versatile species within the Genus *Lactobacillus*. *PLoS One*, *9*(9), Online article E107232.

34. De Simone, C., Vesely, R., Bianchi, S. B., & Jirillo, E., (1993). The role of probiotics in modulation of the immune system in man and in animals. *International Journal of Immunotherapy*, *9*, 23–28.

35. Del Mar Contreras, M., Sevilla, Ma A., Monroy-Ruiz, J., Amigo, L., Gomez-Sala, B., Molina, E., Ramos, M., & Recio, I., (2011). Food-grade production of an antihypertensive casein hydrolysate and resistance of active peptides to drying and storage. *International Dairy Journal*, *21*, 470–476.

36. Diaz-Lopez, A., Bullo, M., Martinez-Gonzalez, M. A., Corella, D., Estruch, R., Fito, M., Gomez-Gracia, E., Fiol, M., de la Corte, F. J. G., & Ros, E., (2016). Dairy product consumption and risk of type 2 diabetes in an elderly Spanish Mediterranean population at high cardiovascular risk. *European Journal of Nutrition*, *55*, 349–360.

37. Dilmi-Bouras, A., (2006). Assimilation (*in vitro*) of cholesterol by yogurt bacteria. *Annals of Agricultural and Environmental Medicine*, *13*, 49–53.

38. Durrieu, C., Degraeve, P., Chappaz, S., & Martial-Gros, A., (2006). Immunomodulating effects of water-soluble extracts of traditional French Alps cheese on a human T-lymphocyte cell line. *International Dairy Journal*, *16*, 1505–1514.

39. Fadaei, V., (2012). Milk Proteins-derived antibacterial peptides as novel functional food ingredients. *Annals of Biological Research*, *3*(5), 2520–2526.

40. Farvin, K. H. S., Baron, C. P., Nielsen, N. S., & Jacobsen, C., (2010). Antioxidant activity of yogurt peptides: Part 1-*in vitro* assays and evaluation in x-3 enriched milk. *Food Chemistry*, *123*, 1081–1089.

41. Farvin, K. H. S., Baron, C. P., Nielsen, N. S., Otte, J., & Jacobsen, C., (2010). Antioxidant activity of yogurt peptides: Part 2–Characterization of peptide fractions. *Food Chemistry*, *123*, 1090–1097.

42. Fiat, A. M., Levy-Toledano, S., Caen, J. P., & Jolles, P., (1989). Biologically active peptides of casein and lactotransferrin implicated in platelet function. *Journal of Dairy Research*, *56*, 351–355.

43. Fitzgerald, R. J., & Murray, B. A., (2006). Bioactive peptide and lactic fermentations. *International Journal of Dairy Technology*, *59*, 118.

44. Flambard, B., & Johansen, E., (2007). Developing a functional dairy product: From research on *Lactobacillus helveticus* to industrial application of Cardi-04TM in novel

antihypertensive drink yogurts. In: *Functional Dairy Products* (Vol. 2, pp. 506–520). CRC Press LLC, Boca Raton, FL.

45. Frenkel, K., (1992). Carcinogen-mediated oxidant formation and DNA damage. *Pharmacology & Therapeutics, 53*, 127–166.

46. Gautier, S. F., Pouliot, Y., & Saint-Sauveur, D., (2006). Immunomodulatory peptides obtained by the enzymatic hydrolysis of whey proteins. *International Dairy Journal, 16*, 1315–1323.

47. Gill, I., Lopez-Fandino, R., Jorba, X., & Vulfson, E. N., (1996). Biologically active peptides and enzymatic approaches to their production. *Enzyme and Microbial Technology, 18*, 162–183.

48. Gjorgievski, N., Tomovska, J., Dimitrovska, G., Makarijoski, B., & Shariati, M. A., (2014). Determination of the antioxidant activity in yogurt. *Journal of Hygienic Engineering and Design, 6*, 88–92.

49. Gobbetti, M., Minervini, F., & Rizzello, C. G., (2004). Angiotensin I-converting enzyme-inhibitory and antimicrobial bioactive peptides. *International Journal of Dairy Technology, 57*, 173–188.

50. Gomez-Ruiz, J. A., Ramos, M., & Recio, I., (2002). Angiotensin converting enzyme-inhibitory peptides in Manchego cheeses manufactured with different starter cultures. *International Dairy Journal, 12*(8), 697–706.

51. Gonzalez-Cordova, A. F., Torres-Llanez, M. J., Rodriguez-Figueroa, J. C., Espinoza-De-Los-Monteros, J. J., Garcia, H. S., & Vallejo-Cordoba, B., (2011). Angiotensin converting enzyme inhibitory activity in milks fermented by Lactobacillus strains. *CYTA—Journal of Food, 9*, 146–151.

52. Gonzalez-Gonzalez, C., Gibson, T., & Jauregi, P., (2013). Novel probiotic-fermented milk with angiotensin I-converting enzyme inhibitory peptides produced by *Bifidobacterium bifidum* MF 20/5. *International Journal of Food Microbiology, 167*, 131–137.

53. Gu, Q., Zhang, C., Song, D., Li, P., & Zhu, X., (2015). Enhancing vitamin B_{12} content in soy-yogurt by *Lactobacillus reuteri*. *International Journal of Food Microbiology, 206*, 56–59.

54. Halliwell, B., & Whiteman, M., (2004). Measuring reactive species and oxidative damage *in vivo* and in cell culture: How should you do it and what do the results mean? *British Journal of Pharmacology, 142*, 231–255.

55. Hanson, A. D., & Roje, S., (2001). One-carbon metabolism in higher plants. *Annual Review of Plant Physiology and Plant Molecular Biology, 52*, 119–137.

56. Hartmann, R., & Meisel, H., (2007). Food-derived peptides with biological activity: From research to food applications. *Current Opinion in Biotechnology, 18*, 163–169.

57. Hartmann, R., Wal, J. M., Bernard, H., & Pentzien, A. K., (2007). Cytotoxic and allergenic potential of bioactive proteins and peptides. *Current Pharmaceutical Design, 13*(9), 897–920.

58. Hata, I., Higashiyama, S., & Otani, H., (1998). Identification of a phosphopeptide in bovine αs1-casein digest as a factor influencing proliferation and immunoglobulin production in lymphocyte cultures. *Journal of Dairy Science, 65*, 569–578.

59. Hayes, M., Ross, R. P., Fitzgerald, G. F., Stanton, C., & Hill, C., (2006). Casein-derived antimicrobial peptides generated by *Lactobacillus acidophilus* DPC6026. *Applied and Environmental Microbiology, 72*, 2260–2264.

60. Hernandez-Ledesma, B., Beatriz, M., Lourdes, A., Mercedes, R., & Isidra, R., (2005). Identification of antioxidant and ACE-inhibitory peptides in fermented milk. *Journal of the Science of Food and Agriculture, 85*, 1041–1048.

61. Hernandez-Ledesma, B., Contreras, M., & Recio, I., (2011). Antihypertensive peptides: Production, bioavailability and incorporation to foods. *Advances in Colloid and Interface Science, 165*, 23–35.

62. Hernandez-Ledesma, B., Davalos, A., Bartolome, B., & Amigo, L., (2005). Preparation of antioxidant enzymatic hydrolysates from α-lactalbumin and β-lactoglobulin. Identification of active peptides by HPLC-MS/MS. *Journal of Agricultural and Food Chemistry, 53*(3), 588–593.

63. Holasova, M., Fiedlerova, V., Roubal, P., & Pechacova, P. M., (2005). Possibility of increasing natural folate content in fermented milk products by fermentation and fruit component addition. *Czech Journal of Food Sciences, 23*, 196–201.

64. http://.aps.unmc.edu/AP/main.php (Accessed on 20 July 2019).

65. Hugenschmidt, S., Schwenninger, S. M., Gnehm, N., & Lacroix, C., (2010). Screening of a natural biodiversity of lactic and propionic acid bacteria for folate and vitamin B12 production in supplemented whey permeate. *International Dairy Journal, 20*, 852–857.

66. Hugenschmidt, S., Schwenninger, S. M., & Lacroix, C., (2011). Concurrent high production of natural folate and vitamin B12 using a co-culture process with *Lactobacillus plantarum* SM39 and *Propionibacterium freudenreichii* DF13. *Process Biochemistry, 46*, 1063–1070.

67. Iyer, R., Tomar, S., Kapila, S., Mani, J., & Singh, R., (2010). Probiotic properties of folate producing *Streptococcus thermophilus* strains. *Food Research International, 43*, 103–110.

68. Jakala, P., & Vapaatalo, H., (2010). Antihypertensive peptides from milk proteins. *Pharmaceuticals, 3*, 251–272.

69. Jarmolowska, B., (2012). The influence of storage on contents of selected antagonist and agonist opioid peptides in fermented milk drinks. *Milchwissenschaft, 67*(2), 130–135.

70. Jarmolowska, B., Kostyra, E., Krawczuk, S., & Kostyra, H., (1999). ß-casomorphin- 7 isolated from Brie cheese. *Journal of the Science of Food and Agriculture, 79*, 1788–1792.

71. Jones, P. J., (2002). Clinical nutrition: 7. Functional foods—more than just nutrition. *Canadian Medical Association Journal, 166*(12), 1555–1563.

72. Kampa, M., Loukas, S., Hatzoglou, A., Martin, P., Martin, P. M., & Castanas, E., (1996). Identification of a novel opioid peptide (Tyr-Val-Pro-Phe-Pro) derived from human alpha S1 casein (alpha S1-casomorphin, and alpha S1-casomorphin amide). *Biochemical Journal, 319*, 903–908.

73. Kawahara, T., Aruga, K., & Otani, H., (2005). Characterization of casein phosphopeptides from fermented milk products. *Journal of Nutritional Science and Vitaminology, 51*, 377–381.

74. Kayser, H., & Meisel, H., (1996). Stimulation of human peripheral blood lymphocytes by bioactive peptides derived from bovine milk proteins. *FEBS Letter, 383*, 18–20.

75. Kerry, A., Jackson, B. S., & Dennis, A., (2001). Lactose maldigestion calcium intake and osteoporosis in Africa-, Asian- and Hispanic-Americans. *Journal of the American College of Nutrition, 20*, S198–S207.

76. Kilpi, E. E. R., Kahala, M. M., Steele, J. L., Pihlanto, A. M., & Joutsjoki, V. V., (2007). Angiotensin I-converting enzyme inhibitory activity in milk fermented by wild-type

and peptidase-deletion derivatives of *Lactobacillus helveticus* CNRZ32. *International Dairy Journal, 17*, 976–984.

77. Kim, S., Park, S., & Choue, R., (2010). Effects of fermented milk peptides supplement on blood pressure and vascular function in spontaneously hypertensive rats. *Food Science and Biotechnology, 19*, 1409–1413.

78. Kitazawa, H., Yonezawa, K., Tohno, M., Shimosato, T., Kawai, Y., Saito, T., & Wang, J. M., (2007). Enzymatic digestion of the milk protein beta-casein releases potent chemotactic peptide(s) for monocytes and macrophages. *International Immunopharmacology, 7*, 1150–1159.

79. Kneifel, W., Kaufmann, M., Fleischer, A., & Ulberth, F., (1992). Screening of commercially available mesophilic diary starter culture: Biochemical, sensory and morphological properties. *Journal of Dairy Science, 75*, 3158–3266.

80. Korhonen, H., & Pihlanto, A., (2006). Bioactive peptides: Production and functionality. *International Dairy Journal, 16*(9), 945–960.

81. Korhonen, H., & Pihlanto, A., (2003). Food-derived bioactive peptides—opportunities for designing future foods. *Current Pharmaceutical Design, 9*, 1297–1308.

82. Kriss, J. L., Ramakrishnan, U., Beauregard, J. L., Phadke, V. K., Stein, A. D., Rivera, J. A., & Omer, S. B., (2017). Yogurt consumption during pregnancy and preterm delivery in Mexican women: A prospective analysis of interaction with maternal overweight status. *Maternal and Child Nutrition, Online Article e-12522*, 1–8.

83. Kunda, P. B., Benavente, F., Catala-Clariana, S., Gimenez, E., Barbosa, J., & Sanz-Nebot, V., (2012). Identification of bioactive peptides in a functional yogurt by micro liquid chromatography time-of-flight mass spectrometry assisted by retention time prediction. *Journal of Chromatography A, 1229*, 121–128.

84. Lahov, E., & Regelson, W., (1996). Antibacterial and immunostimulating casein-derived substances from milk: Caseicidin, isracidin peptides. *Food and Chemical Toxicology, 34*(1), 131–145.

85. Laino, J. E., Juarez Del Valle, M., Savoy de Giori, G., & LeBlanc, J. G. J., (2013). Development of a high folate concentration yogurt naturally bio-enriched using selected lactic acid bacteria. *LWT Food Science and Technology, 54*, 1–5.

86. Laino, J. E., LeBlanc, J. G., & Savoy De Giori, G., (2012). Production of natural folates by lactic acid bacteria starter cultures isolated from artisanal Argentinean yogurts. *Canadian Journal of Microbiology, 58*, 581–588.

87. LeBlanc, J. G., Laino, J. E., Del Valle, M. J., Vannini, V., Van Sinderen, D., Taranto, M. P., De Valdez, G. F., De Giori, G. S., & Sesma, F., (2011). B-group vitamin production by lactic acid bacteria--current knowledge and potential applications. *Journal of Applied Microbiology, 111*(6), 1297–1309.

88. Levkovich, T., Poutahidis, T., Smillie, C., Varian, B. J., & Ibrahim, Y. M., (2013). Probiotic bacteria induce a 'glow of health.' *PLoS One, 8*, e53867, 1–12.

89. Li, P., Gu, Q., Yang, L., Yu, Y., & Wang, Y., (2017). Characterization of extracellular vitamin B12 producing *Lactobacillus plantarum* strains and assessment of the probiotic potentials. *Food Chemistry, 234*, 494–501.

90. Li, S., & Shah, N. P., (2016). Characterization, anti-inflammatory and antiproliferative activities of natural and sulfonated exo-polysaccharides from *Streptococcus thermophilus* ASCC 1275. *Journal of Food Science, 81*(5), M1167–M1176.

91. Lin, M. Y., & Chang, F. J., (2000). Antioxidative effect of intestinal bacteria *Bifidobacterium longum* ATCC 15708 and *Lactobacillus acidophilus* ATCC 4356. *Digestive Diseases and Sciences*, *45*, 1617–1622.

92. Lin, M. Y., & Yen, C. L., (1999). Reactive oxygen species and lipid peroxidation product-scavenging ability of yogurt organisms. *Journal of Dairy Science*, *82*(8), 1629–1633.

93. Lin, M. Y., & Young, C. M., (2000). Folate levels in cultures of lactic acid bacteria. *International Dairy Journal*, *10*, 409–413.

94. Lin, M., & Yen, C., (1999). Antioxidative ability of lactic acid bacteria. *Journal of Agricultural and Food Chemistry*, *47*, 1460–1466.

95. Lopez-Exposito, I., Quiros, A., Amigo, L., & Recio, I., (2007). Casein hydrolysates as a source of antimicrobial, antioxidant and antihypertensive peptides. *Le Lait*, *87*, 241–249.

96. Lorenzen, P. C., & Meisel, H., (2005). Influence of trypsin action in yogurt milk on the release of caseinophosphopeptide-rich fractions and physical properties of the fermented products. *International Journal of Dairy Technology*, *58*, 119–124.

97. Lourens–Hattingh, A., & Viljoen, B. C., (2001). Yogurt as probiotic carrier food. *International Dairy Journal*, *11*, 1–17.

98. Maeno, M., Yamamoto, N., & Takano, T., (1996). Identification of an antihypertensive peptide from casein hydrolysate produced by a proteinase from *Lactobacillus helveticus* Cp790. *Journal of Dairy Science*, *79*(8), 1316–1321.

99. Majumder, K., & Wu, J., (2015). Molecular targets of antihypertensive peptides: Understanding the mechanisms of action based on the pathophysiology of hypertension. *International Journal of Molecular Sciences*, *16*, 256–283.

100. Makino, S., Ikegami, S., Kume, A., Horiuchi, H., Sasaki, H., & Orii, N., (2010). Reducing the risk of infection in the elderly by dietary intake of yogurt fermented with *Lactobacillus delbrueckii* ssp. *bulgaricus* OLL1073R-1. *British Journal of Nutrition*, *104*, 998–1006.

101. Marette, S., Roosen, J., Blanchemanche, S., & Feinblatt-Meleze, E., (2010). Functional food, uncertainty and consumers' choices: A lab experiment with enriched yogurts for lowering cholesterol. *Food Policy*, *35*(5), 419–428.

102. Marques, C., Amorim, M., Odila, J., Estevez, M., Moura, D., Calhau, C., & Pinheiro, H., (2012). Bioactive peptides–Are there more antihypertensive mechanisms beyond ace inhibition? *Current Pharmaceutical Design*, *18*, 4706–4713.

103. Maruyama, S., Mitachi, H., Awaya, J., Kurono, M., Tomizuka, N., & Suzuki, H., (1987). Angiotnsin I-converting enzyme inhibitor activity of the C-terminal hexapeptide of αs1-casein. *Agricultural and Biological Chemistry*, *51*, 2557–2561.

104. Maruyama, S., & Suzuki, H., (1982). A peptide inhibitor of angiotensin I-converting enzyme in the tryptic hydrolysate of casein. *Agricultural and Biological Chemistry*, *46*, 1393–1394.

105. Matar, C., LeBlanc, J. G., Martin, L., & Perdigon, G., (2003). Biologically active peptides released in fermented milk: Role and functions. In: Farnworth, E. D., (ed.), *Handbook of Fermented Functional Foods, Functional Foods and Nutraceuticals Series* (pp. 177–201). CRC Press, Boca Raton, FL.

106. Mazza, G., (1998). *Functional Food, Biochemical and Processing Aspects* (pp. 357–374). Taylor and Francis Group LLC, Roca Raton.

107. Megalemou, K., Sioriki, E., Lordan, R., Dermiki, M., Nasopoulou, C., & Zabetakis, I., (2017). Evaluation of sensory and *in vitro* antithrombotic properties of traditional Greek yogurts derived from different types of milk. *Heliyon 3, e00227*, 1–18.

108. Meisel, H., (2004). Multifunctional peptides encrypted in milk proteins. *Biofactors, 21,* 55–61.

109. Meisel, H., & FitzGerald, R. J., (2003). Biofunctional peptides from milk proteins: mineral binding and cytomodulatory effects. *Current Pharmaceutical Design, 9,* 1289–1295.

110. Meisel, H., & Frister, H., (1989). Chemical characterization of bioactive peptides from *in vivo* digests of casein. *Journal of Dairy Research, 56,* 343–349.

111. Meisel, H., & Schlimme, E., (1994). Inhibitors of angiotensin converting- enzyme derived from bovine casein (casokinins). In: Brantl, V., & Teschemacher, H., (eds.), *b-Casomorphins and Related Peptides: Recent Developments* (pp. 27–33). VCH, Weinheim–Germany.

112. Meydani, S. N., & Ha, W. K., (2000). Immunologic effects of yogurt. *American Journal of Clinical Nutrition, 71,* 861–872.

113. Migliore-Samour, D., Floch, F., & Jolles, P., (1989). Biologically active casein peptides implicated in immunomodulation. *Journal of Dairy Research, 56,* 357–362.

114. Migliore-Samour, D., & Jolles, P., (1988). Casein, a prohormone with an immunomodulating role for the newborn? *Experientia, 44,* 188–193.

115. Minervini, F., Algaron, F., Rizzello, C. G., Fox, P. F., Monnet, V., & Gobbetti, M., (2003). Angiotensin-I-converting enzyme inhibitory and antibacterial peptides from *lactobacillus helveticus* PR4 proteinase-hydrolyzed caseins of milk from six species. *Applied and Environmental Microbiology, 69,* 5297–5305.

116. Mozaffarian, D., Hao, T., Rimm, E. B., Willett, W. C., & Hu, F. B., (2011). Changes in diet and lifestyle and long-term weight gain in women and men. *The New England Journal of Medicine, 364,* 2392–2404.

117. Nabavi, S., Rafraf, M., Somi, M., Homayouni-Rad, A., & Asghari-Jafarabadi, M., (2015). Probiotic yogurt improves body mass index and fasting insulin levels without affecting serum leptin and adiponectin levels in non-alcoholic fatty liver disease (NAFLD). *Journal of Functional Foods, 18,* 684 691.

118. Nakamura, Y., Masuda, O., & Takano, T., (1996). Decrease of tissue angiotensin I-converting enzyme activity upon feeding sour milk in spontaneously hypertensive rats. *Bioscience, Biotechnology and Biochemistry, 60,* 488–489.

119. Nakamura, Y., Yamamoto, N., Sakai, K., Okubo, A., Yamazaki, S., & Takano, T., (1995). Purification and characterization of angiotensin-I-converting enzyme inhibitors from sour milk. *Journal of Dairy Science, 78*(4), 777–783.

120. Nguyen, D. D., Solah, V. A., Johnson, S. K., Charrois, J. W. A., & Busetti, F., (2014). Isotope dilution liquid chromatography tandem mass spectrometry for simultaneous identification and quantification of beta-casomorphin 5 and beta-casomorphin 7 in yogurt. *Food Chemistry, 146,* 345–352.

121. Padghan, P. V., Mann, B., & Hati, S., (2018). Purification and characterization of antioxidative peptides derived from fermented milk (*lassi*) by lactic cultures. *International Journal of Peptide Research and Therapeutics, 24*(2), 235–249.

122. Papadimitriou, C. G., Vafopoulou-Mastrojiannaki, A., Silva, S. V., Gomes, A. M., Malcata, F. X., & Alichanidis, E., (2007). Identification of peptides in traditional and probiotic sheep milk yogurt with angiotensin I-converting enzyme (ACE)-inhibitory activity. *Food Chemistry, 105*(2), 647–656.

123. Ibeagha-Awemu, E.M.; Liu, J.-R.; Zhao, X. Bioactive Components in Yogurt Products. Chapter 9; In Bioactive Components in Milk and Dairy Products, Park, Y.W., Ed., Wiley-Blackwell, New York; 2009; pages 235–250.

124. Parker, F., Migliore-Samour, D., Floch, F., Zerial, A., Werner, G. H., Jolles, J., Casaretto, M., Zahn, H., & Jolles, P., (1984). Immunostimulating hexapeptide from human casein: Amino acid sequence, synthesis and biological properties. *European Journal of Biochemistry, 145*, 677–682.

125. Pattorn, S., Horimoto, Y., Hongsprabhas, P., & Yada, R. Y., (2012). Influence of aggregation on the antioxidative capacity of milk peptides. *International Dairy Journal, 25*, 3–9.

126. Paul, M., & Somkuti, G. A., (2009). Degradation of milk-based bioactive peptides by yogurt fermentation bacteria. *Letters in Applied Microbiology, 49*(3), 345–350.

127. Perdigon, G., De Moreno de Leblanc, A., Valdez, J., & Rachid, M., (2002). Role of yogurt in the prevention of colon cancer. *European Journal of Clinical Nutrition, 56*(3), S65–S68.

128. Perna, A., Intaglietta, I., Simonetti, A., & Gambacorta, E., (2013). Effect of genetic type and casein haplotype on antioxidant activity of yogurts during storage. *Journal of Dairy Science, 96*, 3435–3441.

129. Perpetuo, E. A., Juliano, L., & Lebrun, I., (2003). Biochemical and pharmacological aspects of two bradykinin-potentiating peptides obtained from tryptic hydrolysis of casein. *Journal of Protein Chemistry, 22*(7/8), 601–606.

130. Piaia, M., Antoine, J. M., Guardia, J. A. M., Leplingard, A., & Wijnkoop, I. L., (2003). Assessment of the benefits of live yogurt: Methods and markers for *in vivo* studies of the physiological effects of yogurt cultures. *Microbial Ecology in Health and Disease, 15*, 79–87.

131. Pierce, N. F., Cray, J. W. C., Kaper, J. B., & Mekalanos, J. J., (1988). Determination of immunogenicity and mechanisms of protection by virulent and mutant *Vibrio cholerae* 01 in rabbits. *Infection and Immunity, 56*, 142–148.

132. Pigeon, R. M., Cuesta, E. P., & Gilliland, S. E., (2002). Binding of free bile acids by cells of yogurt starter culture bacteria. *Journal of Dairy Science, 85*(11), 2705–2710.

133. Pihlanto, A., (2006). Antioxidative peptides derived from milk proteins. *International Dairy Journal, 16*, 1306–1314.

134. Pihlanto, A., Virtanen, T., & Korhonen, H., (2010). Angiotensin I converting enzyme (ACE) inhibitory activity and antihypertensive effect of fermented milk. *International Dairy Journal, 10*, 3–10.

135. Pihlanto-Leppala, A., Rokka, T., & Korhonen, H., (1998). Angiotensin I-converting enzyme inhibitory peptides derived from bovine milk proteins. *International Dairy Journal, 8*(4), 325–331.

136. Plaisancie, P., Claustre, J., Estienne, M., Henry, G., Boutrou, R., Paquet, A., & Leonil, J., (2013). A novel bioactive peptide from yogurts modulates expression of the gel-forming MUC2 mucin as well as population of goblet cells and Paneth cells along the small intestine. *Journal of Nutritional Biochemistry, 24*, 213–221.

137. Puri, P., Mahapatra, S. C., Bijlani, R. L., Prasad, H. K., & Nath, I., (1994). Feed efficiency and splenic lymphocyte proliferation response in yogurt- and milk-fed mice. *International Journal of Food Sciences and Nutrition, 45*, 231–235.

138. Quiros, A., Ramos, M., Muguerza, B., Delgado, M., Miguel, M., Aleixandre, A., & Recio, I., (2007). Identification of novel antihypertensive peptides in milk fermented with *Enterococcus faecalis*. *International Dairy Journal, 17*, 33–41.

139. Ramchandran, L., & Shah, N. P., (2009). Effect of exopolysaccharides and inulin on the proteolytic, angiotensin-I-converting enzyme- and alpha-glucosidase-inhibitory

activities as well as on textural and rheological properties of low-fat yogurt during refrigerated storage. *Dairy Science & Technology, 89*, 583–560.

140. Recio, I., & Visser, S., (1999). Identification of two distinct antibacterial domains within the sequence of bovine as2-casein. *Biochimica et Biophysica Acta, 1428*(2/3), 314–326.

141. Rival, S. G., Fornaroli, S., Boeriu, C. G., & Wichers, H. J., (2001). Caseins and casein hydrolysates. 1. Lipoxygenase inhibitory properties. *Journal of Agricultural and Food Chemistry, 49*, 287–294.

142. Rizzello, C. G., Losito, I., Gobbetti, M., Carbonara, T., De Bari, M. D., & Zambonin, P. G., (2005). Antibacterial activities of peptides from the water-soluble extracts of Italian cheese varieties. *Journal of Dairy Science, 88*(7), 2348–2360.

143. Rodriguez-Figueroa, J. C., Gonzalez-Cordova, A. F., Astiazaran-Gacia, H., & Vallejo-Cordoba, B., (2013). Hypotensive and heart rate-lowering effects in rats receiving milk fermented by specific *Lactococcus lactis* strains. *British Journal of Nutrition, 109*, 827–833.

144. Rodriguez-Figueroa, J. C., Gonzalez-Cordova, A. F., Astiazaran-Garcia, H., Hernandez-Mendoza, A., & Vallejo-Cordoba, B., (2013). Antihypertensive and hypolipidemic effect of milk fermented by specific *Lactococcus lactis* strains. *Journal of Dairy Science, 96*, 4094–4099.

145. Rodriguez-Figueroa, J. C., Gonzalez-Cordova, A. F., Torres-Llanez, M. J., Garcia, H. S., & Vallejo-Cordoba, B., (2012). Novel angiotensin I-converting enzyme inhibitory peptides produced in fermented milk by specific wild *Lactococcus lactis* strains. *Journal of Dairy Science, 95*, 5536–5543.

146. Rodriguez-Figueroa, J. C., Reyes-Diaz, R., Gonzalez-Cordova, A. F., Troncoso-Rojas, R., Vargas-Arispuro, I., & Vallejo-Cordoba, B., (2010). Angiotensin-converting enzyme inhibitory activity of milk fermented by wild and industrial *Lactococcus lactis* strains. *Journal of Dairy Science, 93*, 5032–5038.

147. Ruiz-Gimenez, P., Salom, J. B., Marcos, J. F., Valles, S., Martinez-Maqueda, D., Recio, I., Torregrosa, G., Alborch, E., & Manzanares, P., (2012). Antihypertensive effect of a bovine lactoferrin pepsin hydrolysate: Identification of novel active peptides. *Food Chemistry, 131*, 266–273.

148. Sadat, L., Cakir-Kiefer, C., N'Negue, M. A., Gaillard, J. L., Girardet, J. M., & Miclo, L., (2011). Isolation and identification of antioxidative peptides from bovine α-lactalbumin. *International Dairy Journal, 21*(4), 214–221.

149. Sandre, C., Gleizes, A., Forestier, F., Gorges-Kergot, R., Chilmonczyk, S., Leonil, J., Moreau, M. C., & Labarre, C., (2001). A peptide derived from bovine b-casein modulates functional properties of bone marrow-derived macrophages from germfree and human flora-associated mice. *The Journal of Nutrition, 131*(11), 2936–2942.

150. Shakerian, M., Razavi, S. H., Ziai, S. A., Khodaiyan, F., Yarmand, M. S., & Moayedi, A., (2015). Proteolytic and ACE-inhibitory activities of probiotic yogurt containing non-viable bacteria as affected by different levels of fat, inulin and starter culture. *Journal of Food Science and Technology, 52*(4), 2428–2433.

151. Shori, A. B., & Baba, A. S., (2013). Antioxidant activity and inhibition of key enzymes linked to type-2 diabetes and hypertension by Azadirachta indica-yogurt. *Journal of Saudi Chemical Society, 17*, 295–301.

152. Sieuwerts, S., De Bok, F. A. M., Hugenholtz, J., & Van Hylckama, V. J. E. T., (2008). Unraveling microbial interactions in food fermentations: From classical to genomics approaches. *Applied and Environmental Microbiology, 6*, 4997–5007.

153. Sipola, M., Finckenberg, P., Korpela, R., Vapaatalo, H., & Nurminen, M., (2002). Effect of long-term intake of milk products on blood pressure in hypertensive rats. *Journal of Dairy Research, 69*, 103–111.

154. Sipola, M., Finckenberg, P., Santisteban, J., Korpela, R., Vapaatalo, H., & Nurminen, M., (2001). Long term intake of milk peptides attenuates development of hypertension in spontaneously hypertensive rats. *Journal of Physiology and Pharmacology, 52*, 745–754.

155. Sipola, M., Finckenberg, P., Vapaatalo, H., Pihlanto-Leppala, A., Korhonen, H., Korpela, R., & Nurminen, M.-L., (2002). α-Lactorphin and β-lactorphin improve arterial function in spontaneously hypertensive rats. *Life Sciences, 71*(11), 1245–1253.

156. Stuknyte, M., De Noni, I., Gugliemetti, S., Minuzzo, M., & Mora, D., (2011). Potental immunomodulatory activity of bovine casein hydrolysates produced after digestion with proteinase of lactic acid bacteria. *International Dairy Journal, 21*, 763–769.

157. Sun, L., Finegann, C. M., Kish-Catalone, T., Blumenthal, R., GarzinoDemo, P., La Terra Maggiore, G. M., Berrone, S., Kleinman, C., Abdelwahab, S., Lu, W., & Garzino-Demo, A., (2005). Human beta-defensins suppress human immunodeficiency virus infection: Potential role in mucosal protection. *Journal of Virology, 79*, 14318–14329.

158. Sybesma, W., Starrenburg. M., Tijsseling, L., Hoefnagel, M. H., & Hugenholtz, J., (2003). Effects of cultivation conditions on folate production by lactic acid bacteria. *Applied and Environmental Microbiology, 69*, 4542–4548.

159. Tang, Z., Yin, Y., Zhang, Y., Huang, R., Sun, Z., Li, T., Chu, W., Kong, X., Li, L., Geng, M., & Tu, Q., (2009). Effects of dietary supplementation with an expressed fusion peptide bovine lactoferricin-lactoferrampin on performance, immune function and intestinal mucosal morphology in piglets weaned at age 21 d. *British Journal of Nutrition, 101*, 998–1005.

160. Tauzin, J., Miclo, L., & Gaillard, J. L., (2002). Angiotensin-converting enzyme inhibitory peptides from tryptic hydrolysate of bovine as2-casein. *FEBS Letters, 531*(2), 369–374.

161. The Dairy Council, (2013). *The Nutritional Composition of Dairy Products* (p. 112). Dairy Council, London.

162. Thibessard, A., Borges, F., Fernandez, A., Gintz, B., Decaris, B., & Leblond-Bourget, N., (2004). Identification of *Streptococcus thermophilus* CNRZ368 genes involved in defense against superoxide stress. *Applied and Environmental Microbiology, 70*, 2220–2229.

163. Tidona, F., Criscione, A., Guastella, A. M., Zuccaro, A., Bordonaro, S., & Marletta, D., (2009). Bioactive peptides in dairy products, *Italian Journal of Animal Science, 8*, 315–340.

164. Tsompo, A., Romanowski, A., Banda, L., Lavoie, J. C., Jenssen, H., & Friel, J. K., (2011). Novel anti-oxidative peptides from enzymatic digestion of human milk. *Food Chemistry, 126*, 1138–1143.

165. Udenigwe, C., & Mohan, A., (2014). Mechanisms of food protein-derived antihypertensive peptides other than ACE inhibition. *Journal of Functional Foods, 8C*, 45–52.

166. Ueta, E., Tanida, T., & Osaki, T., (2001). A novel bovine lactoferrin peptide, FKCRRWQWRM, suppresses Candida cell growth and activates neutrophils. *Journal of Peptide Research, 57*, 240–249.

167. Van Boven, A., Tan, P. S. T., & Konings, W. M., (1986). Purification and characterization of a dipeptidase from *Streptococcus cremoris* Wg2. *Netherlands Milk and Dairy Journal, l40*, 117–127.

168. Van't Hof, W., Veerman, E. C. I., Helmerhorst, E. J., & Amerongen, A. V. N., (2001). Antimicrobial peptides: Properties and applicability. *Biological Chemistry, 382,* 597–619.

169. Vassou, D., Bakogeorgou, E., Kampa, M., Dimitriou, H., Hatzoglou, A., & Castanas, E., (2008). Opioids modulate constitutive B-lymphocyte secretion. *International Immunopharmacology, 8,* 634–644.

170. Water, J. V., Keen, C. L., & Gershwin, M. E., (1999). The influence of chronic yogurt consumption on immunity. *Journal of Nutrition, 129,* 1492S–1495S.

171. WHO, (2011). *Enfermedades Cardiovasculares (Heart Diseases).* http://www.who.int/mediacentre/factsheets/fs317/es/index.html (Accessed on 20 July 2019).

172. Yamamoto, N., Akino, A., & Takano, T., (1994). Antihypertensive effects of different kinds of fermented milk in spontaneously hypertensive rats. *Bioscience, Biotechnology, and Biochemistry, 58,* 776–778.

173. Yamamoto, N., Maeno, M., & Takano, T., (1999). Purification and characterization of an antihypertensive peptide from a yogurt-like product fermented by *Lactobacillus helveticus* CPN4. *Journal of Dairy Science, 82,* 1388–1393.

174. Ziarno, M., (2009). *In vitro* cholesterol uptake by *Lactobacillus delbrueckii* subsp. *bulgaricus* isolates. *Acta Scientiarum Polonorum Technologia Alimentaria, 8*(2), 21–32.

175. Zucht, H. D., Raida, M., Adermann, K., Magert, H. J., & Forssman, W. G., (1995). Casocidin-I: A casein as2-derived peptide exhibits antibacterial activity. *FEBS Letters, 372*(2/3), 185–188.

CHAPTER 8

ANTIBIOTIC RESISTANT PATHOGENS IN MILK AND MILK PRODUCTS

AMI PATEL, FALGUNI PATRA, and NIHIR SHAH

ABSTRACT

The food chain is one of the most common routes for the spread of antibiotic resistance among microorganisms. Especially, milk and milk products are widely consumed and preferred by all age groups. However, both beneficial and harmful microorganisms have been found to possess resistance to major antibiotics. This chapter focuses on the historical development of antibiotic resistance in microorganisms; mechanisms of horizontal transfer of specific genes; the current status of antibiotic resistance among pathogenic bacteria in dairy products; and preventive measures to control the problem.

8.1 INTRODUCTION

Before the innovation of antibiotics, 97% of endocarditis patients died [23], up to 40% cases of pneumonia caused by *Streptococcus pneumonia* became fatal [6], and mortality rate in patients with *Staphylococcus aureus* bacteremia reached as high as 80% [43]. Antibiotics also had a major role in the advancement of increasingly complex surgery, chemotherapy, and organ transplantation. Unfortunately, antibiotic resistance to common antibiotics against majority of the pathogenic bacteria has increased tremendously throughout the world [16].

The accomplishments of contemporary medicines are at greater risk due to the emergence of antibiotic resistance. With the generation of antibiotics resistant pathogens, surgeries such as cesarean sections, organ transplantations, chemotherapy, etc. have become risky without effective antibiotics and other therapies for the prevention and treatment of infections. In the US, each year, at least 2 million people become infected with common antibiotic-resistant pathogens, and approximately 23,000 patients die due to these infections [12]. When the first-line antibiotics are not working for the treatment of any infection, more expensive medicines must be used, which prolong the length of disease and treatment ultimately leading to financial pressure on families and societies due to increase in medical bills.

The development of antibiotic resistance is a natural evolutionary response to antibiotic exposure. The mechanism of antibiotic resistance has evolved, including the production of enzymes to inactivate or modify the antimicrobials, preventing the entry of or exporting the antibiotics out or altering antibiotic target. Selective pressure with the use, misuse, or excessive use of antibiotics is regarded as the most important factor for the emergence of drug resistance, which enables the antimicrobial-resistant microorganisms with inherent resistance or newly acquired mutations or resistance genes to survive and proliferate [4]. Antibiotic resistance is also developed through the exchange of a new resistance gene between organisms by conjugation, transformation, and transduction [36].

It has been reported that infants are fast colonized by *Enterobacteriaceae* after birth, irrespective of the factor, whether they are breastfed or not [36]. In an environment, which is free from external antimicrobial selection pressure, resistant, and non-resistant species exist together in a stable balance.

Antimicrobials are one of the most frequently prescribed, and 50% of those are considered non-essential, which eventually leads to the emergence of antimicrobial resistance in pathogens. Antibiotic-resistant bacteria have been isolated from almost every possible source including soil, drinking water, sea, various food products even in the Antarctica region [79]. Resistances against synthetic antimicrobials have also emerged, e.g., fluoroquinolones resistance in *Escherichia coli* isolated from patients in Europe is now at 10–40% [6]. The 2015 report shows that drug resistance is continuously increasing for most of the bacteria and antibiotics under observation [19]. In the EU, it was reported that there was an increase in carbapenem resistance from 6.2% in 2012 to 8.1% in 2015 in *Klebsiella pneumonia*. The carbapenems and polymyxins (e.g., colistin) groups of antibiotics are last-line antibiotics for the treatment of patients infected with other multidrug-resistant pathogens [19]. Even a yogurt sample from California was found to have colistin-resistant bacteria [82].

CDC has also reported that one infection out of five resistant infections is due to the antibiotic-resistant strain of *Salmonella* and *Campylobacter* from food and animals sources [12]. Antibiotic-resistant animal pathogens have also been found to be transferred to humans. When diseased animals are treated with antibiotics, drug-resistant pathogens emerge. The animals are then slaughtered and processed for food, and the antimicrobial-resistant pathogens from the animal can contaminate meat or other food products.

Drug-resistant pathogens may be spread to the environment through animal feces, which subsequently contaminate fruits and vegetables, and drinking water. The probable transmission of extended-spectrum β-lactamase (ESBL) and AmpC-β-lactamase genes on plasmids and of *E. coli* clones to human beings from livestock through food chain has also been supported [48]. Milk and milk products can be contaminated with pathogenic microorganisms, mainly due to poor processing, handling, and unhygienic environment. As milk is highly perishable in nature, it can act as a vehicle for the transmission of a range of food-borne diseases, particularly in countries, where stringent hygienic regulations are not mandatory.

This chapter focuses on: the historical development of antibiotic resistance among bacteria; antibiotic resistance mechanisms and related detection methods; the current status of the prevalence of antibiotic resistance among pathogenic bacteria in dairy-based food products; and remedial strategies to reduce or overcome the issue.

8.2 HISTORY OF DEVELOPMENT OF ANTIBIOTIC RESISTANCE

Adaptation is very commonly observed phenomenon in living cells that may lead to develop resistance against adverse conditions. Hence, resistant bacteria have always been around and existed long before antibiotics were employed therapeutically in the history of mankind. Right after the beginning of the use of penicillin, several strains of *Staphylococcus* appeared to resistant against it, and currently, about 80% of *Staphylococcus* strains do not respond to treatment of antibiotics [5]. Chloramphenicol, streptomycin, and tetracycline were discovered in the 1940s and early 1950s. Soon in 1953, a strain of *Shigella* was appeared to resist these antibiotics.

During 1940–50s, a single antibiotic, such as streptomycin, was effective against bacteria causing tuberculosis (T.B.) that is now no longer able to treat the disease. Additionally, multi-drug resistant T.B. strains had emerged, and thus, tuberculosis is one of the chief causes of death under the infectious disease category. The resistant strains of gonorrhea emerged in the 1970s; and since 1990s, the development of bacteria that resist all known antibiotics called 'true superbugs' was scrutinized. Vancomycin is one of the effective antibiotics from the last decade that hits bacteria on many fronts. However, some enterococcal strains have been discovered recently that showed resistant against vancomycin [36, 50].

8.3 ANTIBIOTIC RESISTANCE IN PATHOGENIC BACTERIA

There is a close association between the quantities of antimicrobials being used and the rate of development of resistance against them. In human and animal medicine, irrational use of antibiotics and related drugs is believed to be the foremost reason of the antibiotic resistance problem. Another significant fact is related with the vigorous use of antimicrobial agents in animal husbandry and livestock farms that has led to contamination of resistant bacteria in the food chain [17, 91]. Several reports state that presence of antibiotic residues in milk [58, 74, 92] because of antimicrobial drugs are used to prevent, control, and cure infections (such as mastitis), and moreover, to enhance animal growth and feed efficiency.

Although clinical bacterial isolates have been considered as key culprits contributing to the emergence and transfer of resistance, yet the role of food-borne commensal and/or pathogenic organisms have not received considerable focus, and thus research concerning antibiotic resistance in normal food chain bacteria is enough. Food chain can serve as the main source for the transmission of antibiotic-resistant microflora among the animals and humans [15, 93, 98].

Milk, curd, cheese, and other dairy products are routinely consumed in the diet, which makes them to serve significantly in our daily diet. Therefore, if pathogenic bacteria present in these products acquire antibiotic-resistant, it may cause serious infections and other illnesses in consumer's that might be difficult to cure or treat then after. Moreover, since fermented dairy products are often consumed without any heat treatment; thus such products may serve as a source or vehicle of antibiotic-resistant bacteria, and represent a direct link between the indigenous microflora of animal and human gastrointestinal (GI) tract [56]. Clinical investigations have documented the persistence of antibiotic-resistant strains in the human gut even in the absence of selective pressure, indicating that drug exposure induces long-term alterations within complex microbial communities [41].

Enough literature is available on the occurrence of antibiotic residues in milk and milk products. Nevertheless, there is less emphasis on the predominance of antibiotic resistances among foodborne pathogens present in milk and dairy products. Thus, the authors of this chapter have tried to cover and discuss the significant findings related to the prevalence of antibiotic resistance among pathogenic bacteria, which have been more frequently isolated from milk and milk products.

8.4 MECHANISM OF ANTIBIOTIC RESISTANCE IN FOODBORNE PATHOGENS

A single strain of any bacteria may own a number of mechanisms for antibiotic resistance. The resistance mechanism can be categorized into two basic types: biochemical and genetic; which mechanism exists in the particular bacterial cell depends on the antibiotic, its site of action (target site), and the bacterial species itself. Further, weather resistance is associated with a chromosome (through mutation) or by a plasmid is necessary to investigate.

While the acquisition of resistance may differ between types of bacteria, many times, resistance is created by a few common mechanisms [17, 56], such as:

- **Inactivation of Antibiotic:** Direct inactivation of the active antibiotic;
- **Target Modification:** Modification of the susceptibility against the antibiotic by modification of the target site;
- **Efflux Pumps and Outer Membrane (OM) Permeability Changes:** Reduction of the concentration of drug without modification of the compound itself; or
- **Target Bypass:** Some bacteria become refractory to specific antibiotics by bypassing the inactivation of a given enzyme [17, 56].

The majority of these mechanisms have been established or studied in enterococcal species; conversely, there have not been definite studies dealing with such mechanisms in other foodborne pathogenic bacteria.

The naturally occurring mechanism of antibiotic resistance involves mutations or horizontal transfer of resistant genes between the bacteria [17]. In microorganisms, the evolution of antibiotic resistance is enhanced by horizontal transfer of resistance genes over species and genus borders by means of conjugative plasmids, transposons, the possession of integrons and insertion elements; and also mediated through virulent and temperate bacteriophages [14, 15].

The development of antibiotic resistance problem, particularly against tetracycline and/or macrolide, is mainly due to the transfer of R-plasmid through conjugation among bacteria [88]. The transfer led to the development of several resistant genes against the specific antibiotic(s) in the same bacteria or different bacteria and thus creating widespread resistance [35]. The antimicrobial resistance in bacteria may be due to the production of beta-lactamase enzyme, inactivating enzyme chloramphenicol acetyltransferase, decrease uptake of drug, plasmid-mediated transferable multidrug resistance (MDR), changes in the porin molecules in OM, or changes in the target organ such as penicillin-binding proteins [21, 94].

8.5 METHODS FOR TESTING OF ANTIBIOTIC SUSCEPTIBILITY

Antimicrobial susceptibility testing can be performed using different phenotypic methods, in which agar dilution and broth microdilution are regular methods based on CLSI (Clinical and Laboratory Standards Institute: formerly NCCLS, National Committee on Clinical Laboratory Standards). Other widely used methods are: agar/disc diffusion method, agar gradient method and commercial methods, such as for instance E-test (AbBiomerieux, Sweden), with a predefined gradient of antibiotic concentrations on a plastic strip [70]. Besides determining the phenotypic antibiotic resistance, the genotypic detection of specific genes responsible for resistance is also essential. These include various polymerase chain reaction (PCR)-based techniques, DNA hybridization, plasmid profiling, and DNA microarray. In recent years, few rapid assays based on MALDI-TOF mass spectrometry in combination with PCR/electrospray ionization mass spectrometry or mini-sequencing has also been developed [102].

The situation is apparent when the phenotypic and genotypic resistance patterns are in agreement. However, a phenotypically resistant bacterium strain may be genotypically "susceptible." This is usually due to the fact that appropriate genes are not included in the test patterns, or there exists unknown resistance genes. Tetracycline, for example, has more than 40 different genes conferring antibiotic resistance, and a number of tetracycline resistance genes continue to increase [77].

8.6 ANTIBIOTIC RESISTANCE AMONG DIFFERENT PATHOGENS

8.6.1 STAPHYLOCOCCUS AUREUS

S. aureus is a foodborne pathogen known to be responsible for disease outbreaks associated with the consumption of various food products. Milk and milk products, including raw milk, pasteurized milk, butter, cheese, ice cream, and SMP are also well-known sources of food poisoning caused by S. aureus. The S. aureus strains are known to cause gastroenteritis due to the production of highly heat-stable enterotoxins. The organism is easily killed in milk by pasteurization temperature, but the heat-stable staphylococcal enterotoxins are not easily inactivated in foods during cooking.

S. aureus is one of the human pathogen, which has adopted itself and gained resistance to each new antibiotic developed initially with penicillin and methicillin, latest against linezolid and daptomycin. The antibiotic resistance mechanisms in *S. aureus* include enzymatic inactivation of the antibiotic, e.g., with penicillinase and aminoglycoside-modification enzymes; modification of the target, e.g., penicillin-binding protein 2a of methicillin-resistant *S. aureus* and D-Ala-D-Lac of peptidoglycan precursors of vancomycin-resistant strains, efflux pumps for fluoroquinolones and tetracycline and trapping of the antibiotics for vancomycin and possibly daptomycin. *S. aureus* also acquired complex genetic arrays like staphylococcal chromosomal cassette-*mec* elements (SCC*mec*) or the *van A* operon through horizontal gene transfer (HGT). The resistance to the most recent antibiotics like fluoroquinolones, linezolid, and daptomycin has been emerged through spontaneous mutations and positive selection [54]. A number of reports have confirmed the presence of antibiotic-resistant *S. aureus* in milk and dairy products.

Gundogan et al., [30] isolated 110 *S. aureus* from 180 samples of raw milk, pasteurized milk and ice cream; and they found that all the isolates were highly resistant to penicillin G, methicillin, and bacitracin, and few strains were also erythromycin resistant. In another study with 160 *S. aureus* strains, 164 were isolated from 1634 food samples in Italy, six isolates (bovine milk 4 strain, pecorino cheese 1 strain and mozzarella cheese I strain) were *mecA (for methicillin resistance) positive [59]. The same researcher also reported that* 68.8% analyzed isolates demonstrated antimicrobial resistance properties against at least one of the antibiotics tested. In a study with 150 bulk milk tank samples from Minnesota revealed 84% samples were positive for methicillin-susceptible *S. aureus* (MSSA), and 4% were MRSA. From these 150 samples, 93 MSSA and 2 MRSA were isolated; and antibiotic-resistant profiling showed that 21 isolates were resistance to a single antibiotic class, 13 isolate to two antibiotic classes and 5 isolate to ≥3 antibiotics classes; thus indication multidrug-resistant. The two MRSA isolates showed resistance to β-lactams, cephalosporins, and lincosamides [32].

In another study, 36 MRSA were isolated and characterized from bulk tank milk (BTM) samples from German dairy herds, and it was found that all the isolates were resistant to tetracycline, and 58%, 52%, 36%, and 27% isolates were resistant to clindamycin, erythromycin, quinupristin/dalfopristin, and kanamycin, respectively [52].

Thaker et al., [92] isolated 10 *S. aureus* from milk (100), curd (30), and pedha (30) and they found that all the isolates were resistant to penicillin-G; 40% of the isolates resistant to ampicillin, 20% resistant to oxytetracycline and oxacillin, and 10% resistant to streptomycin and gentamicin. Later, Gundogan and Avci [29] surveyed 150 samples of raw milk, white cheese and ice cream from dairy-processing plants in Turkey, and they found that 56% raw milk, 48% cheese and 36% ice cream samples were contaminated with *S. aureus*. The isolated *S. aureus* demonstrated resistance to ampicillin, penicillin, tetracycline, erythromycin, gentamicin, and trimethoprim/sulfamethoxazole.

MRSA (*mecC* positive) were detected on 10 bovine bulk milk tank samples out of 465 dairy farms samples tested in England and Wales [71]. MRSA strains were also isolated from ice cream (*mecA* positive) in Turkey [28], traditional goat or sheep's milk cheese in Iran [84] and Swiss bovine raw milk cheese (*spa* type t127) [38]. MRSA (3 isolates) were found in 3 samples out of 565 samples of raw milk, thermized milk, curd, yogurt, cheese under test. The isolates were positive for SCC*mec* type Iva, *spa* type t127, clonal complex 1, and sequence type 1 [10].

Jamali et al., [39] screened 1930 samples of raw milk and 720 samples of dairy products from Iran, and they found that 328 samples were contaminated with *S. aureus* out of which 53 samples were confirmed for MRSA. The resistance pattern of isolated *S. aureus* was in the following order: 56.1% tetracycline, 47.3% penicillin G, 16.2% oxacillin, 11.9% lincomycin, 11.3% clindamycin, 7.9% erythromycin, 5.8% streptomycin, 5.5% cefoxitin, 4% kanamycin, 3.7% chloramphenicol, and 2.1% gentamicin, respectively. They also reported a higher incidence of *blaZ* (46%) and *tetM* (34.8%) resistance genes in isolated *S. aureus*. The 162sheep and goat BTM samples from Southern Italy were tested for the presence of MRSA; and out of 162 samples, 2 samples were found to be positive for MRSA (characterized by Multi-Locus Sequence Typing (MLST), *spa*, and SCC*mec* typing) [11].

Al-Ashmawy et al., [3] also detected 106 MRSA positive samples (53%) of raw milk and dairy products like yogurt, Ice cream, Kareish cheese and Damietta cheese in Egypt, out of 200 samples under test. The MRSA strains were highly resistant to penicillin-G (87.9%), and antibiotic resistance pattern of cloxacillin, tetracycline, and amoxicillin were 75.9%, 65.2%, and 55.6%, respectively. Similarly, Parisi et al., [69] studied the prevalence of *S. aureus* (coagulase producing) and MRSA in BTM samples collected in Southern Italy. They analyzed 486 BTM samples and reported that 12 samples (2.5%) were positive for MRSA. All the tested isolates were resistant to tetracycline, ampicillin, and oxacillin.

Basanisi et al., [7] isolated 484 *S. aureus* strains from milk and milk products (3760 samples) in Southern Italy. They reported that 40 isolated strains (8.3%) were MRSA, which were validated by various methods like MLST, *spa*-typing, SCC*mec* typing, Panton-Valentine Leukocidin (PVL) genes, Staphylococcal enterotoxins genes, and also by the ability of the strains to form a biofilm. Most recently, Papadopoulos et al., [68] isolated *S. aureus* from 36 BTM samples and 19 dairy products in Greece and they found that 99.6% of the isolates were resistant to at least one antibiotics studied and 13.3% of the isolates were multi-drug resistant showing resistance to three or more antibiotic classes.

8.6.2   *LISTERIA MONOCYTOGENES*

L. monocytogenes is considered one of the emerging foodborne pathogens and is reported to cause sporadic outbreaks and epidemics due to the consumption of contaminated food products such as ready-to-eat foods, meat, poultry, milk, cheeses made from unpasteurized milk and other dairy-based products. *L. monocytogenes* is known to be responsible for encephalitis, meningitis, septicemia, abortion, premature birth, and stillbirth in humans [86]. The mortality rate due to *L. monocytogenes* infection ranges from 20–30%, but fatality rate can be as high as 75% in immunocompromised persons, fetuses, infants, and in aged people [89]. *L. monocytogenes* have been found to be naturally resistant to fosfomycin, cephalosporins, aztreonam, pipemidic acid, dalfopristin/quinupristin, and sulfamethoxazole. However, the emergence of resistance strain to other groups of antibiotics has been reported due to the improper use of antibiotics in humans and animals.

Harakeh et al., [31] isolated *L. monocytogenes* from traditional dairy-based Lebanese food products like *Baladi* cheese, *Shankleesh, Kishk*. They reported that out of all isolates tested, 93%, 90%, 60%, and 26.66% were resistant to oxacillin, penicillin, ampicillin, and vancomycin, respectively. In 2012, Osaili et al., [62] collected 350 Brined white cheese samples such as *halloumi, akkawi, shellal, boiled, and pasteurized* from the local market in Jordan and they found that 11.1% samples were contaminated with *L. monocytogenes*. Isolated *L. monocytogenes* strains showed drug resistance towards clindamycin, fosfomycin, and oxacillin.

Rodas-Suarez et al., [78] isolated *L. monocytogenes* strains form dry milk samples collected in Mexico, and the strains showed multidrug resistance to ampicillin, erythromycin, tetracycline, dicloxacillin, and trimethoprim-sulfamethoxazole in the range of 9%-14%. The 18 strains of *L. monocytogenes* were also isolated from 446 BTM raw milk samples in Iran and were screened for antibiotic resistance. The resistance pattern of the strains was in the order of tetracycline (72.2%), penicillin (61.1%), chloramphenicol (27.8%), amoxicillin-clavulanic acid (22.2%); and 71.4% of the *L. monocytogenes* were multidrug-resistant. Internalin genes *inlA, inlC,* and *inlJ* were identified in the tested *L. monocytogenes* isolates [40].

Garedew et al., [24] analyzed 384 food samples and 24 (6.25%) samples were found to be contaminated with *L. monocytogenes. The isolated L. monocytogenes* from positive samples of ice cream, unpasteurized milk, cake, pizza, etc. were also studied for antibiotic susceptibility. The tested isolates demonstrated high resistance to penicillin, tetracycline, nalidixic acid, and chloramphenicol; among all the isolates, 4 strains were multi-drug resistant. All *L. monocytogenes* isolates showed sensitivity towards vancomycin, gentamicin, amoxicillin, cloxacillin, cephalothin, and sulfamethoxazole.

Similarly, Osman et al., [64] analyzed 203 udder milk samples from cow and buffalo in Egypt for listeria, and they found 3 (1.4%) samples positive for *L. monocytogenes.* The isolated *L. monocytogenes* were found to be resistant to 9 out of 27 antibiotics tested. In Turkey, a study was conducted with 210 samples of milk and milk products like ice cream, butter, white cheese, *kashar* cheese, farm cheese, *cokelek,* and *kuymak* for the presence of *L. monocytogenes* and the 8.2% in dairy products and 5% milk samples were found positive for *L. monocytogenes.* Also, 36.5% of the isolated *L. monocytogenes* were multidrug-resistant, and 15.3% of isolates were resistant to at least one antibiotic. Out of all the tested strains, 34.6% were resistant to tetracycline; and resistant to penicillin G and chloramphenicol was @ 23% and 25%, respectively [45]. Gohar et al., [26] screened 125 milk and dairy products and reported that 13.6% samples were contaminated with *L. monocytogenes.* The isolated *L. monocytogenes* demonstrated resistant towards penicillin, amoxicillin, and ampicillin.

Recently, Sharma et al., [85] screened 457 raw milk samples procured from 15 major cities in Rajasthan, India for the presence of *L. monocytogenes* and they found 5 samples positive for *L. monocytogenes.* The study also assessed virulence potential of the strains and found positive for Listeriolysin O (LLO) *inlA, inlC, inlB (two strains)* and phosphatidylinositol-specific phospholipase C (PI-PLC) activity. Antibiotic susceptibility test showed high resistance to penicillin G, azithromycin, piperacillin, oxacillin, ceftriaxone, ampicillin, amoxicillin-clavulanic acid, and nalidixic acid; and to a lesser extent to ceftazidime, linezolid, tetracycline, teicoplanin, chloramphenicol, and rifampicin [85].

Escolar et al., [18] identified seven *L. monocytogenes* strains from meat and dairy products and tested for antimicrobial susceptibility against nine antimicrobials using real-time PCR targeting the antimicrobial resistance genes like *tet M, tet L, mef A, msr A, erm A, erm B, lnu A* and *lnu B*. MDR was identified in four *L. monocytogenes* and resistance to clindamycin was the most common resistance phenotype. Most recently, a study with samples of milking equipment, hand swabs, and raw milk (total 300 samples collected from four dairy farms) in Egypt found 69 samples positive for *L. monocytogenes*. The researchers studied antibiotic resistance pattern (14 antibiotics) of the highly pathogenic *L. monocytogenes* isolates and observations were in the order of tetracycline 81%, clindamycin 81%, rifampicin 71.4%, gentamycin 66.7%, daptomycin 66.7% and doxycycline 66.7%. Out of all the pathogenic strains of *L. monocytogenes,* 37 (88%) isolates were multidrug-resistance [89].

8.6.3   SALMONELLA AND SHIGELLA

Both the *Salmonella and Shigella* comprise of Gram-negative short rods, which are catalase (CAT) positive and anaerobic and usually motile by *peritrichous flagella*. They are sensitive to low pH, get readily destroyed by acid and heat treatment; but if proteins and protectants are present, they can resist freezing and drying. Salmonella is pathogenic to humans, animals, and birds; and responsible to cause typhoid fever, enteric fever, gastroenteritis, and septicemia while Shigella spp. generally causes dysentery (also called shigellosis) in humans. Salmonella spp. are normally found in raw milks, but usually absent in heat-treated dairy products. However, improper sanitary practices may lead to show their presence in finished products, i.e., post pasteurization contamination and therefore *Salmonella* and *Shigella* have been associated with several foodborne outbreaks. Few such studies have been discussed in this section.

Khan et al. [46] analyzed antimicrobial-resistant of bacteria associated with different milk and milk products like raw milk, packaged milk, curd, khoa, and paneer. All the isolates (mainly belonging *Salmonella* spp. and *Shigella* spp.) were resistant to ampicillin, tetracycline, chloramphenicol, gentamycin, cotrimoxazole, ceftriaxone, vancomycin, methicillin, imipenem, and meropenem. In one of the studies performed in India, *Salmonella* spp. isolated for the samples of raw milk were able to be resistant against penicillin and erythromycin, while they were intermediate sensitive to chloramphenicol, and sensitive against tetracycline and streptomycin [67].

Out of 100 raw milk samples in Ethiopia, 20 were positive for *Salmonella* [13]. The isolated species of salmonella were not sensitive to two or more antibiotics. Isolated strains were most resistant to Nalidixic acid (80%) while were most susceptible to ciprofloxacin (95%) among all tested antimicrobials. In a separate study in Slovakia, 67 samples of milk, cheese, and related milk products were collected to determine antibiotic resistance of Salmonella species [34]. A high resistance was detected in all samples towards streptomycin, ampicillin, tetracycline, and chloramphenicol.

8.6.4   ESCHERICHIA COLI

E. coli constitute one of the important parts of the normal gastrointestinal tract (GIT) microflora of humans and other mammals. Generally, within a few hours after birth GIT of the newborn baby are colonized by *E. coli*. Most of the *E. coli* are harmless and coexist in a symbiotic relationship with their human host. These harmless commensal *E. coli* strains do not cause disease in healthy humans unless the normal GUT barriers are broken or in immunocompromised hosts. Still, some serovar of several highly adapted *E. coli* have gained specific virulence attributes making them pathogenic and able to infect healthy individuals. Generally, pathogenic *E. coli* strains are well known to cause enteric/diarrheal disease, urinary tract infections (UTIs) and sepsis/meningitis [42].

With the development of antibiotic-resistant strains, the treatment of *E. coli* infection has also become increasingly complicated, as most of the first-generation antimicrobial agents are of no use. Also, multidrug-resistant ESBL producing strains have made things more difficult. Antibiotic resistance in *E. coli* has also been reported throughout the world. The *E. coli* strains were identified from well-known Lebanese dairy products like *Shankleesh*, *Baladi*, and *Kishk* and were studied for susceptibility to 10 common antibiotics. All the isolates were highly resistant to at least one antibiotic [81].

Zanella et al., [101] examined 260 milk samples in Brazil for the presence of coliforms and reported that 77% of samples were contaminated with *E. coli*. The isolated *E. coli* strains were resistant to ampicillin, cephalothin, and tetracycline. Diarrheagenic *E. coli* (DEC) strains were also identified from beverages, and food items in Mexico and dairy products were found to be the major source of DEC (2.8%) among all food and beverages samples. DEC strains (66%) were found to be resistant to at least one frequently recommended antibiotic [9]. In a study with 150 samples of ice cream, raw milk, and white cheese from dairy-processing plants in Turkey, 56% ice cream, 74% raw milk, and 60% cheese samples were positive for *E. coli* and 2% of cheese samples were infected with *E. coli* O157. The isolated *E. coli* were resistance to ampicillin, penicillin, tetracycline, erythromycin, gentamicin, trimethoprim/sulfamethoxazole, chloramphenicol, and ciprofloxacin [38].

Rasheed et al., [75] screened 30 raw buffalo milk samples in Hyderabad, India, for the presence of *E. coli* strains, and they reported that 13 samples were contaminated with *E. coli*. Two isolated *E. coli* out of 13 were resistant to various antibiotics like ampicillin, amoxicillin, amoxiclav, cefotaxime, chloramphenicol, ciprofloxacin, ofloxacin, streptomycin, and tetracycline. However, the resistant strains were a non-ESBL producer.

Ahmed and Shimamoto [2] studied 54 Shiga toxin-producing *E. coli* O157:H7 strains from 1600 samples of dairy and meat products collected from different street vendors, butchers, retail markets, and slaughter-houses in Egypt for MDR. They reported that 57.4% isolates (34) were MDR to at least three classes of antimicrobials. Majority of the *E. coli* O157:H7 strains were resistant to kanamycin (96.8%), followed by spectinomycin (93.6%), ampicillin (90.3%), streptomycin (87.1%), and tetracycline (80.6%).

8.6.5 ENTEROCOCCI

Enterococci consist of Gram-positive anaerobic cocci, which frequently occur in pairs or short chains; and on the basis of physical characteristics alone, it is difficult to distinguish from streptococci. However, enterococci can tolerate in range of adverse environmental conditions, including extreme temperature (10–45°C), pH (4.5–10.0), and high concentration of sodium chloride [22]. *E. faecalis* (90–95%) and *E. faecium* are examples of enterococcal species occurring as commensal organisms in the oral cavity, GIT, and vaginal tract of humans and other animals; while other species like *E. raffinosus, E. casseliflavus,* and *E. gallinarum* are commonly associated with important clinical infections such as UTIs, bacteremia, diverticulitis, bacterial endocarditis, and meningitis. From dairy prospective, inherent enterococci in milk may function as natural starter cultures for the production of a number of cheese types, particularly artisan cheeses produced from both raw and pasteurized milk in several European countries [25].

One of the important features of enterococci species is their intrinsic antibiotic resistance at a high level. Several enterococcal species are resistant to β-lactam antibiotics intrinsically such as penicillins, carbapenems, cephalosporins, and many aminoglycosides generally employed to cure infections caused by Gram-positive bacteria. Currently, the emerging resistance of enterococci to vancomycin, streptogramins, and teicoplanin is of particular interest among researchers. Vancomycin is one of the key antibiotics applied for enterococci infections. However, the presence of *van* genes in vancomycin-resistant in bacteria and approximately nine types of vancomycin-resistance has been demonstrated in Enterococci. Eight of these types (i.e., *van*A, *van*B, *van*C, *van*D, *van*E, *van*G, *van*L, *van*M, and *van*N) are responsible for this resistance; amongst all *van*A and *van*B are clinically most significant [37].

Further, *van*A is the most frequently occurring genotype in vancomycin-resistant Enterococci (VRE) worldwide associated with the transfer of high-level vancomycin resistance from Enterococci, particularly, vancomycin-resistant *Ent. faecium* (VREF) to *Staph. Aureus* [8, 33]. Enterococci have noteworthy aptitude to acquire novel antimicrobial resistance mechanisms and by way of conjugation to transfer resistance determinants [1]. During the last few decades, predominantly virulent *Enterococci* strains–resistant to vancomycin, known as vancomycin-resistant *Enterococcus,* or VRE- have appeared in nosocomial infections of hospitalized patients. The development of antibiotic resistance in enterococci represents the biggest hazard to the enterococcal infections treatment in humans [47].

In one of the investigations, it was established that *Enterococcus faecium* (32.61%) and *E. faecalis* (21.74%) were most frequently isolated species from milk, cheese, and other food products particularly with higher resistance against streptomycin and gentamicin [27]. For vancomycin, identical types of groups of *van*A of enterococci genes from hospital patients, pet food, and people in the community, fecal samples of animals, and water samples have been identified. It is obvious that these resistance genes may contaminate mankind through the food chain [47].

Researchers at the University of Sao Paulo, Brazil stated that 52.5% of the raw milk, pasteurized milk, cheese, vegetables, and meat samples had shown the presence of enterococci; the meat and the cheese being the most contaminated. In that, *E. faecium* was the main species among other enterococcal species. *E. faecalis* was resistant against tetracycline, gentamicin, and erythromycin while three strains of *E. faecium* were resistant against vancomycin, and related virulence genes were also detected in these strains [27].

In a separate investigation, a group of researchers in Turkey assessed antibiotic resistance patterns of enterococci from various milk products procured from local markets and observed that out of 200 samples, 50%

(100 samples) had enterococci contamination at high levels; and that to the greater resistance was noticed in cream cheese samples [49].

It was observed that out of 135 isolates from 200 samples of milk (24 raw milk and 6 cheese samples), meat products 100 (74.1%), 91 (67.4%), 84 (62.2%) and 53 (39.25%) were resistance to tetracycline, synercid, erythromycin, and ampicillin, respectively. Nitrofurantoin showed lower antibiotic resistance (1.5%) [37]. In one of the previous studies, all dairy product isolates belonging to enterococcal species and *E. coli* were multi-resistant (100%) to tetracycline, streptomycin, enrofloxacin, and compound sulphonamides [60].

Antibiotic resistance of 68 strains of *Enterobacteriaceae* family and *Enterococcus* spp. isolated from poultry and livestock meat, pasteurized dairy products acquired from the retail in Moscow–Russia was investigated by Korotkevich and his co-workers [51]. In general, 38% of *Enterobacteriaceae* strains and 40% of *Enterococcus* spp., isolated from meat products were resistant to tetracycline and doxycycline, and 21 and 33%–from dairy products, respectively; 26% of milk isolates and 54% of meat isolates were resistant to ampicillin. Among the *Enterobacteriaceae* strains, 26% of dairy isolates and 38% of meat isolates were highly resistant to tetracycline (MIC ranged from 8 to 120 mg/kg), while among *Enterococcus* spp. about 17–40% were resistant to tetracycline.

The occurrence of antibiotic resistance among enterococci and *E. coli* isolates of fresh Slovak cheese *bryndza* was recently studied [96]. Enterococci were maximally resistance against tetracycline (29.73%), and there was no resistance towards vancomycin (0%). None of the enterococci isolates was resistant to all six antibiotics employed, and two antibiotics (32.43%) were highly MDR. About nineteen (42.22%) *E. coli* isolates were appeared to produce ESBL.

8.6.6   *CAMPYLOBACTER*

Campylobacter is a Gram-negative, comma-shaped bacteria, which are found as a normal inhabitant of the intestinal tract of animals and birds; excreted in large numbers in feces. Foods of animal origin get contaminated with fecal material following unhygienic practices; further, these bacteria get transferred to humans due to the consumption of uncooked or partially cooked foods. Campylobacteriosis is an infectious disease caused by *Campylobacter jejuni*.

In a study by Rahimi et al. [73], 9.3%samples (13 of 552) were found to be contaminated with *Campylobacter* species; the highest occurrence was observed in raw cow milk (6.3%), followed by traditional cheese (5%), and butter (4%) samples. *C. jejuni* (76.92%) was the most prevalent than *C. coli* (23.08%). All thirteen isolates were sensitive to gentamicin, chloramphenicol, and amoxicillin whereas resistance to nalidixic acid (46%), tetracycline (38%), and ciprofloxacin (31%) as determined by using the disk diffusion assay. In another experiment, species of *Campylobacter* was isolated from seven (4.6%) samples among 150 bulk milk samples [99]. All seven tested samples were susceptible to chloramphenicol, erythromycin, gentamicin, ampicillin, doxycycline (28.5%); tetracycline (14.2%) and ciprofloxacin (14.2%).

In Tanzania, a group of researchers investigated the dominance, antimicrobial resistance, and genetic determinants of isolated *Campylobacter* from raw milk and beef carcasses. The antimicrobial susceptibility was tested using the disk diffusion assay and microdilution method. Researchers found highest resistance against ampicillin (63% and 94.1%) followed by streptomycin (35.2% and 84.3%), erythromycin (53.7% and 70.6%), gentamicin (0% and 150.7%), tetracycline (18.5% and 17.7%), and ciprofloxacin (9.3% and 11.8%), respectively. Further, authors also detected genetic determinants including bla_{OXA-61} (52.6% and 28.1%), *cmeB* (26.3% and 31.3%), *tet(O)* (26.3% and 31.3%), and *aph-3-1* (5.3% and 3.0%) in *C. coli* and *C. jejuni* [44].

Out of 240 samples of cow milk, buffalo milk, ice-cream, cheese, and paneer, *Campylobacter spp.* were ascertained in seven (2.91%) samples of raw milk, whereas it was not detected in any tested milk product. Majority of the isolates showed resistance to nalidixic acid, tetracycline, and ciprofloxacin. Three virulence genes *cad*F, *cdt*B, and *flg*R were found in all isolates, while one isolate was positive for *iam*A gene, and *fla* gene was present in six isolates [57].

In one of the bulk studies, out of 590 samples, 141 (23.9%) samples were positive for *Campylobacters*. *Campylobacter* spp. that were isolated in 40.8% (106/260), 14% (28/200), and 8.7% (7/80) of poultry meat, red meat, and milk samples, respectively [72]. Antimicrobial susceptibility test indicated a high frequency of resistance to ciprofloxacin, tetracycline, and nalidixic acid among the isolates. There was a weak correlation between antibiotics resistance and occurrence of the pathogen genes; and prevalence of *waa*C, *cia*B, and *pld*A genes were 91.7%, 86.7%, and 80.8%, respectively. In contrast, no campylobacter were noticed in any of the 80 milk samples collected from the animal farms [66]. Few other studies also reported similar results in raw milk samples [87, 97].

8.6.7 BACILLUS

Bacillus consist of Gram-positive, rig rod-shaped and strictly aerobic or facultative bacteria. Most of them are spoilage causing, while few like *B. cereus* may cause health problems like food poisoning.

Yibar et al. [100] determined the occurrence of *B. cereus* in a variety of full-fat milk and cheese samples and they assessed production the HBL (hemolysin BL) and NHE (non-hemolytic enterotoxin) as well as resistance to numerous antibiotics. *B. cereus* isolates were susceptible to erythromycin, oleondamycin, and streptomycin, while all others were resistant to penicillin G. In addition, 84.2% (n = 16) of isolates demonstrated resistance against multiple antibiotics. In a recent investigation in Ghana, *B. cereus sensulato* isolates from 89 samples of different milk and milk products were resistant to β-lactam antibiotics such as ampicillin (98%) [65].

Organji et al. [61] analyzed 110 different samples of dairy products (including raw milk, long life pasteurized milk, yogurt, infant powdered milk formulas), raw rice, and they observed that all *B. cereus* isolates were resistant to penicillin G, but vulnerable to erythromycin, vancomycin, and clindamycin. The antibiotic resistance profile of *Bacillus* species isolated from 392 milk samples revealed that the all isolates were resistant to methicillin, followed by 91.40%, 80.54%, 54.75%, 51.13%, and 50.67% resistant against penicillin G, oxacillin, cefixime, cefaclor, ampicillin, respectively; while all species were susceptible to vancomycin [80].

In another investigation, the prevalence of *B. cereus* group was analyzed in 230 samples of milk products in India [53]. The occurrence of *B. cereus* in ice cream, cheese, milk powder, and milk was comparatively high (33%–55%), and it was noticed that all the 144 *B. cereus* isolates were multidrug-resistant. In the Czech Republic, dairy products available in commercially were investigated for the prevalence of antibiotic resistance in *B. cereus* [83]. About 31% of dairy products showed the presence of *B. cereus* and nearly all isolates of *B. cereus* displayed low sensitivity against ampicillin, cephalothin, and oxacillin.

8.6.8 CLOSTRIDIUM

The genus *Clostridium* contains spore-forming Gram-positive anaerobic bacillus, which is broadly scattered in the environment including soil, food, sewage, and dust; and is found in human's and animal's intestinal tracts. One of the species, *Cl. Perfringens* is responsible to cause mild food poisoning to life-threatening disease like gas gangrene in humans; while in domestic livestock, it causes a wide range of enteric diseases.

Several researchers have detected their presence in milk and milk products and further evaluated the prevalence of antibiotic resistance in *Cl. perfringens*. Singh and Bist [88] evaluated pasteurized milks and dairy products samples, including ice cream and shrikhand to detect antibiotic resistance of *Clostridium perfringens*. Authors found that ampicillin, penicillin, tetracycline, co-trimoxazole, and cloaxacillin showed higher resistance, whereas erythromycin and gentamicin displayed intermediate sensitivity to *Cl. perfringens* isolates.

Recently, a group of researchers evaluated 150 samples of raw milk, cheese, and other milk products to detect and measure the prevalence of *Cl. perfringens*. In raw milk, milk powder, Kariesh cheese, Damietta cheese, and Ras cheese, the prevalence of *Cl. perfringens* was 20, 20, 36, 60, and 60%, respectively. Each and every isolate showed resistant against colistin and ampicillin. Furthermore, 91.8% and 75.5% were resistant to lincomycin and erythromycin while for other antimicrobials, the resistance varied from 73 to 18.36%, minimum for vancomycin [63]. In the previous study, isolates of *Cl. perfringens* from infant formula had shown resistance against penicillin, ampicillin, sulfonamides, tetracycline, gentamicin, and cephalosporins [95].

8.6.9   OTHER PATHOGENIC BACTERIA

Apart from the above discussed foodborne bacteria, several researchers had investigated antibiotic resistance among other not much common pathogens like *Klebsiella*. *Klebsiella* spp. isolated from samples of raw milk showing resistance against chloramphenicol, penicillin, and erythromycin while intermediate sensitive towards streptomycin and tetracycline [67]. In another study, erythromycin resistance was observed in more than 90% of *Klebsiella* spp. isolated from raw milk. These isolates were observed multidrug-resistant profile against antibiotics erythromycin, streptomycin, chloramphenicol, trimethoprim, and gentamicin; conversely, isolates were sensitive towards ciprofloxacin, tetracyclin, norfloxacin, imipenem, and nalidixic acid [90]. In the same study, all *E. coli* isolates were appeared to resist erythromycin, whereas 86% showed ceftriaxone and amoxycillin resistance.

Reta et al. [76] identified and measured the antibiotic resistance of raw milk contaminants. The identified bacteria and their prevalence rates were *E. coli* 70 (58%), *Staph. aureus* 29 (24.2%), *Shigella spp.* 21 (17.5%), *Proteus spp.* 9 (7.5%) and *Salmonella spp.* 4 (3.3%). About 50% isolates of *Salmonella spp.* were highly resistant to amoxicillin, whereas *Shigella spp.* was resistant to ampicillin (38.1%). High antibiotic resistance for *E. coli* isolates were observed for doxycycline (42.3%) and ampicillin (30%). *S. aureus* were resistant to penicillin G (93.1%), tetracycline (69%), vancomycin (6.9%) and rifampicin (3.4%). Overall, the authors observed MDR in 55.2% of the total isolates [76].

The antibiotic resistance of pathogens (including *E. coli, Klebsiella* spp., *Enterobacter* spp., *Proteus vulgaris, Salmonella typhi, Enterococcus faecalis, Staphylococcus aureus* and *Staph. epidermidis*) isolated from 8 different types of branded and unbranded milk in Accra, Ghana were analyzed by Mahami and his co-workers [55]. Authors noticed that all isolates (100%) were multi-resistant to ampicillin, tetracycline, chloramphenicol, gentamycin, cotrimoxazole, ceturoxime, and cefotaxime. Among these, locally produced unpasteurized cow milk accounted for the highest (26.42%) of resistant strains while imported skimmed milk accounted for the least (3.77%). The locally-produced pasteurized cow milk, imported whole milk, soya milk, and powdered milk showed 20.75%, 11.32%, 22.64%, and 15.09% resistance microbes, respectively [55].

In one of the previous approach, *Enterobacter* spp. and *E. coli,* isolated from traditional milk products khoya and burfi established a higher degree of sensitivity against ampiclox, tetracycline, septran, and amikin. However, *Enterobacter* spp., *Klebsiella* spp., and *E. coli* were appeared to be resistance against Urixin [20]. Outcomes of such studies suggest that locally prepared traditional milk products might be a possible source of foodborne bacteria, which further poses a considerable clinical risk to consumers through disproportionate use of different antibiotics against these bacteria.

8.7 SUMMARY

Various bacteria have been detected to possess antibiotic resistance, and the responsible genes are also noticeable in resistant phenotypes strains. The possible transfer of these antibiotic-resistant genes from one bacterium to another poses a danger to food safety. Presence of antibiotic-resistant strains in local milk and milk products proposes the obligation of hygienic practices throughout handling, processing plus post-processing of the raw milk to improve and enhance the overall microbiological quality and hence safety of raw milk. The preventive measures and control strategies necessitate the use of epidemiological and behavioral approaches, as well as the research methodologies intended at basic mechanisms of antimicrobial resistance.

Further, the monitoring of resistant bacterial strains is desired to eradicate or at least to diminish this universal problem.

KEYWORDS

- antibiotic resistance
- foodborne pathogens
- *listeria monocytogenes*
- milk products
- raw milk
- *salmonella*
- *shigella*
- *staphylococcus aureus*

REFERENCES

1. Aarestrup, F. M., (2006). *Antimicrobial Resistance in Bacteria of Animal Origin* (pp. 19–27). ASM Press, Washington DC.
2. Ahmed, A. M., & Shimamoto, T., (2015). Molecular analysis of multidrug resistance in Shiga toxin-producing Escherichia coli O157:H7 isolated from meat and dairy products. *International Journal of Food Microbiology, 16*(193), 68–73.
3. Al-Ashmawy, M. A., Sallam, K. I., Abd-Elghany, S. M., Elhadidy, M., & Tamura, T., (2016). Foodborne pathogens and disease: Prevalence, molecular characterization, and antimicrobial susceptibility of methicillin-resistant *Staphylococcus aureus* isolated from milk and dairy products. *Foodborne Pathogens and Disease, 13*(3), 156–162.
4. Aminov, R. I., (2009). The role of antibiotics and antibiotic resistance in nature. *Environmental Microbiology, 11,* 2970–2988.
5. Ammor, M. S., Florez, A. B., & Mayo, B., (2007). Antibiotic resistance in non-enterococcal lactic acid bacteria and bifidobacteria. *Food Microbiology, 24,* 559–570.
6. Bartlett, J. G., & Mundy, L. M., (1995). Community-acquired pneumonia. *The New England Journal of Medicine, 333,* 1618–1624.
7. Basanisi, M. G., La Bella, G., Nobili, G., Franconieri, I., & La Salandra, G., (2017). Genotyping of methicillin-resistant Staphylococcus aureus (MRSA) isolated from milk and dairy products in South Italy. *Food Microbiology, 62,* 141–146.
8. Batistao, D. W. D. F., Gontijo-Filho, P. P., Conceiçao, N., Oliveira, A. G. D., & Ribas, R. M., (2012). Risk factors for vancomycin-resistant enterococci colonization in critically ill patients. *Memórias do Instituto Oswaldo Cruz* (Memories of Oswaldo Cruz Institute), *107* (1), 57–63.

9. Canizalez-Roman, A., Gonzalez-Nunez, E., Vidal, J. E., Flores-Villaseñor, H., & León-Sicairos, N., (2013). Prevalence and antibiotic resistance profiles of diarrheagenic *Escherichia coli* strains isolated from food items in northwestern Mexico. *International Journal of Food Microbiology, 164*(1), 36–45.

10. Carfora, V., Caprioli, A., Marri, N., & Sagrafoli, D., (2015). Enterotoxin genes, enterotoxin production, and methicillin resistance in *Staphylococcus aureus* isolated from milk and dairy products in Central Italy. *International Dairy Journal, 42,* 12–15.

11. Caruso, M., LaTorre, L., Santagada, G., & Fraccalvieri, R., (2016). Methicillin-resistant *Staphylococcus aureus* (MRSA) in sheep and goat bulk tank milk from Southern Italy. *Small Ruminant Research, 135,* 26–31.

12. Anonymous, (2017). *Antibiotic / Antimicrobial Resistance (AR/AMR).* https://www.cdc.gov/drugresistance/index.html/ (Accessed on 20 July 2019).

13. Dabassa, A., & Bacha, K., (2012). The prevalence and antibiogram of *Salmonella* and *Shigella* isolated from Abattoir-Jimma Town, Southwestern Ethiopia. *International Journal of Pharmaceutical and Biological Research, 3*(4), 143–148.

14. Davies, J., (1994). Inactivation of Antibiotics and the dissemination of resistance genes. *Science, 264*(5157), 375–382.

15. Devirgiliis, C., Barile, S., & Perozzi, G., (2011). Antibiotic resistance determinants in the interplay between food and gut microbiota. *Genes & Nutrition, 6,* 275–284.

16. DeWaal, C. S. J. D., & Grooters, S. V. M. P. H., (2013). Antibiotic resistance in foodborne pathogens. *Center for Science in the Public Interest,* 1–22.

17. Dzidic, S., Suskovic, J., & Kos, B., (2008). Antibiotic resistance mechanisms in Bacteria: Biochemical and genetic aspects. *Food Technology and Biotechnology, 46*(1), 11–21.

18. Escolar, C., Diego, G., María, C. R. G., Pilar, C., & Antonio, H., (2017). Antimicrobial resistance profiles of *Listeria monocytogenes* and *Listeria innocua* isolated from ready-to-eat products of animal origin in Spain. *Foodborne Pathogens and Disease, 14*(6), 357–363.

19. European Centre for Disease Prevention and Control (ECDC), (2016). Last-line antibiotics are failing. *Science Daily.* <www.sciencedaily.com/releases/2016/11/161118104821.htm> (Accessed on 20 July 2019).

20. Farzana, K., Akhtar, S., & Jabeen, F., (2009). Prevalence and antibiotic resistance of bacteria in two ethnic milk based products. *Pakistan Journal of Botany, 41*(2), 935–943.

21. Finegold, S. M., (1989). Mechanisms of resistance in anaerobic bacteria and new developments in testing. *Diagnostic Microbiology and Infectious Disease, 12,* 117S–120S.

22. Fisher, K., & Phillips, C., (2009). The ecology, epidemiology and virulence of Enterococcus. *Microbiology, 155*(6), 1749–1757.

23. Friedman, N. D., Temkin, E., & Carmeli, Y., (2016). The negative impact of antibiotic resistance. *Clinical Microbiology and Infection, 22,* 416–422.

24. Garedew, L., Taddese, A., & Biru, T., (2015). Prevalence and antimicrobial susceptibility profile of listeria species from ready-to-eat foods of animal origin in Gondar Town, Ethiopia. *BMC Microbiology, 15*(1), 100–105.

25. Giraffa, G., (2003). Functionality of enterococci in dairy products. *International Journal of Food Microbiology, 88*(2/3), 215–222.

26. Gohar, S., Abbas, G., Sajid, S., Sarfraz, M., Ali, S., Ashraf, M., Aslam, R., & Yaseen, K., (2017). Prevalence and antimicrobial resistance of *Listeria monocytogenes* isolated from raw milk and dairy products. *Matrix Science Medica, 1*(1), 10–14.

27. Gomes, B. C., Esteves, C. T., & Palazzo, I. C., (2008). Prevalence and characterization of Enterococcus spp. isolated from Brazilian foods. *Food Microbiology, 25*(5), 668–675.

28. Gucukoglu, A., Cadirci, O., Terzi, G., Kevenk, T. O., & Alisarli, M., (2013). Determination of enterotoxigenic and methicillin resistant *Staphylococcus aureus* in ice cream. *Journal of Food Science, 78,* 738–741.

29. Gundogan, N., & Avci, E., (2014). Occurrence and antibiotic resistance of *Escherichia coli, Staphylococcus aureus* and *Bacillus cereus* in raw milk and dairy products in Turkey. *International Journal of Dairy Technology, 67,* 562–569.

30. Gundogan, N., Citak, S., & Turan, E., (2006). Slime production, DNase activity and antibiotic resistance of *Staphylococcus aureus* isolated from raw milk, pasteurized milk and ice cream samples. *Food Control, 17*(5), 389–392.

31. Harakeh, S., Saleh, I., Zouhairi, O., Baydoun, E., Barbour, E., & Alwan, N., (2009). Antimicrobial resistance of *Listeria monocytogenes* isolated from dairy-based food products. *Science of Total Environment, 407*(13), 4022–4027.

32. Haran, K. P., Godden, S. M., Boxrud, D., Jawahir, S., Bender, J. B., & Sreevatsan, S., (2012). Prevalence and characterization of *Staphylococcus aureus*, including methicillin-resistant *Staphylococcus aureus*, isolated from bulk tank milk from Minnesota dairy farms. *Journal of Clinical Microbiology, 50*(3), 688–695.

33. Hegstad, K., Giske, C. G., Haldorsen, B., & Matuschek, E., (2014). Performance of the EUCAST disk diffusion method, the CLSI agar screen method, and the Vitek 2 automated antimicrobial susceptibility testing method for detection of clinical isolates of enterococci with low-and medium-level VanB-type vancomycin resistance: A multicenter study. *Journal of Clinical Microbiology, 52*(5), 1582–1589.

34. Hleba, L., Kacániová, M., Pochop, J., & Lejková, J., (2011). Antibiotic resistance of *Enterobacteriaceae* genera and *Salmonella* spp., *Salmonella enterica ser. typhimurium* and *enteritidis* isolated from milk, cheese and other dairy products from conventional farm in Slovakia. *The Journal of Microbiology, Biotechnology and Food Sciences, 1*(1), 1–20.

35. Holgel, C. S., Harms, K. S., Schwaiger, K., & Johann, B., (2010). Resistance to linezolid in a porcine *Cl. perfringens* strain carrying a mutation in the rplD gene encoding the Ribosomal Protein L4. *Antimicrobial Agents and Chemotherapy, 54,* 1351–1353.

36. Holmes, A. H., Moore, L. S. P., & Sundsfjord, A., (2016). Understanding the mechanisms and drivers of antimicrobial resistance. *The Lancet, 387*(10014), 176–187.

37. Hosseini, S. M., Zeyni, B., Rastyani, S., & Jafari, R., (2016). Presence of virulence factors and antibiotic resistances in *Enterococcus* spp. collected from dairy products and meat. *Der Pharmacia Letter, 8*(4), 138–145.

38. Hummerjohann, J., Naskova, J., & Baumgartner, A., (2014). Enterotoxin-producing *Staphylococcus aureus* genotype B as a major contaminant in Swiss raw milk cheese. *Journal of Dairy Science, 97,* 1305–1312.

39. Jamali, H., Paydar, M., Radmehr, B., & Ismail, S., (2015). Prevalence and antimicrobial resistance of *Staphylococcus aureus* isolated from raw milk and dairy products. *Food Control, 54,* 383–388.

40. Jamali, H., Radmehr, B., & Thong, K. L., (2013). Prevalence, characterization, and antimicrobial resistance of *Listeria* species and *Listeria monocytogenes* isolates from raw milk in farm bulk tanks. *Food Control, 34*(1), 121–125.

41. Jernberg, C., Lofmark, S., Edlund, C., & Jansson, J. K., (2010). Long-term impacts of antibiotic exposure on the human intestinal microbiota. *Microbiology, 156,* 3216–3223.

42. Kaper, J. B., Nataro, J. P., & Mobley, H. L., (2004). Pathogenic *Escherichia coli. Nature Reviews Microbiology, 2*(2), 123–140.

43. Karchmer, A. W., (1991). *Staphylococcus aureus* and Vancomycin: The Sequel. *Annals of Internal Medicine, 115,* 739–741.
44. Kashoma, I. P., Kassem, I. I., & John, J., (2016). Prevalence and antimicrobial resistance of campylobacter isolated from dressed beef carcasses and raw milk in Tanzania. *Microbial Drug Resistance, 22*(1), 40–52.
45. Kevenk, T. O., & Gulel, G. T., (2016). Prevalence, antimicrobial resistance and serotype distribution of *Listeria monocytogenes* isolated from raw milk and dairy products. *Journal of Food Safety, 36,* 11–18.
46. Khan, J., Musaddiq, M., & Budhlani, G. N., (2014). Antibiotic resistant of pathogenic bacteria isolated from milk and milk products in Akola city of Maharashtra. *Indian Journal of Life Sciences, 4*(1), 59–62.
47. Klare, I. Konstabel, C., Badstübner, D., Werner, G., & Witte, W., (2003). Occurrence and spread of antibiotic resistances in *Enterococcus faecium*. *International Journal of Food Microbiology, 88*(2), 269–290.
48. Kluytmans, J. A., Overdevest, I. T., & Willemsen, I., (2013). Extended spectrum β-lactamase-producing *Escherichia coli* from retail chicken meat and humans: Comparison of strains, plasmids, resistance genes, and virulence factors. *Clinical Infectious Disease, 56*(4), 478–487.
49. Koluman, A., Akan, L. S., & Çakiroglu, F. P., (2009). Occurrence and antimicrobial resistance of enterococci in retail foods. *Food Control, 20*(3), 281–283.
50. Korhonen, J. M., Danielsen, M., & Mayo, B., (2008). Antimicrobial susceptibility and proposed microbiological cut-off values of lactobacilli by phenotypic determination. *International Journal of Probiotics and Prebiotics, 3,* 257–268.
51. Korotkevich, Y. V., (2016). Antibiotic resistance analysis of *Enterococcus spp.* and *Enterobacteriaceae spp.* isolated from food. *Voprosy Pitaniia, 85*(2), 5–13.
52. Kreausukon, K., Fetsch, A., & Kraushaar, B., (2012). Prevalence, antimicrobial resistance, and molecular characterization of methicillin-resistant *Staphylococcus aureus* from bulk tank milk of dairy herds. *Journal of Dairy Science, 95*(8), 4382–4388.
53. Kumari, S., & Sarkar, P. K., (2014). Prevalence and characterization of *Bacillus cereus* group from various marketed dairy products in India. *Dairy Science & Technology, 95*(5), 483–497.
54. Lowy, F. D., (2003). Antimicrobial resistance: The example of *Staphylococcus aureus*. *Journal of Clinical Investigation, 111,* 1265–1273.
55. Mahami, T., Odonkor, S., Yaro, M., & Adu-Gyamfi, A., (2011). Prevalence of antibiotic resistant bacteria in milk sold in Accra. *International Research Journal of Microbiology, 2*(4), 126–132.
56. Mathur, S., & Singh, R., (2005). Antibiotic resistance in food lactic acid bacteria—a review. *International Journal of Food Microbiology, 105,* 281–295.
57. Modi, S., Brahmbhatt, M. N., Chatur, Y. A., & Nayak, J. B., (2015). Prevalence of campylobacter species in milk and milk products, their virulence gene profile and anti-bio gram. *Veterinary World, 8*(1), 1–8.
58. Nikolic, N., Mirecki, S., & Blagojevic, M., (2011). Presence of inhibitory substances in raw milk in the area of Montenegro. *Mljekarstvo, 61*(2), 182–187.
59. Normanno, G., Corrente, M., La Salandra, G., & Dambrosio, A., (2007). Methicillin resistant *Staphylococcus aureus* (MRSA) in foods of animal origin product in Italy. *International Journal of Food Microbiology, 117*(2), 219–222.

60. Nováková, I., Kačániová, M., & Arpášová, H., (2010). Antibiotic resistance of enterococci and coliform bacteria in dairy products from commercial farms. *Scientific Papers Animal Science and Biotechnologies, 43*(1), 307–309.

61. Organji, S. R., Abulreesh, H. H., Elbanna, K., Osman, G. E. H., & Khider, M., (2015). Occurrence and characterization of toxigenic *Bacillus cereus* in food and infant feces. *Asian Pacific Journal of Tropical Biomedicine, 5*(7), 515–520.

62. Osaili, T. M., Al-Nabulsi, A. A., & Taha, M. H., (2012). Occurrence and antimicrobial susceptibility of *Listeria monocytogenes* isolated from brined white cheese in Jordan. *Journal of Food Science, 77*(9), M528–M532.

63. Osama, R., Khalifa, M., Marwa, Al-Toukhy, & Al-Ashmawy, M., (2015). Prevalence and antimicrobial resistance of *Clostridium perfringens* in milk and dairy products. *World Journal of Dairy & Food Sciences, 10*(2), 141–146.

64. Osman, K. M., Samir, A., & Abo-Shama, U. H., (2016). Determination of virulence and antibiotic resistance pattern of biofilm producing *Listeria* species isolated from retail raw milk. *BMC Microbiology, 16*, 263.

65. Owusu-Kwarteng, J., Wuni, A., & Akabanda, F., (2017). Prevalence, virulence factor genes and antibiotic resistance of *Bacillus cereus* isolated from dairy farms and traditional dairy products. *BMC Microbiology, 17*(1), 65–70.

66. Pallavi, A., (2014). Prevalence and antibiotic resistance pattern of *Campylobacter* species in foods of animal origin. *Veterinary World, 7*(9), 681–684.

67. Pant, R., Nirwal, S., & Rai, N., (2013). Prevalence of antibiotic resistant bacteria and analysis of microbial quality of raw milk samples collected from different regions of Dehradun. *Prevalence, 5*(2), 804–810.

68. Papadopoulos, P., Papadopoulos, T., & Angelidis, A. S., (2018). Prevalence of *Staphylococcus aureus* and of methicillin-resistant *S. aureus* (MRSA) along the production chain of dairy products in north-western Greece. *Food Microbiology, 69*, 43–50.

69. Parisi, A., Caruso, M., Normanno, G., & Latorre, L., (2016). Prevalence, antimicrobial susceptibility and molecular typing of Methicillin-Resistant *Staphylococcus aureus* (MRSA) in bulk tank milk from southern Italy. *Food Microbiology, 58*, 36–42.

70. Patel, A. R., Shah, N. P., & Prajapati, J. B., (2012). Antibiotic resistance profile of lactic acid bacteria and their implications in food chain. *World Journal of Dairy & Food Science, 7*(2), 202–211.

71. Paterson, G. K., Morgan, F. J., Harrison, E. M., & Peacock, S. J., (2014). Prevalence and properties of mecC methicillin resistant *Staphylococcus aureus* (MRSA) in bovine bulk tank milk in Great Britain. *Journal of Antimicrobial Chemotherapy, 69*(3), 598–602.

72. Raeisi, M., Khoshbakht, R., Ghaemi, E. A., & Bayani, M., (2017). Antimicrobial resistance and virulence-associated genes of *Campylobacter* spp. Isolated from raw milk, fish, poultry, and red meat. *Microbial Drug Resistance, 23*(7), 925–933.

73. Rahimi, E., Sepehri, S., & Momtaz, H., (2013). Prevalence of campylobacter species in milk and dairy products in Iran. *Revue De Medecine Veterinaire, 164*(5), 283–288.

74. Rama, A., Lucatello, L., Benetti, C., Galina, G., & Bajraktari, D., (2017). Assessment of antibacterial drug residues in milk for consumption in Kosovo. *Journal of Food and Drug Analysis, 25*(3), 525–532.

75. Rasheed, M. U., Thajuddin, N., Ahamed, P., Teklemariam, Z., & Jamil, K., (2014). Antimicrobial drug resistance in strains of Escherichia coli isolated from food sources.

Revista do Instituto de Medicina Tropical de São Paulo (Magazine of the Tropical Medicine Institute of São Paulo), 56(4), 341–346.

76. Reta, M. A., Bereda, T. W., & Alemu, A. N., (2016). Bacterial contaminations of raw cow's milk consumed at Jigjiga City of Somali Regional State, Eastern Ethiopia. *International Journal of Food Contamination, 3*(1), 4–8.

77. Roberts, M. C., (2005). Update on acquired tetracycline resistance genes. *FEMS Microbiology Letters, 245*, 195–203.

78. Rodas-Suárez, O. R., Quiñones-Ramírez, E. I., Fernández, F. J., & Vázquez-Salinas, C., (2013). *Listeria monocytogenes* strains isolated from dry milk samples in Mexico: Occurrence and antibiotic sensitivity. *Journal of Environmental Health, 76*(2), 32–37.

79. Rubin, J. E., Ekanayake, S., & Fernando, C., (2014). Carbapenemase producing organism in food. *Emerging Infectious Diseases, 20*(7), 1264–1265.

80. Sadashiv, S. O., & Kaliwal, B. B., (2014). Isolation, characterization and antibiotic resistance of Bacillus sps. from bovine mastitis in the region of north Karnataka, India. *International Journal of Current Microbiology & Applied Science, 3*(4), 360–373.

81. Saleh, I., Zouhairi, O., Alwan, N., Hawi, A., Barbour, E., & Harakeh, S., (2009). Antimicrobial resistance and pathogenicity of *Escherichia coli* isolated from common dairy products in the Lebanon. *Annals of Tropical Medicine and Parasitology, 103*(1), 39–52.

82. Sanchez, B., Bayangos, T., Rocha, K., et al., (2017). Levels of antibiotic resistant bacteria in ready-to-eat foods. *Poster Presented at ASM Microbe* (p. 1). New Orleans, Louisiana.

83. Schlegelova, J., Babak, V., Brychta, J., Klimova, E., & Napravnikova, E., (2003). The prevalence of and resistance to antimicrobial agents of *Bacillus cereus* isolates from foodstuffs. *Veterinarni Medicina-UZPI (Czech Republic), 11,* 331–338.

84. Shanehbandi, D., Baradaran, B., Sadigh-Eteghad, S., & Zarredar, H., (2014). Occurrence of methicillin resistant and enterotoxigenic *Staphylococcus aureus* in traditional cheeses in the north west of Iran. *ISRN Microbiology*, 1–5.

85. Sharma, S., Sharma, V., Dahiya, D. K., Khan, A., Mathur, M., & Sharma, A., (2017). Prevalence, virulence potential, and antibiotic susceptibility profile of *Listeria monocytogenes* isolated from bovine raw milk samples obtained from Rajasthan, India. *Foodborne Pathogens and Diseases, 14*(3), 132–140.

86. Silk, B. J., Date, K. A., Jackson, K. A., & Pouillot, R., (2012). Invasive listeriosis in the foodborne diseases active surveillance network (FoodNet), 2004–2009: Further targeted prevention needed for higher-risk groups. *Clinical Infectious Disease, 54*(5), S396–S404.

87. Singh, H., Rathore, R. S., Singh, S., & Cheema, P. S., (2011). Comparative analysis of cultural isolation and PCR based assay for detection of *Campylobacter jejuni* in food and fecal samples. *Brazilian Journal of Microbiology, 42,* 181–186.

88. Singh, R. V., & Bist, B., (2013). Antimicrobial profile of *Clostridium perfringens* isolates from dairy products. *Journal of Animal Research, 3*(2), 147–151.

89. Tahoun, A. B. M. B., Abou-Elez, R. M. M., & Abdelfatah, E. N., (2017). *Listeria monocytogenes* in raw milk, milking equipment and dairy workers: Molecular characterization and antimicrobial resistance patterns. *Journal of Global Antimicrobial Resistance, 10*, 264–270.

90. Tasnim, U. T., & Islam, M. T., (2015). Pathogenic and drug resistant bacteria in raw milk of Jessore city: A potential food safety threat. *Bangladesh Journal of Veterinary Medicine, 13*(1), 71–78.

91. Teale, C. J., (2002). Antimicrobial resistance and the food chain. *Journal of Applied Microbiology, 92*, 85S–89S.
92. Thaker, H. C., Brahmbhatt, M. N., & Nayak, J. B., (2013). Isolation and identification of *Staphylococcus aureus* from milk and milk products and their drug resistance patterns in Anand, Gujarat. *Veterinary World, 6*(1), 10–13.
93. Trombete, F. M., Dos Santos, R. R., & Souza, A. L., (2014). Antibiotic residues in Brazilian milk: A review of studies published in recent years. *Revista Chilena de Nutrición, 41*(2), 191–197.
94. Vankerckhoven, V., Huys, G., Vancanneyt, M., & Vael, C., (2008). Biosafety assessment of probiotics used for human consumption: Recommendations from the EU-PROSAFE project. *Trends in Food Science & Technology, 19*(2), 102–114.
95. Voidarou, C., Apostolidis, P., Skoufos, I., & Vassos, D., (2005). Hygienic quality of some infant formula milk. Abstract In: *The European Congress of Clinical Microbiology and Infectious Diseases* (p. 2). Copenhagen –Denmark.
96. Vrabec, M., Lovayová, V., Dudriková, K., Gallo, J., & Dudriková, E., (2015). Antibiotic resistance and prevalence of *Enterococcus* spp. and *Escherichia coli* isolated from Bryndza cheese. *Italian Journal of Animal Science, 14*(4), 609–614.
97. Wegmuller, B., Luthy, J., & Candrian, U., (1993). Direct polymerase chain reaction detection of *Campylobacter jejuni* and *Campylobacter coli* in raw milk and dairy products. *Applied and Environmental Microbiology, 59*(7), 2161–2165.
98. Witte, W., (1998). Medical consequences of antibiotic use in agriculture. *Science, 279,* 996–997.
99. Wysok, B., Wiszniewska-Łaszczych, A., Uradziński, J., & Szteyn, J., (2011). Prevalence and antimicrobial resistance of Campylobacter in raw milk in the selected areas of Poland. *Polish Journal of Veterinary Sciences, 14*(3), 473–477.
100. Yıbar, A., Çetinkaya, F., Soyutemiz, E., & Yaman, G., (2017). Prevalence, enterotoxin production and antibiotic resistance of *Bacillus cereus* isolated from milk and cheese. *Kafkas Universitesi Veteriner Fakultesi Dergisi* (Kafkas University Faculty of Veterinary Medicine Journal), 23(4), 635–642.
101. Zanella, G. N., Mikcha, J. M. G., & Bando, A. E., (2010). Occurrence and antibiotic resistance of coliform bacteria and antimicrobial residues in pasteurized cow's milk from Brazil. *Journal of Food Protection, 73*(9), 1684–1687.
102. Zboromyrska, Y., Ferrer-Navarro, M., Marco, F., & Vila, J., (2014). Detection of antibacterial resistance by MALDI-TOF mass spectrometry. *Revista Espanola de Quimioterapia* (Spanish Magazine of Chemotherapy), 27, 87–92.

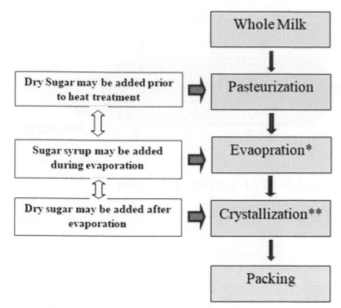

Dry Sugar may be added prior to heat treatment	⇨	Pasteurization
Sugar syrup may be added during evaporation	⇨	Evaopration*
Dry sugar may be added after evaporation	⇨	Crystallization**

Whole Milk

Packing

* Water is evaporated to give desired solid content
** To ensure excess sugar forms lactose crystals, fine seed crystals are added during cooling
** Crystallization at 30 degree C, Quantity of fine seeds: 375-500gms/ 1000 kg and Size of seed <200 microns

FIGURE 4.1   Traditional process for the manufacture of sweetened condensed milk.

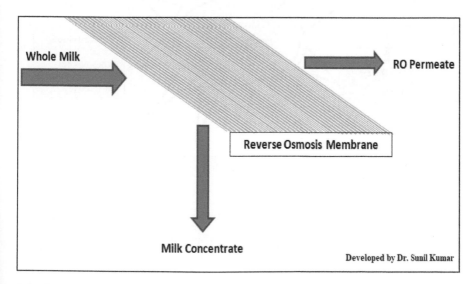

Whole Milk

RO Permeate

Reverse Osmosis Membrane

Milk Concentrate

Developed by Dr. Sunil Kumar

FIGURE 4.2   Principle of reverse osmosis.

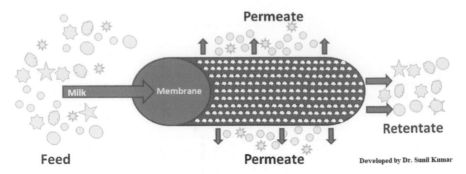

Principle of Membrane Filtration

FIGURE 4.3 Principle of microfiltration.

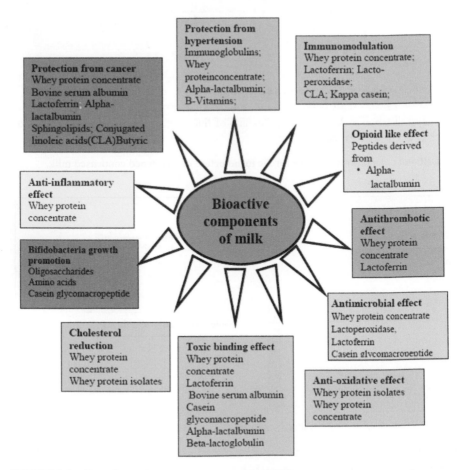

FIGURE 6.2 Functions of bioactive components of milk.

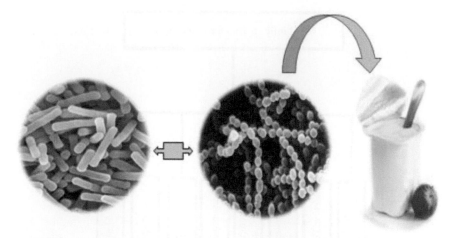

FIGURE 7.1   Yogurt cultures *Lactobacillus delbrueckii* ssp. *bulgaricus* and *Streptococcus thermophilus*.

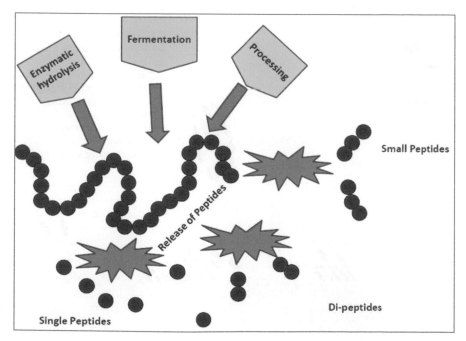

FIGURE 7.2   Methods to release bioactive peptides from native protein.

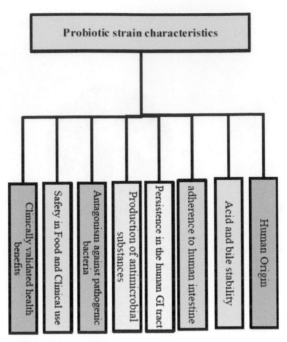

FIGURE 15.1   Characteristics of the probiotic strain.

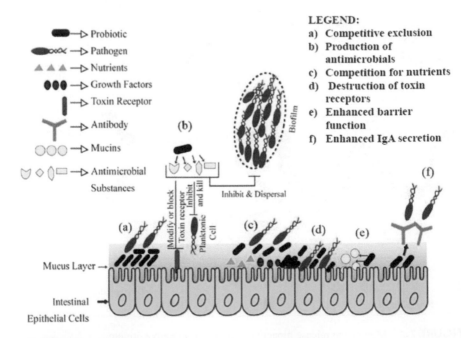

FIGURE 15.2   Mechanism of probiotic action.

PART IV
Quality and Safety of Milk Products

CHAPTER 9

QUALITY AND SAFETY MANAGEMENT IN THE DAIRY INDUSTRY

KUNAL M. GAWAI and V. SREEJA

ABSTRACT

Food safety and quality assurance are one of the important pillars of any management systems, and the applicability of these systems in the dairy industry is integral when the focus is on customer safety. The use of food safety and quality assurance in dairy plants is very important to eliminate chemical and microbiological hazards. Strict regulatory law implementation in the dairy industries and long-term planning is required to achieve and ensure the quality and safety of dairy products. For the said purpose, implementation of pre-requisite programs like training of personnel, good manufacturing practices (GMP), etc. are needed. Operational pre-requisite programs and risk analysis need to be established for the effective applicability of HACCP that determines physical, chemical, and microbiological hazards in the dairy industry. Food Safety systems based on HACCP approaches, integrated with ISO 9000, 22000, 14000 and OHSOS systems look for hazards or any factors that may harm product safety and implement control measures.

9.1   INTRODUCTION

Milk and milk products are easy targets for adulteration. Hence the dairy industry needs to continuously monitor the quality and safety aspects of milk and milk products at all levels of its production, processing, marketing, and consumption. Quality and safety of milk and milk products are mainly governed by the possibility of occurrence of physical, chemical, or microbiological hazards. These hazards may be detrimental to a population who may get exposed to these risks due to the consumption of non-conforming products [50].

Globalization of food trade has further highlighted the need to strengthen the quality and safety of foods. Traditional sampling and analysis programs are insufficient to provide a satisfactory level of protection by most of the countries. With the continuous increase in the international food trade, an effective measures need to be put in place in the production facility to guarantee the quality and safety of food from exporting countries beside certified laboratory's analyses report. Best in class food quality assurance systems are needed at each stage of the product manufacturing and supply chain and in each allied area of the food industry for ensuring complete food, i.e., in terms of quality as well as safety. At one end policy regulator has taken the accountability of establishing the standards, rules, and regulations and enforcement of programs necessary for food quality control and safety, while the industry is also countable to implement quality assurance systems with hope for compliance of the prescribed standards and legal obligations [48].

Modern-day approaches of dairy industries require adequate quality and food safety practices, which are important for food security. Dairy industries have been transforming itself with adaptation and implementation of the most effective and efficient food safety monitoring systems to achieve complete elimination of hazards and better quality of food produced. Among them, the Hazard Analysis and Critical Control Points (HACCP) system have been considered as one of the most successful ways to certify the high quality and safety food [13, 30].

The *Codex General Principles of Food Hygiene* provide a concrete framework, which is needed for ensuring food hygiene and should be used in combination with each specific code of practice when it is most appropriate. These principles compile to follow strict hygienic controls in the food production system, starting from primary raw material selection up to final product preparation [48]. The main objective of HACCP is to identify Critical Control Points (CCP) before a problem has occurred. Simultaneously, it elevates the food safety at each stage in the production process with the establishment of control measures. It is important to note that when HACCP is used with other quality management programs like Good Manufacturing Practices (GMP) and Good Hygiene Practices (GHP), the success ratio increases significantly [40].

Implementation of the quality management system (QMS) helps the dairy companies to continually improve its activities. The International Organization for Standardization (ISO) has published more than twenty-one thousands of standards, which covered nearly all features of technology and production. The first version of the series of ISO 9000 standards was published in the year 1987, describing the requirements for QMS for organizations and enterprises [42].

The ISO introduced the Food Safety Management System (FSMS) in 2005. The FSMS set interrelated or interacting elements for organizations to establish policy, objectives, and methodologies to achieve maximum food safety irrespective of their size and shape. The FSMS focused to control and guide an organization for devoting maximum intention towards achieving the food safety [30]. HACCP mediated quality programs (i.e., ISO 22000 and ISO 14000) have been widely implemented in the dairy industry [34, 50].

The food industry in the present competitive environment is highly regulated by local, national, and international laws relating to food safety. It governs markets, captures, shares, and profit of the industry. More importantly, consumers are very careful and have started taking extra precautions in making any buying decision relating to raw or processed foods. They usually feel more confident by visualization of certification mark like evidence of quality management during the manufacturing of food and beverages, in addition to supervision by food regulators.

This chapter focuses on the need of quality and safety assurance practices in the dairy industry, concerns for food safety and risk management, and QMS in the dairy industry.

9.2 WHAT IS QUALITY?

Goetsch and Davis [19] described "*Quality as an active condition related with products, services, public, method, and surroundings that meet up or surpasses potential and help to produce superior standards.*" Van den Berg and Delsing [54] focus on essential circumstances and its relation among resource that provide business house actually delivering products with the intention to satisfy and fulfill the need of customers. Evans and Lindsay [17] emphasized on the definition of quality concept by describing different criteria. These criteria targeted end-user and manufacturing conditions, based on judgments and product category.

Companies can achieve higher customer satisfaction level, increase the sale of products, sustain competition, strive to gain more profit as well as can acquire maximum market share through only assured high-quality products.

Food manufacturers need to adhere strictly to the quality parameters to ensure the safe production of food products [4]. To reach to that level, a systematic approach of Total Quality Management (TQM) is required. TQM is a management system and also an integrated philosophy, which improves the competitiveness of firms. Quality generally implies "Quality Assurance and Quality Control" of which, Bolton [8] gave distinctive meaning.

- **Quality Assurance:** It is the confidence-building process among consumers by providing them a product and/or service, which is superior in all attributes through deliberated and organized actions.
- **Quality Control:** It is an activity that is used to complete requirements for the quality.

Thus quality is the most competitive strategic instrument in businesses and hence has got noticed as an important major factor for developing products and services. This later on gave the additional benefit of sustained development of businesses, which enabled them for developing high-quality products and services, and helped to increase the company's profit and competitive capacity [7].

9.3   QUALITY ASSURANCE IN DAIRY INDUSTRY

In the past few decades, many scientists, technologists, and managers have contributed largely for the good quality produce. Due to this, quality has become a significant concern in the dairy and food industry.

FAO/WHO have defined guiding principles for implementation of FSMS emphasizing on basic requirements for hygiene and application of Hazard Analysis Critical Control Point (HACCP) [10]. National Food Control Systems are deliberated to ensure the supply of a safe food and to promote the better health of consumers. Food legislation around the world is targeting food businesses to have hazard analysis of their product and introduction of necessary actions to complete assurance of safe food production.

The term "Quality assurance" expresses the actual control, assessment, and reviewing of an established system for a food processing operation. The main objective of quality assurance is to provide confidence to the management and customers. Consumers anticipate uniform quality products with complete safety. This is replicated steadily with the participation of regulatory and advisory organizations in food quality practices for implementation [20].

Quality assurance systems are planned in such a way to ensure customers that each time whenever they purchased the same product, idea land uniform product is delivered. These systems performed an important task in the establishment of safe and sound quality attributes, which otherwise are difficult to get noticed. Generally, the success of quality assurance systems is certified by valuable consumers [9, 22].

Food safety assurance comprises the entire supply chain from raw ingredients to the final product; and includes primarily control of microbiological, chemical, and physical hazards [5]. These food quality assurance systems are necessary at each stage of the food chain to ensure the quality and safety of food. To meet the terms the mandatory legal requirements, the industry has the responsibility to implement a quality assurance program successfully. Generally, the mandatory standards are being enforced through laws, regulations, and they represent the minimum quality of standards. However, the voluntary standards display a consumer image and may become a brand or identity mark of a product quality. Food safety and quality assurance systems can be applied in the food industry in many ways, such as:

- Adaptation of voluntary international quality assurance standards, ISO 9000;
- National level assurance systems applicable at farms;
- Proprietary quality assurance systems, such as those maintained by the large retail food chains.

No matters where these quality assurance systems are applied, they have two general characteristics [22]: (1) Proper documentation of processes and practices followed; and (2) Auditing and certification from certified auditors.

9.4   DAIRY SAFETY: ISSUES, MEASURES AND IMPORTANCE FOR CONTROL

Food safety concerns are noticeably high particularly relating to the irregularity and unpredictability in food safety systems. One of the best examples is milk adulteration while procurement in India [6]. Approximately 15% from the total 37% of processed milk is utilized on a small scale, and retail dairy businesses for converting it into various Indian dairy products [26, 27] and these products are highly perishable and are being packaged without the use of any aseptic packaging conditions [15]. Also, it is difficult to map out the possible source of milk procurement in such segments, and reason for this condition is obviously related with failure to comply safety and legitimate transparent standards due to deficient funds, resources, and expertise [21, 26].

In the present arena, safety measures have been given more focus based on previous experiences, as developing countries still have been identified as suffering from lack of good agricultural, manufacturing, and hygiene practices. To get rid of this, worldwide, food industries have started using Good Agricultural Practices (GAPs), Hazard Analysis of Critical Control Points (HACCPs) and ISO as a point of reference of food quality assurance.

Variations in information between consumers and producers complicate the problem of identifying the quality that may lead towards uncertainty [1]. Those practices are in core triangle revolution of food quality system in food safety management, which includes focus on: food safety fundamentals; certified HACCP Food Safety Plans; and comprehensive Food safety and QMS [41].

9.5   QUALITY MANAGEMENT SYSTEM (QMS)

QMS ensure all aspects of a business that are working efficiently and cost-effectively. QMS will provide a competitive advantage that can increase marketing and fetch sales opportunities, and in this way, it will help a company to expand business with new customers as well as retaining long relations with existing business partners.

9.5.1   TOTAL QUALITY MANAGEMENT (TQM)

The food industry sector is supporting TQM as the main line of attack to ensure the best service or product quality [8]. TQM systems approach a manufacturing organization to fulfill its needs to ensure product quality, which it is manufacturing. It involves each member of an organization in the fulfillment of management objectives to make food safe and wholesome, boost customer delight, confidence, and simultaneously point out various means of continuous improvement.

Basically, TQM methodology needs to be established at all levels to fulfill the product processing requirements, which can be reflected exactly from the business quality report indicating its operational process status [37]. TQM schemes endorsing HACCP and its documental recording form an important framework formulated for a process which can be communicated effectively by the ways that can be demonstrated and audited [32].

QMS such as ISO 9001 and ISO 22000, which are most suited to the needs of the dairy industry, have widespread international acceptance [35].

TQM is a management tool (a sort of quality assurance program is popular in the business community). TQM approaches can be applied to all of the dairy processing operations. These are helpful to the dairy industry to gain a better execution and knowledge of the entire production process. These strategies employ written records, monitors ways, establishing quality objective, and gathering of useful knowledge to detect variations in process that would affect final quality [38].

A number of different concepts for quality control have been proposed by many quality *gurus*. Among them, the most established in the context of food quality and food safety are GMP, HACCP, and ISO Systems [39, 46]. All these concepts are intended to control food safety and have one thing in common that they are based upon deployment of basic hygiene practices and encompassing preventive measures for its reduction or elimination. Compliance with basic hygiene requirements, the existence of an established hygiene concept, availability of effective controlling and reliable monitoring are pre-conditions before HACCP can be implemented in an establishment. GMP, GHP, and sanitation standard operational procedures also count among these pre-requisites [2, 51, 50] and can be regarded as evolutionary precursors of HACCP and towards approaches using food safety risk analysis [47].

9.5.2 RISK ANALYSIS

Risk analysis is quantifying risk and locating levels at which casual agents should be controlled to assure food safety. It is a structured and formalized approach, which has three major components: risk assessment, risk management, and risk communication. Microbiological risk analysis protocols are becoming a key component in deciding the intensity level of end-user safety [32]. HACCP system, which is appropriately incorporated into TQM, is generally the most preferred risk management tool.

9.5.3 HAZARD ANALYSIS AND CRITICAL CONTROL POINT (HACCP)

The HACCP system is recognized globally as an organized and hazard protective methodology to eliminate the presence of all types of hazards (biological, chemical, and physical) [11, 12]. HACCP is applicable in the food sector to ascertain the quality of food products, which is based on two main objectives: hazard analysis and determining such points in the production process, at which chances of hazard occurrence have increased [5, 49]. HACCP has the main concern of the safety of the products through risk identification and risk management in the production process itself [44]. It is pro-actively working and emphasizing on the prevention of food hazard rather than the detection of defects in final prepared food products [11, 43].

In the 1960s, HACCP was developed by Pillsbury Company and the United States Army Laboratories at Natick, and the National Aeronautics and Space Administration (NASA–USA) collaboratively, for safe food production specifically meant for manned space flights [25, 52]. During later years, the *Codex Alimentarius Commission* (CAC) was adopted and developed the present concept of HACCP as the food safety management tool for advice on protection of end-users under *Sanitary and Phytosanitary Measures (1994) granted at the Uruguay round of GATT* negotiations [14]. As such, HACCP is used as a reference point in international trade disputes, and it is increasingly protected in national legislation by many countries [43]. At present, HACCP, and its principles (Table 9.1) are the foundation of frameworks, on which most food quality and safety assurance systems have been established. Making HACCP mandatory in some agricultural sectors, including dairy and food sector, can be beneficial for developing countries.

TABLE 9.1 Seven Principles of HACCP

Principle	Scope
Principle 1	Identify the potential hazard/s associated with food production involved in all stages from raw material, in-process products, final products, and product in the supply chain just before consumption.
	Evaluate the possibility of an incident of the hazard/s and identify the possible protective measures.
Principle 2	Decide the most suited processes that can be placed to control and eliminate the hazard/s or at least minimize its possibility of happening, i.e., identify Critical Control Point (CCP).
Principle 3	Establish limits to ascertain that each CCP is under control.
Principle 4	Set up a monitoring system to ensure that each CCP is under control.
	It must be established both by testing procedures or recording relevant observations.
Principle 5	Ascertain the corrective action against any particular CCP if not under control and has exceeded the permissible limit.
Principle 6	Establish a method for verification to confirm that the HACCP system is working efficiently.
Principle 7	Systematic documentation and record maintenance of all concern procedures.

The HACCP procedure is normally the point of attack at food safety management (safety from harmful pathogens and/or toxic metabolites derived from them), but advancing towards broader quality management concept, it can efficiently be applied to various hazards arising like microbiological spoilage, extraneous matters or pesticide contamination. It is better to conduct a HACCP program to minimize tempering by hazard/s.

However, an experienced team might opt to cover the wide range of hazards, depending on: (i) the competency in producing and maintaining a composite HACCP plan; and (ii) the systematic approach to incorporate it into the local quality plan and in the system [32]. HACCP's foresightedness is seen as more cost-effective than testing a final product for defect and then working out either by destroying or reworking it [8, 28, 33]. HACCP is guided by seven principles, which are outlined in Table 9.1.

9.5.4 STANDARDS UNDER ISO-9000 SERIES

ISO provides standards to achieve uniformity in quality and to eliminate technical complexity to trade food products throughout the world [43]. The ISO-9000 is a family of standards that are wide-ranging and has support from almost all organizations that desire to put into practice a QMS. ISO-9000 family of standards can be implemented by any sector of the industry, including the food industry. Large and small companies that are part of the production process or even just only packaging food products have also implemented ISO-9000 and obtained third party certification. To provide healthy and safe food, it is necessary that producers of food must fulfill quality along with safety approach to succeed in the market [16]. The fundamental nature of an ISO-9000-based quality system is that all activities and operations must be well-maintained in procedures, which must be pursued by a guaranteed apparent obligation of responsibilities and authorities.

The three standards in the ISO 9000 family are:

- ISO 9000, which focuses on fundamentals and vocabulary of standards: The rationale of ISO-9000 is to provide a positive reception of the basic principles of QMS and give details on the meaning of the vocabulary used in the family of standards;
- ISO 9001, which focuses on requirements for system establishment: ISO-9001 is meant to offer requirements to the organization if they want to fulfill customer and applicable regulatory requirements; and

- ISO 9004, which focuses on guidelines for performance improvements at organization: ISO 9004 provides guidance for improving the competency, efficacy, and in general performance of an organization.

ISO 9001 and ISO 9004 both have been established as adjusting standards, which are complementary to each other. They have a common structure but need to be implemented independently [53].

The well-known ISO-9001 was published first by ISO (www.iso.org) in 1987 as a key tool to sustain and help in promoting the global expansion of business. The ISO management Standard all over the world has achieved great international visibility with around 1,643,523 valid certificates were recorded across nine standards compared to 1,520,368 in 2015 [31].

Companies may also adopt HACCP and/or similar systems as their quality assurance system singly or in combinations. Out of them, many have been using ISO-9000 and HACCP both, incorporated food safety as well as QMS. Applying HACCP within an ISO-9000 QMS can result in food safety along with the quality assurance that is more effective than applying ISO 9001 or HACCP separately. The emphasis of both systems is to prevent hazard in food rather than correction of problems, recall, and clarification for a deficiency after they have occurred. A company implementing a HACCP system does not have an obligation to have ISO-9001certification, but it is desirable, rather positive for any organization [43]. HACCP act as a specialized tool in controlling hazards and provide a hurdle against its proliferation on food. In this way, safe food is assured with creating a number of hurdles to tackle the growth of microbes effectively. These hurdles may include Good Manufacturing Procedures (GMP), Good Handling Practices (GHP), etc. as pre-requisite programs together with HACCP system [32].

9.5.5 ISO-22000 STANDARD

ISO 22000 provides an effective framework for the development, implementation, and continual improvement of an FSMS. Getting this certification, a company can win the trust of their customers ascertaining that they have an FSMS, and it is in place. It is the first standard applicable to organizations directly or indirectly involved in the food supply chain, including suppliers of raw materials and service providers such as cleaning, equipment manufacturers, and logistics companies [18]. The ISO 22000 adopts a flexible approach allowing an organization to implement and get accreditation of management systems standards to assure both quality and safety.

Manufacturing of food products with the speculated standard is a prime key step in achieving goals, assure consumer safety, and build confidence [55].

The ISO 9001 module can be incorporated with existing ISO-22000 to get combined ISO 22000 and ISO 9001 certification for both quality and safety. The format, which ISO 22000 uses, is the same like other QMS standards. This makes integration more real and logical, which is helpful for an organization to develop a transparent system covering all aspects related to production targeting quality and safety of food [3].

ISO-22000 is the first international standard, which further defines HACCP's role in FSMS. This unique standard encompasses the needs of the consumer and market by timely and simplified processes without making adjustments with other quality and safety management systems. Though it has combined multiple principles, yet approaches and functions under ISO 22000 are simpler to learn, apply, and acquire. Interestingly like other quality programs, it does not follow a prescriptive checklist approach. This makes it more efficient prime certification marking tool than previous combinations of national and other standards [29].

9.5.6   OCCUPATIONAL HEALTH AND SAFETY MANAGEMENT SYSTEMS (OHSAS): BRITISH STANDARD

In a present market scenario, customers are quite aware of what they are purchasing. Now, they have wider interest apart from just pricing and discounts availed from the products. To sustain under such conditions, companies have to assure that their businesses are managed professionally and correctly. Similarly, they must be reliable service provider without a long history of industrial mishaps and incidents. An effective Health & Safety Management System can be helpful: in minimizing burden; improve business efficiency and achieve cost savings, which ultimately improves the company's image. Along with this, it builds a framework to compliance with all regulatory requirements.

Organizations have abided with legal and public responsibility, while establishing the occupational health and safety of all employees working in factory premises. It requires a perfect occupational health and safety management system (OHSAS) to keep away the hazards and risks, which otherwise may be harmful to an employee. OHSAS-18001:2007 has been suitable to all kinds of organizations. Its standard objective is to enable organizations to identify, control, eliminate, or at least minimize occupational hazards and lower down associated risks through OHSAS performance improvement.

OHSAS-18001 has helped organizations to fulfill their health and safety obligations in competent and successful behaviors possible potential hazards within an organization. Though it is not the ISO Standard, yet it is developed in such a way so that must be mutually compatible with the ISO-9001 (QMS) and ISO-14001 (Environmental Management System (EMS)) standards. It has some prominent benefits like gain competitive advantages in the global market, enhance health and safety working environment, reputation and brand building of the company, help continual improvement in performance, morale-boosting of organizations to improve the working environment with prime focus on health and safety for the employees [23].

9.5.7 ENVIRONMENTAL MANAGEMENT SYSTEM (EMS) (ISO–14000)

The EMS supports the organization with objectives to identify the environmental aspects, to control the occurrence of environmental impacts through the activity of organization; improving the use of natural resources to promote sustainability. The ISO-14001 is different from other standards and certifications with no criteria for demarcation from other management systems. This provides all the requirements for the establishment of an EMS subsequently by the Plan, Do, Check, and Act cycle, a TQM tool. It develops and implements strategies to encompass legal requirements and information about significant environmental aspects; thus, it allows the sustainable environmental protection and the pollution prevention with the aim to accomplish the socioeconomic needs [37].

9.6 STATUS OF QUALITY MANAGEMENT SYSTEMS (QMS) IN THE DAIRY INDUSTRY

TQM implementation in the Indian dairy industry started way back in 1994. These TQM programs were then explored to other stakeholders of business partners, including the farmer, who produces raw material, a wholesale distributor located in a bigger city and up to the production factories. Later on, the dairy industry implemented QMS, an internationally recognized quality, and safety standard. The company Amul has certified to be a maiden dairy organization in India to get accredited with the certification of ISO 2200:2005 and ISO-9001 for its production. Amul then achieved the bigger milestone and an example to all others in dairy sectors with the accreditation of its village Dairy Co-operative Societies with ISO 9001:2000. The dairy plants manufacturing AMUL branded products are certified by Agricultural and Processed Food Exports Development Authority (APEDA) and thus now can export their branded dairy products to international markets. Their plants are also periodically under-going audit for various Indian Statutory Bodies, International statutory Bodies, and Quality Management Agencies Hygiene and QMS by various agencies like Export Inspection Agency (EIA).

It must be noteworthy that Amul Asia's largest cooperative organization feathered the hat by certifying its primary village cooperative societies in the rural sector with ISO 9001:2000 certification. It is the first in India and possibly premier in the world. The factors, which help in the retention of the freshness of milk being received at dairy-plant, are mainly focus on the health of animal and hygienic conditions of premises, housekeeping activities, installation of bulk milk coolers in villages, prompt management of village co-operative societies, facility for rapid raw milk transportation to the dairy plant and dedicated milk handling at reception dock. To maintain the uniform quality routinely throughout the year, training programs are planned for all those involved in the milk procurement chain. It also needs laboratories well equipped with state of the art equipments and well-trained professionals to sustain this development in production facility [24].

Currently, the Indian dairy sector contributes to about 18% of world milk production [45]. Over the years, Indian dairy sector has accepted the challenges associated with the continuous upgradation of quality and safety standards and emerged as the highest milk producing country in the world and exports its various milk products to many countries.

9.7 QUALITY SYSTEM IMPLEMENTATION: BENEFITS AND CHALLENGES

The external business environment and subsequently anything, which relates to the adoption and possible implementation of quality systems in small food enterprises, is influenced by the government's policy. The type and intensity of measures included in public policy have been influenced by an interest in human health and quality of life, social welfare activities, employment status, and competitiveness in maintaining socio-economic lifestyle. Actually, these policies related issues are impacted by the implementation of various quality systems in business. The governments considering this have serious reasons to tackle the issues and contribute towards the diffusion and large scale establishment of quality systems in small food enterprises [35].

The government initiated a national or sector standards, which make easier to certify a small food enterprise and in many countries have been actually promoting business in this way. This approach has been extensively taken up by the agro-food business and tourism sectors on the initiative of government-supervised and by private institutions [36]. By such standards, small food enterprises can shape the basic elements of quality management, which they need to implement in their organization to fulfill the basic needs of their customers and cumulative cost saving can be subsequently achieved. The benefits of such food safety system are that the organization can credit of completing formal commitment done in terms of consumer's happiness about their product. The implementation of the quality system can also enhance the ability of small food enterprises to identify and promote other companies allied in the supply chain to supply good grade raw material to them and make easy their trade further. It also helps enterprises to achieve a level of quality, which can be transitional that will initiate the development [35, 36].

The adaption of a quality system, however, also has some disadvantages, including more effort in initial stages with high capital investment for the development and implementation of the quality system. The safety and food system, which is unknown to customers, need encouragement for partners to invest and at the same time may lead to loose market opportunities if an improper selection of quality assurance system, which is not known to the customers [35].

Food safety and quality assurance are one of the important pillars of any management systems, and the applicability of these systems in the dairy industry is integral when the focus is on customer safety. The use of food safety and quality assurance in dairy plants is important to eliminate chemical and microbiological hazards.

Strict regulatory law implementation in the dairy industries and long term planning is required to achieve and ensure the quality and safety of dairy products. Implementation of such management systems at all stages of production, processing, marketing is a must. For the said purpose, implementation of pre-requisite programs like training of personnel, GMP, etc. are needed. Operational pre-requisite programs and risk analysis need to be established for the effective applicability of HACCP that determine physical, chemical, and microbiological hazards in the dairy industry. Milk Safety systems based on HACCP approaches, integrated with ISO 9000, 22000, 14000 and OHSOS systems, look for hazards, or anything that could harm product safety and also implements control measures. Further, the dairy industry needs to adapt and modify to the timely changes, which are being made in the safety management systems to remain in the domestic market and compete more effectively in the world market by reducing barriers to international trade.

9.8 SUMMARY

It has become essential to streamline quality parameters across the globe for every product. Such demand can be met only through ensuring the safety and wholesomeness of milk products. The deficiency in quality relates to safety and overall sensory attributes, which can result in personal illness. Food can get contaminated at any time and point of operation in factory premises or while in the consumer's house just before serving.

Many food industries are still not in a position to provide consistently quality safe food. The solution lies in adaption of specific protective approaches about food safety that engages the organization in checking and monitoring each step of the processing, categorize the essential procedures on the basic of risk associated and making sure that each parameter must remain constant.

The only solution to achieve this is to set up and retain food safety and quality assurance systems. Dairy industry uses various systems for quality and safety management, such as ISO: 9001 for Quality Management, ISO: 22000 for Food Safety Management, ISO: 14001 for Environment Management, TQM, HACCP, etc. These systems are much appreciated, but need basic establishment and infrastructure with some monetary investment. These standards are subjected to revision to meet the quality and food safety aspects, and any industry who wants to sustain in business and to grow in open market across the world should meet these standards, and the dairy industry is no exception.

KEYWORDS

- *Codex Alimentarius* **Commission**
- **critical control point**
- **environmental management system (EMS)**
- **food safety management system (FSMS)**
- **good hygiene practices (GHP)**
- **good manufacturing procedures (GMP)**
- **hazard analysis and critical control point (HACCP)**
- **total quality management (TQM)**

REFERENCES

1. Akerloff, G., (1970). The market for lemons: Quality uncertainty and the market mechanism. *Quarterly Journal of Economics, 84*, 488–500.
2. Anonymous, (2016). *Hong Kong: The Facts, Food, Food and Environmental Hygiene Department, Environmental Hygiene Department.* https://www.hk.gov/en/about/abouthk/factsheets/docs/f%26e_hygiene.pdf (Accessed on 20 July 2019).
3. Arouma, I., (2006). The impact of food regulation on the food supply chain. *Food Toxicology, 221*(1), 119–127.
4. Arvanitoyannis, I., & Mavropoulos, A., (2000). Implementation of the hazard analysis critical control point (HACCP) system to Kasseri/Kefalotiri and Anevato cheese production lines. *Food Control, 11*(1), 31–40.
5. Batt, C. A., (2016). *Food Safety Assurance: Reference Module in Food Science* (p. 342). Elsevier Pub., New York. https://doi.org/10.1016/B978–0–08–100596–5.03442–9 (Accessed on 20 July 2019).
6. Biswas, A. K., & Hartley, K., (2015). India's food safety crisis is indicative of bureaucratic failure. https://thediplomat.com/2015/09/india-and-food-safety/ (Accessed on 20 July 2019).
7. Boateng-Okrah, E., & Fening, F. A., (2012). TQM implementation: A case of a mining company in Ghana. *Benchmarking: An International Journal, 19*(6), 743–759.
8. Bolton, F. J., (1998). Quality assurance in food microbiology–a novel approach. *International Journal of Food Microbiology, 45,* 7–11.
9. Botonaki, A., Polymeros, K., Tsakiridou, E., & Mattas, K., (2006). The role of food quality certification on consumers' food choices. *British Food Journal, 108*(2), 77–90.
10. CAC (*Codex Alimentarius* Commission), (2001). *Food Hygiene Basic Texts* (2nd edn., p. 112). *Codex Alimentarius* Commission Publication, Food and Agriculture Organization/World Health Organization, Rome, Italy.
11. Cannas, J., & Noordhuizen, J., (2008). Consumer safety and HACCP like quality risk management programs on dairy farms: The role of veterinarians. *The Open Veterinary Science Journal, 2,* 37–49.
12. Chountalas, P., Tsarouchas, D., & Lagodimos, A., (2009). Standardized food safety management: The case of industrial yogurt. *British Food Journal, 111*(9), 897–914.
13. *Codex Alimentarius* Commission (CAC), (1996). *Hazard Analysis and Critical Control Point (HACCP) System and Guidelines for its Application* (p. 108). Report of the 29th session of the Codex Committee on Food Hygiene, Alinorm. 97/13A, Appendix II. *CAC*, FAO, Rome.
14. *Codex Alimentarius* Commission (CAC), (1993). *Procedural Manual* (p. 99). Joint FAO and WHO Foods Standards Program, CAC–FAO, Rome.
15. Dabbene, F., Gay, P., & Tortia, C., (2014). Traceability issues in food supply chain management: A review. *Biosystems Engineering, 120*, 65–80.
16. Djordjevic, D., Cockalo, D., & Bogetic, S., (2011). An analysis of the HACCP system implementation: The factor of improving competitiveness in Serbian companies. *African Journal of Agricultural Research, 6,* 515–520.
17. Evans, J. R., & Lindsay, W. M., (1996). *The Management and Control of Quality* (3rd edn., p. 245). West Publication, Minneapolis.
18. Faergemand, J., (2008). The ISO-22000 series: Global standards for safe food supply chains. *ISO Management Systems* (pp. 1–39). Rome. www.iso.org/ims (Accessed on 20 July 2019).

19. Goetsch, D. L., & Davis, S. B., (2010). *Quality Management for Organizational Excellence: Introduction to Total Quality* (p. 110). Prentice Hall, Pearson, NJ.

20. Gould, W. A., & Gould, R. W., (1988). *Total Quality Assurance for the Food Industries* (p. 84). CTI Publications Inc., Baltimore, Maryland.

21. Gupta, P. R., (2007). *Dairy India* (p. 543). Dairy India Yearbook Pub., New Delhi.

22. Holleran, E., Bredahl, M. E., & Zaibet, L., (1999). Private Incentives for adopting food safety and quality assurance. *Food Policy, 24*, 669–683.

23. https://www.siemens.com/content/dam/internet/siemensc.com/global/company/sustainability/downloads/ehs-policy-managementsystem.pdf (Accessed on 20 July 2019).

24. http://www.amuldairy.com/index.php/cd-programmes/quality-movement (Accessed on 20 July 2019).

25. Hulebak, K. L., & Schlosser, W., (2002). Hazard analysis and critical control point (HACCP): history and conceptual overview. *Risk Anal, 22*(3), 547–552.

26. IAI–vision 2020: 1st white paper document for Indian dairy industry, (2011). *First International Symposium on Future of Indian Dairy Industry* (p. 32). NDRI, Karnal, Haryana.

27. IBEF, (2012). *Food Processing and Market and Opportunities*. Aranca, https://www.ibef.org/download/Food_Processing_270608.pdf (Accessed on 20 July 2019).

28. ICMSF (International Commission on Microbiological Safety of Foods), (1998). Microorganisms in foods. In: *Application of the Hazard Analysis Critical Control Point (HACCP) to Ensure Microbiological Safety and Quality* (pp. 89–114). Blackwell Scientific Publications, London, UK.

29. ISO, (2005). *ISO-22000: Food Safety Management Systems–Requirements for Any Organization in the Food Chain* (p. 78). ISO Publication, Switzerland.

30. ISO, (2007). *Food Safety Management Systems Requirements for Bodies Providing Audit and Certification of Food Safety Management Systems* (p. 51). Report No. 2007ISO/TS 22003, First Edition, ISO Publication, Switzerland.

31. ISO, (2018). *ISO Survey 2016*. https://www.iso.org/the-iso-survey.html (Accessed on 20 July 2019).

32. Jervis, D., (2002). Application of process control. In: *Dairy Microbiology Handbook* (3rd edn., pp. 593–654). Wiley-Interscience Pub., New York.

33. Joint FAO/WHO Codex Alimentarius Commission, (1993). Guidelines for the application of hazard analysis critical control point (HACCP) system. In: *Training Consideration for the Application of the HACCP System to Food Processing and Manufacturing* (pp. 17, 18). Report No. WHO/ FNU/FOS/93. 31993 by Codex Alimentarius Commission, World Health Organization, Geneva.

34. Karaman, A. D, Cobanoglu, F, Tunalioglu, R., & Ova, G., (2012). Barriers and benefits of the implementation of food safety management systems among the Turkish dairy industry: A case study. *Food Control, 25,* 732–739.

35. Karipidis, P., Athanassiadis, K., Aggelopoulos, S., & Giompliakis, E., (2009). Factors affecting the adoption of quality assurance systems in small food enterprises. *Food Control, 20*, 93–98.

36. Karla, M. P., Carvalho, Picchi, F., Camarini, G., & Chamon, E. M. Q. O., (2015). Benefits in the implementation of safety, health, environmental and quality integrated system. *IACSIT International Journal of Engineering and Technology, 7*(4), 333–338.

37. Kirk, J. H., Sischo, W. C., Klingborg, D. J., Arana, M., Higginbotham, G., Mullinax, D., & Shultz, T., (1999). Dairies adopt TQM to improve milk quality and food safety. *California Agriculture*, 33–35.

38. Knura, S., Gymnich, S., Rembialkowska, E., & Petersen, B., (2006). Agri-food production chain. In: *Safety in the Agri-Food Chain* (pp. 19–65). Academic Publishers, Wageningen.

39. Kok, S., (2009). Application of food safety management systems (ISO-22000/HACCP) in the Turkish poultry industry: A comparison based on enterprise size. *Journal of Food Protection*, *72*(10), 2221–2225.

40. Lamuka. P. O., (2014). Challenges of developing countries in management of food safety. *Encyclopedia of Food Safety*, *4*, 20–26.

41. Lukichev, S., (2016). Romanovich. The quality management system as a key factor for sustainable development of the construction companies. *Procedia Engineering*, *165*, 1717–1721.

42. Luning, P. A., Marcelis, W. J., & Jongen, W. M. F., (2002). *Food Quality Management: Techno-Managerial Approach* (p. 213). Academic Publishers, Wageningen.

43. Mortimore, S., & Wallace, C., (2013). *HACCP: A Practical Approach* (pp. 36–48). Springer Science and Business Media, New York.

44. Nanda-Kumar, T., (2018). *Dairying in India by 2030: Make in Rural India* (p. 19). Key Note Address at Indian Dairy Association 44th Dairy Industry Conference, NDRI, Karnal. http://www.nddb.coop/about/speech/make-in-india/ (Accessed on 20 July 2019).

45. Noordhuizen, J. P. T. M., & Metz, J. H. M., (2005). Quality control on dairy farms with emphasis on public health, food safety, animal health and welfare. *Livestock Production Science*, *94*, 51–59.

46. OIE, World Organization for Animal Health, (2011). Terrestrial animal health code: Chapters 6, 1, In: *The Role of the Veterinary Services in Food Safety*. http://www.oie.int/fileadmin/Home/eng/Health_standards/tahc/2010/en_chapitre_1.6.1.pdf/ (Accessed on 20 July 2019).

47. Orriss, G. D., & Whitehead, A. J., (2000). Hazard analysis and critical control point (HACCP) as a part of an overall quality assurance system in international food trade. *Food Control*, *11*, 345–351.

48. Panfiloiu, M., Firczak, M., Perju, D. M., & Simion, G., (2010). Quality control of ice-cream products using the HACCP method. *Banat's Journal of Biotechnology*, *2*, 61–65.

49. Papademas, P., & Bintsis, T., (2010). Food safety management systems (FSMS) in the dairy industry: Review. *International Journal of Dairy Technology*, *63*(4), 489–503.

50. Sampers, I., Toyofuku, H., Luning, P. A., Uyttendaele, M., & Jacxsens, L., (2012). Semi quantitative study to evaluate the performance of a HACCP-based food safety management system in Japanese milk processing plants. *Food Control*, *23*(1), 227–233.

51. Sperber, W. H., (2005). HACCP does not work from farm to table. *Food Control*, *16*(6), 511–514.

52. Thompson, J., (2001). Role, origins and application of ISO 9000: Chapter 3. In: *ISO-9000 Quality Systems Handbook* (4th edn., pp. 80–115). Completely revised, Butterworth-Heinemann Pub Ltd., London.

53. Van der Berg, M. G., & Delsing, B. M. A., (1999). *Quality of Food* (2nd edn., p. 360). Kluwer Pub., Deventer, The Netherlands.

54. Verano, D., & Ponce, C., (2008). ISO-22000: Security in food chain. *UNE*, *236*, 41–43.

CHAPTER 10

GENETIC IMPROVEMENT OF DAIRY STARTER CULTURES AND LACTIC ACID BACTERIA

LOPAMUDRA HALDAR and SOUMYASHREE SAHA

ABSTRACT

The diversity and multi-facet functions of lactic acid bacteria (LAB) have brought tremendous progress in starter microbiology and fermentation technology. Genetic characterization of LAB and continuous genetic manipulation using recombinant DNA technology have made available many new LAB strains with enhanced industrial utility. With the advancement of molecular biology, ample opportunities exist to do research for the development of new food-grade LAB strains that can expand the diversity of fermented dairy products, improve product quality, reduce manufacturing cost and promote human health benefits in future.

10.1 INTRODUCTION

Fermented dairy products are the natural choice for consumers due to their delicate flavor, taste, and longer shelf-life. Health protecting attributes of fermented milk products is a bonus to the consumers, due to the production of a variety of metabolites by a group of beneficial microorganisms called starter cultures, during fermentation. A variety of cheese and indigenous milk products have gained popularity not only among the consumers, but also a choice of production for dairy industries because fermented milk and milk products have a longer shelf life, high profitability, increased market demand for their delicate taste, flavor, and additional functional attributes beyond nutrition.

In order to sustain a high level of productivity and diversification of fermented dairy products and economic viability of fermentation industry, the requirement of well-characterized, robust starter microorganisms with improved immunity and stress tolerance and capable of enhanced production of useful metabolites, became evident. Thus, a thorough understanding of phenotypic and genetic characteristics as well as metabolic activity of dairy starter cultures drew significant attention of food scientists.

A radiant development of an understanding of microbial genetics opens up avenues for using genetic tools for the authentic characterization of dairy starter cultures, which ensure proper maintenance of starter cultures and their precise utilization. Remarkable progress in the understanding of gene transfer mechanism, comprehensive knowledge of microbial gene expression and genome sequence enables to design appropriate genetic manipulation of dairy starter cultures especially Lactic Acid Bacteria (LAB) for expression of constitutive and inducible genes and thus enhancing the production of desired phenotypic as well as metabolic traits.

Genetic strategy can be adopted to address various problems of starter failure. Besides, genetic, and biochemical mechanisms of production and mode of action of bioactive components by starter and nonstarter organisms must be explored to validate various health claims of probiotic LAB. Understanding of molecular mechanisms of starter activity and their life cycles is necessary for genetic manipulation of the same to ensure improved product quality, minimization of industrial loss, and value-added and safe product delivery to the consumers.

This chapter outlines the present trend of the activities related to genetic improvement of dairy starter cultures; and discusses the technology of LAB.

10.2   GENETIC CHARACTERIZATION OF DAIRY STARTER CULTURES

The dairy starter cultures are mainly comprised of LAB, few yeasts and molds capable of enhancing flavor and texture of the cheese and other fermented milks. LAB represents a diverse group of organisms, functionally related by their attributes of producing lactic acid during homo- or heterofermentative metabolism. They are predominantly Gram-positive, catalase (CAT) negative, microaerophilic, non-sporulatingcocci, coccobacilli or bacilli, nutritionally fastidious, acid-tolerant organisms. LAB is generally unable to synthesize porphyrins, and lack cytochrome or other components of respiratory chains. They have low (<55% mol) G+C content [56].

According to taxonomic classification, most of the LAB belong to the phylum Firmicutes, class Bacilli and order Lactobacillales, while few of them belong to Actinobacteria. Several genera under Lactobacillales order are included as per phylogenetic relationship and biochemical activities, but seven genera, such as Lactobacillus, Lactococcus, Leuconostoc, Oenococcus, Pediococcus, Streptococcus, and Enterococcus, are widely used in fermented dairy and other food industry [69]. Species of Bifidobacterium genus are only LAB in Actinobacteria phylum [56]. LAB is used extensively not only as a starter culture, few of them–being able to produce bacteriocins and other antimicrobials–are also used as biopreservatives [27], while few selected strains are used as probiotics, due to health-promoting attributes of metabolites elaborated by them [36].

The LABs are also used popularly as a mucosal vaccine to deliver vaccine antigens [98, 103]. It is important to identify and decide the taxonomic position of these starter strains for their correct identification and proper use. The conventional methods of phenotypic, biochemical, and chemotaxonomic characterization based on cell wall amino acid composition, fatty acid composition and motility [72] may not be reliable because of variable results under varying environmental and cultural conditions. Besides genus and species, strain-level identification of dairy LAB may not be possible by these classical techniques [23]. Hence, a polyphasic taxonomic approach based on phenotypic, biochemical, and genotypic characterization has been introduced for authentic identification of LAB [92].

In the last three decades, remarkable advancement has taken place for genetic typing and identification of LAB. Various genetic methods for identification and typing of LAB may be broadly classified into [40]:(i) Non-PCR based methods include Plasmid profiling, Probe hybridization, Restriction analysis using pulsed-field gel electrophoresis, ribotyping, etc.; and (ii) PCR based typing methods are Randomly Amplified Polymorphic DNA (RAPD), Amplified rDNA Restriction Analysis (ARDRA), species-specific PCR.

Although genetic characterization and sequencing of variable regions of rRNA operons have been done for many LABS for authentic phylogenetic analysis, yet identification variation in a number of rRNA operon may result into instability in the phylogenetic relationship among closely related species [24]. Few housekeeping genes have been identified as more suitable phylogenetic single-gene markers like *groEL* (encoding heat shock proteins), *rpoB* (β-subunit of RNA polymerase), *recA* (encoding protein for recombination) than the 16S rRNA gene for large-scale analysis [47].

Considering the importance of LAB in the food fermentation industry, an endeavor was made by LAB Genome Sequencing Consortium [56] to sequence the whole genome of non-pathogenic LAB. Whole-genome DNA, DNA-RNA hybridization and GC content analysis led to the delineation of three closely related lineages of Lactobacillales, such as [84]: (1) the Leuco-nostoc group (*Leuconostocmesenteroides* and *Oenococcusoeni*); (2) the Lactobacillus casei-Pediococcus group (*Lb. plantarum*, *Lb. casei*, *Pedio-coccuspentosaceus*, and *Lb. brevis*); and (3) the *Lactobacillus delbrueckii* group (*Lb. delbrueckii*, *Lb. gasseri*, and *Lb. johnsonii*). Streptococci (*Strep-tococcus thermophilus*) and Lactococci (*Lc. lactis* subsp. *lactis* and *Lc. lactis* subsp. *cremoris*) form a separate branch [84].

Whole-genome analysis of *Bifidobacteriumlongum* has revealed the fact that only seven genes, present in *B. longum* have been shared specifically with Lactobacillales [56]. Such common genes exist between Lactobacil-lales and Bifidobacteria, but not in other nonstarter Actinobacteria, thus explaining 'genome cognate' of the LAB phenotype.

10.3   ORGANIZATION OF GENETIC ELEMENTS OF DAIRY STARTER CULTURES

In order to understand the genetic basis of starter activity, the genetic composition of starter cultures must be studied thoroughly. Most of the starter organisms possess number of extrachromosomal DNA in addition to their chromosomal DNA [29]. The organization of different genetic elements and their functional attributes have been revealed for number of starter bacteria. The variability of metabolic activities of LAB is due to the fact that these extra-chromosomal genetic elements can either be lost or gained by horizontal gene transfer (HGT).

10.3.1   BACTERIAL CHROMOSOME

Extensive advancement of DNA based techniques, like pulsed-field gel electrophoresis (PFGE) and DNA sequencing, enable a whole-genome analysis of any bacterium, and thus allow rapid progress in the knowledge of the LAB chromosomal DNA. Bacterial chromosome contains all essential housekeeping genes, and gene encoding essential biochemical activities. In the 1980s, the introduction of the PFGE method brought a major technological breakthrough in the area of whole-genome analysis. This very technique was extensively used to determine the genome size of different strains of LAB [2]. A more rapid method called Arbitrarily Primed Polymerase Chain Reaction (AP-PCR) was used to obtain genomic fingerprints of strains of *Lc. Lactis* [16]. LAB genome fingerprinting has potential industrial usefulness, particularly for selecting a culture ensuring the use of isogenic strains in fermentation.

Considering economic as well as health protecting the importance of LAB, efforts have been started to reveal the whole genome sequence of LAB in order to have more thorough insight for strain development by genetic manipulation. The first complete genome of the LAB group was published on *Lc. Lactis* subsp. *lactis* IL1403 by Bolotin and his co-workers [10]. Although their main focus was to explore the features related to the importance of *Lc. lactis* as a dairy starter, yet several unexpected facts got revealed related to their primary biochemical activities, like biosynthetic pathways for all 20 amino acids, a complete set of late competence genes, five complete prophages, and partial components for aerobic metabolism [11]. This revolutionary finding gave a new insight for the genetic manipulation of starter organisms to achieve efficient fermentation. Extensive sequencing of the genomes of non-pathogenic LAB was announced in 2002 by the LAB Genome Sequencing Consortium, in joint collaboration with Department of Energy–Joint Genome Institute in USA [56].

About 75 genome sequences of industrially important LAB, including different species of *Lactobacilli, Lactococci, Streptococci, Leuconostoc, Bifidobacteria* genera have been published while about 80 genome sequencing projects are in progress [102]. Knowledge of Global gene regulation along with full genome sequence would help understanding of many biochemical mechanisms.

10.3.2   MOBILE GENETIC ELEMENTS (MGE)

Bacteria vary considerably in their metabolic properties, structural phenomena, and life cycle, despite having a single cell with a very small size. One of the important reasons of their variability is accounted for the presence mobile genetic elements (MGE) or 'Mobilome' capable of HGT [8]. Mobilome can be classified into two groups:

- Plasmids and bacteriophage DNA may be grouped together as they are transferred from cell to cell independently, (the intercellular MGEs); and
- The second category of MGE is transmitted integrating into the former group. Transoposons, insertion sequence (IS) and introns are grouped into this category [84]. With no exception, LAB also possesses a variety of transposable elements.

10.3.2.1 PLASMID DNA

LAB possesses a wide range of traits, many of which are encoded by plasmid DNA. Plasmid is a self-replicating, extrachromosomal DNA molecule, small in size but plays an instrumental role in gene transfer, gene expression and genetic recombination of the host. Plasmid DNA mostly owns the covalently closed circular structure, while few bacteria including LAB (*Lb. gasseri*) have linear plasmids too [59]. The role of plasmid may not be indispensable for growth and survival of LAB under a favorable environment, but often they are attributed to overcome many hurdles and to compete better in their ecological niche. Examples of few plasmids encoded traits that potentially contribute to better environmental adaptation of LAB are genes responsible for bacteriocin production, amino acid or sugar transportation, hydrolysis of protein, restriction-modification systems, antibiotic resistance, and bacteriophage resistance [1, 78].

LAB plasmids also carry genes responsible for the degradation of casein, acidification by lactic acid, and production of flavor compounds and exopolysaccharides (EPS) contributing to the desired flavor and texture of the fermentation product and enabling optimal growth of strains in milk [62, 105]. However, in many cases, plasmids remain cryptic, contributing no phenotypic traits. Loss of plasmid during successive propagation of bacteria and variability in plasmid copy number due to HGT are the problems responsible for the inconsistent performance of LAB, resulting into industrial loss in many cases [26]. Safety and compatibility are important considerations for genetic manipulation of LAB, intended for food production. Many cryptic plasmids are preferred to be used as selective marker systems in vector construction, as some plasmid-encoded functions have been discovered in cryptic plasmids originating from *Lactobacillus*, *S. thermophilus*, and *Pediococcus* spp. that are used as selective marker systems in vector construction [81].

Mode of plasmid replication influences remarkably the stability and segregation pattern of plasmid DNA [9]. Circular plasmids undergo mainly either rolling circle replication (RCR) or theta replication, and in some cases strand displacement, while linear plasmids may replicate either by virus-like processes or by protein priming [59].

RCR is the most common type of plasmid replication mechanism in Gram-positive bacteria, including LAB. For RCR, a *rep* gene encodes Rep initiation proteins, a double-stranded origin, *ori*, and a single strand origin, *sso*. Rep protein nicks leading strand at *ori* site thus initiating replication of leading strand. After one replication cycle gets over, Rep releases leading strand by a second nick. Replication is initiated in lagging strand at *sso* site. RCR plasmids are relatively small and are broad host range molecule. RCR plasmids decide plasmid copy number. RCR produces single-strand DNA and thus influence plasmid incompatibility [73].

Theta Replication does not produce any ssDNA. It involves strand separation at specific loci followed by bidirectional DNA replication to synthesize both leading and lagging strands. In contrast to RCR plasmids, theta-type plasmids can be more suitable to construct cloning vector as they replicate by means of a double-stranded rather than a single-stranded replication intermediate, which results in better structural stability, fit for inserting large heterologous DNA fragments [20, 107].

10.3.2.1.1 Diversity in LAB Plasmid

LAB plasmids are extremely diverse in terms of size ranging from 0.87 kb to more than 250 kb, copy number from 1 to more than 100 plasmids per cell, and phenotypes conferred to their hosts. Plasmid DNA plays the most vital functionality in *Lactobacillus*, *Lactococcus*, *Leuconostoc*, *Oenococcus*, *Pediococcus*, *Streptococcus*, *Tetragenococcus*, and *Weissella* genera [26].

Lactobacilli plasmids have shown wide variation in size, copy number, and functionality. In recent days, *Lb. plantarum* strain 16 and *Lb. brevis* KB290 have been reported to have the largest plasmid complement [19]. Further study revealed that all the plasmids present in *Lb. brevis* KB290 have putative replication systems, including *grep ABC* genes for essential plasmid replication to allow co-existence of multiple plasmids. Mega-plasmids of sizes ranging from 120–490 kb are found in *Lb. salivarius*, *Lb. acidophilus*, *Lb. hamsteretc* [22, 53]. A large number of vectors has been designed from Lacobacilli, including shuttle vector [26]. These shuttle vectors provide efficient genetic tools for DNA cloning and heterologous gene expression in LAB.

Most *Lc. Lactis* strains usually contain 4–7 plasmids, which range in size from 0.87 kb to more than 80 kb. The plasmid profile analysis of *Lactococcus* strains has revealed that dairy strains of *Lactococcus* species have more number of plasmids as compared to non-dairy *Lactococcus* species, mostly with a size of less than 10kb [51]. More than 86 lactococcal plasmids have been completely sequenced and about 14 partially sequenced, which are available in public database [http://www.ncbi.nlm.nih.gov] till date. A phylogenetic analysis has been conducted with RCR protein Rep and based on analysis, Lactococcal plasmids have been categorized broadly in two groups [26].

Plasmid is one of the important molecules responsible for HGT resulting into loss or gain of any trait to the host cell. It was noticed that pEOC0, a plasmid possessed by *P. acidilactici* NCIMB 6990, contained a streptomycin resistance gene *aadE* that had 100% identity to an *aadE* gene found in Gram-negative bacterium *Campylobacter jejuni* plasmid. The observation helped to explain how a Gram-negative intestinal pathogen became streptomycin-resistant by gaining the corresponding gene from a Gram-positive food-grade organism through horizontal transfer of plasmid [66].

10.3.2.2 TRANSPOSONS

Transposons are MGE, capable of moving from one location to another in the genome. Transposons are common mobilome for both eukaryotic and prokaryotic organisms and play a significant role in evolution, transgenesis, and mutagenesis. In eukaryotic cells, both RNA transposons (retrotransposons) and DNA transposons are found while prokaryotic cells mainly contain the latter one [64]. Mainly two types of DNA transposons are found in dairy LAB.

  a. **Composite Transposons:** It consists of a protein-coding region flanked by complete identical or non-identical ISs on both sides. Composite transposons may contain more than one protein-coding regions, and the entire stretch of composite transposons transposes as a single unit. Many well-known bacterial composite transposons carry genes for antibiotic resistance or other useful properties. Three of the best known such transposons are: Tn5 (kanamycin resistance), Tn9 (chloramphenicol resistance), and Tn10 (tetracycline resistance). Composite transposons are responsible for antibiotic resistance spread in Enterococci and Streptococci [18].
  b. **Conjugative Transposons:** These are larger and more complex mobile elements as compared to composite transposons. Gram-positive bacteria may have conjugative transposons of 18 to 70 kb. Autonomous replication of broad host range Conjugative transposons of Enterococci *Tn916* has been recently identified [101]. Conjugative transposons are capable of excision from and integration to chromosomal or plasmid DNA. After getting excised from any native structure, the remaining part is converted into plasmid like covalently closed circular DNA, incapable of self-replication, and able to conjugate in single-stranded form into another cell. While

integration is aided by transposon encoded integrase protein, its excision is influenced by transposon's *xis* gene product. A conjugative transposon ICESt1, with an approximate size of 35.5 kb, has been identified in *S. thermophilus*. ICESt1 has been found to have a copy of the lactococcal element IS981, suggesting HGT between *Lc. lactis* and *S. thermophilus* [15].

10.3.2.3   INSERTION SEQUENCES (IS ELEMENTS)

Insertion sequences is a simple transposable element, composed of a short DNA sequence, encoding transposase gene, which is flanked by inverted repeats on both ends. They are relatively small in size (<2.5 kb), phenotypically cryptic, and capable of inserting at multiple sites [84]. The IS elements of LAB range from 0.8 to 1.5 kb in size with 16–40 bp inverted repeats. The first IS element identified in dairy LAB was ISL1 in *Lb. casei*, which was horizontally transferred to virulent bacteriophage φFSV [82]. Many IS elements have been identified from plasmid and chromosome of LAB. In LAB genome, the presence of IS elements in the proximity of plasmid-borne genes encoding important fermentation properties (like citrate utilization, phage resistance, lactose utilization, etc.) suggests that IS may play an important role to help LAB to adapt milk environment [13].

10.3.2.4   INTRONS

Introns, the parts of eukaryotic pre-m-RNA, are spliced immediately after its transcription to produce mature m-RNA. Group I and group II introns act as ribozymes to catalyze self-splicing reaction. The spliced introns then behave like MGE to be inserted into a similar intronless allele. This process called 'homing' is highly flexible; and presence of introns have been identified in t-RNA anticodon part in case of eubacteria and eukaryotic organelles, while phage introns have been noticed in different genes [35].

• The presence of Group I introns in diverse genetic systems (including eukaryotic mitochondria, chloroplasts, bacteriophage of Gram-positive and Gram-negative bacteria and eubacterial genome) suggests a possible evolutionary relationship among eukaryotes and prokaryotes [46]. The mechanism of homing is initiated by the group I intron encoded ds-DNA endonuclease that cleaves on a cognate element of

double-stranded intronless DNA followed by an exonuclease digestion to make a gap on recipient molecule. The gap is filled by an
intron, thus resulting into the simultaneous conversion of adjacent
exon markers degraded by exonuclease activity. Such change in the
exon sequence may alter the functionality of the recombined genome.
Hence, Group I intron mobility may restrict only within multi-copy
genomes like bacteriophage, mitochondria, or chloroplast [65].
Among LAB, Group I intron have been identified in virulent bacteriophage of *S. thermophilus, Lc. Lactis* [58, 35].

- Group II introns are bigger in size as compared to Group I counterpart,
and contain ORFs that code for multi-domain proteins. Group II introns
encode reverse transcriptase (RT) and endonuclease that plays a role
in genetic mobility reaction, while maturase activity is thought to be
responsible for RNA splicing. Group II introns perform a homing activity
in intronless allele through RT-dependent pathway. Pre-m-RNA acts as
a template to form cDNA, which is then spliced by ribozyme activity of
maturase followed by integration of intron to a recipient DNA molecule.
Group II homing may also be performed by the process of reverse
splicing too. In the latter case, the spliced intron may be attached with
a stretch of mRNA, followed by reverse transcription to form cDNA,
which acts as a recombination substrate [95]. In this mechanism, DNA
homology or generalized host recombination function are not required.
Group II intron L1.1trB shows some advantages over Group I intron,
such as [37]:(1) Domain IV of L1.1trB is not essential for splicing or
transposition and thus can carry foreign DNA of more than 1kb size;(2)
Group II introns have relatively relaxed specificity; and (3) Chances of
exon co-conversion during transposition. Owing to these advantages,
Group II intron L1.1trB may be a popular choice for using as a vector
for genetic engineering [37].

10.3.2.5   BACTERIOPHAGE DNA

Bacteriophage, the virus of bacterial cells, plays a crucial role in the economic
viability of cheese and other fermented dairy industries, being one of the major
causes of starter failure. These infectious particles contain single or double-
stranded nucleic acid (DNA or RNA) with varying size within a protein
coat called capsid. According to the International Committee on Taxonomy
of Viruses (ICTV), all the phages infecting dairy starter cultures belong to
Caudovirales order. LAB, infecting phage of Siphoviridae family, have long

noncontractile tails while phages of Podoviridae family possess very short non-contractile tails. Phages belonging to the Myoviridae family have contractile tails with central tube [57]. In recent years, a remarkable advancement in DNA sequencing protocols enables faster progress in phage genome sequencing. Consequently, the huge database has been generated, which immensely helps in explaining LAB-infecting phage biodiversity, taxonomy, and evolution. Presently comparative genome analysis is employed to derive phage taxonomy.

The genome structure of LAB bacteriophage possesses linear and double-stranded characteristics of the DNA molecule, with either cohesive end or presenting terminal redundancy and circular permutation. Essential genes present in phage are genes encoding replicase that exploit host cell replication machinery and genes responsible to produce proteins that insert and package phage DNA into the capsid. Virulent or lytic bacteriophage ORFs are generally transcribed from a common strand and lyse the cell vigorously, while an alternative cluster of genes become functional to allow temperate phage genome to be integrated into the host chromosome silently and replicate with it as a prophage [55]. Temperate bacteriophages are prone to HGT within commensal organisms [7].

As Lactococci is a most popular dairy starter, Lactococcal phages have been extensively studied. Most of the Lactococcal phages can be classified under three groups 936, c2, and P335 [57]. P335shows remarkable DNA homology between lytic and lysogenic counterparts [32]. More than 231 phages have been reported from Lacotobacilli family, mostly existing in lysogenic form [99]. Based on mode of packaging and structural protein composition, S. thermophilus bacteriophages can be divided into two distinct groups: pac-type containing 41, 25 and 13 kDa proteins; and cos-type with 32 and 26 kDa proteins [61].

10.4   STRATEGIES FOR GENETIC IMPROVEMENT OF STARTER CULTURES

Improved quality of starter strain is the primary bio-capital for a well-off fermentation food industry. Wild strains of starter organisms often require a modification to perform optimally for industrial production. Interventions are also required frequently to reduce or eliminate unwanted properties of starter cultures. Knowledge of genome sequences of a large number of LAB, and epoch-making advancement in genetic tools encourage biotechnologists for genetic manipulation of starter organisms to improve their phenotypic and metabolic traits. Genetically modified organisms must be of 'Food Grade' and approved as GRAS (Generally Recognized as Safe) by FDA–USA. Though no legal definition of 'Food grade' has been laid officially, food-grade GMO can contain DNA from the same species, or genes from other homologous GRAS food microorganisms. In both the cases, the use of antibiotic resistance genes as selectable markers is not allowed [50]. Hence, conventional recombinant DNA (rDNA) technology is no longer in use for making any industrial strain. Rather, random mutagenesis and natural genetic exchange methods are relied upon.

Before discussing about various genetic improvement approaches, the natural methods of gene alteration and genetic exchange processes are overviewed in this section.

10.4.1   MUTATION

Mutation is an important factor for evolutionary divergence of microbial starter cultures. Natural mutation occurs at low rate until organisms are exposed to any mutagenic agents. Dairy starter cultures, especially LAB, have shown marked differences in their physiological and metabolic parameters due to alteration in their genetic makeup as a result of evolutionary advancement. Reduction in genome size of *Lactobacillales* from its ancestral Bacilli is one of the important examples of genetic divergence [56].

Genome sequence analysis suggested that gene loss was predominantly due to adaptive mutation of organisms into nutrient-rich food environment rich in protein and carbohydrate, like milk matrix. Presence of pseudogenes is another important reason of genome size reduction of dairy LAB. Yogurt cultures *Lb. bulgaricus* and *S. thermophilus* are reported to have 270 and 182 pseudogenes, respectively, indicating special adaptation to those in nutrient-rich milk media [78]. As mutation results into random alteration, selection of mutant of interest may be difficult. This method may be time-consuming and may result adversely, if any deleterious mutation takes place spontaneously.

10.4.2   NATURAL GENE TRANSFER METHOD: HORIZONTAL GENE TRANSFER (HGT)

The mosaic structures of genetic material in bacterial genomes imply a constant flow of genetic information. Although they do not undergo sexual reproduction, yet bacterial cells perform processes in which genetic material from one cell or environment can be incorporated into another cell forming recombinants. This HGT includes: conjugation, which requires the cell to cell contact, transduction–bacteriophage–facilitated transfer of genetic information, transformation, and protoplast fusion. Usually, the genes to be transferred are part of MGE, which are presented in this chapter.

10.4.2.1   TRANSDUCTION

Transduction is bacteriophage mediated natural gene transfer process among closely related bacterial species. The process is accomplished while part of a bacterial genome, packed into its phage during phage replication. This gene transfer method was first reported in *Salmonella* spp. by Norton Zinder and Joshua Lederberg. Transduction of two important chromosomes linked attributes like tryptophan biosynthesis and streptomycin resistance markers in *Lc. lactis* by its virulent phage was the first testimony of such mechanism taking place in LAB, according to Sandine and his co-workers [77]. Initially, this natural gene transfer mechanism was attempted to improve the metabolic quality of few *Lc. Lactis* strains by incorporating industrially important plasmid linked traits encoding lactose fermenting ability (Lac$^+$) and proteinase activity (Prt$^+$) into chromosomal DNA and in many others. However, this process could not gain wide popularity for genetic improvement because HGT by this mechanism can only be restricted within a narrow host range. More competent gene transfer methods were also explored simultaneously later.

10.4.2.2   CONJUGATION

Conjugation, a natural form of gene transfer between any two related or unrelated microbial species through physical contact between live donor and recipient cell, is one of the most important mechanisms of evolution and diversity in the microbial world [3]. Usually, conjugation machinery involves plasmids and conjugative transposons. Plasmids are categorized based on their genetic organization and mode of transfer. The conjugative plasmids, having a larger size (>30kb), contain all requirements for self-transmission, while mobilizable plasmids (being smaller in size (15 kb)) contain only essential requirement for conjugation, i.e., *oriT* [25], and require *trans*acting gene products from a conjugative element as a donation. Another type of plasmids is there, which lacks even *oriT*, and thus, it is transferred by co-integration with another conjugative element [21].

Since its discovery by Joshua Lederberg and Edward Tatum in 1946, this natural gene transfer process has been studied extensively as an efficient tool for genetic improvement of microbial starter cultures, mainly because no regulatory or ethical agency can curb the use of microorganisms, genetically modified by natural process. Probably, it is the first successful attempt to introduce a conjugative lactococcal plasmid pTRK 2030, encoding restriction/modification and abortive phage infection defense mechanisms, into cheddar cheese starter bacteria [76], and was enough to encourage in the dairy industry.

Natural conjugation mechanism in *Lc. lactis* has been efficient to enhance many other industrially important attributes like lactose and citrate utilization, bacteriocin production [70]. Conjugation has been a more efficient technique for gene transfer by introducing an interspecific and intergeneric wide host range, erythromycin, and lincomycin resistance plasmid pAMβ1 into a variety of bacterial species including *Lactococcai*, *Lactobacilli*, and *Streptococci* [93]. Antibiotic resistance and virulence factors, in the case of pathogenic bacteria, are also transferred by the process of conjugation [67].

Actual DNA translocation process differs considerably between Gram-positive and Gram-negative bacteria during conjugation. In Gram-negative bacteria, stable mating pair formation during conjugation takes place by cell-cell contact through sex pili formation by donor cell. Gram-positive bacteria are unable to produce sex pili, perhaps due to its more rigid peptidoglycan layer in the cell wall; hence, they require other mechanisms for stable mating pair formation.

Different models explaining the strategy of DNA translocation into Gram-positive bacteria have been summarized by Goessweiner-Mohr et al. [41]. Almost all Gram-positive bacteria possess multi-protein complex, known as type IV secretion system (T4SS) across the gram-positive cell envelope, while *Streptomycetes* spp. may have different mechanisms, resembling bacterial cell division or spore formation mechanism [45]. A third mechanism, sex pheromone-responsive DNA transfer system, has been identified in *Enterococcus faecalis* [30].

Among several plasmids present in *E. faecalis*, pAD1 and pCF10 are well studied and found to have hemolysin and tetracycline resistance property. Pheromone induced mating pair formation by these conjugative plasmids occur through protein-protein interaction involving plasmid-encoded aggregation substance (AS) on donor cells and binding substance on the recipients [31]. It has been reported that *Lc. Lactis*ML3 and 712 also secrete protein, CluA, homologous to enterococcal surface proteins involved in conjugation, and takes part in mating pair formation [89]. CluA, Lactococcal sex factor has the potential ability for intergeneric conjugation between Lactococci and other LAB. Transfer of non-conjugative DNA can also occur in LAB by Lactococcal sex factor through conduction, while plasmid co-integration can take place through ISS1 elements present on lactose plasmid [1].

The mechanism of conjugation especially in LAB has been considered as one of the safest and promising technique for genetic improvement of food-grade organisms.

10.4.2.3 TRANSFORMATION

Transformation is another lateral gene transfer process by which a piece of foreign DNA is received, integrated, and expressed by a cell at a condition called competence [52]. The transformed foreign DNA may be a part of a chromosomal DNA, plasmid DNA, or phage DNA. The state of 'competence' is achieved, when a set of genes in recipient cell produces enough competence stimulating peptides to enable cell for allowing foreign DNA to enter the cell. *Lc. lactis* have a complete set of competence gene [10], while many LAB may lack it.

Electroporation technique is applied to make a voltage gradient across membrane, reducing its integrity and thus allowing foreign DNA particle to enter the cell. The process of electro-transformation has been applied for strain improvement of LAB for achieving successful gene transfer [52]. Many research studies have been conducted to improve the process of electroporation too. As transformation efficiency in LAB is quite low (only 10^4/ μg of exogenous DNA) as compared to *E. coli* (10^8/μg of exogenous DNA), introduction of single-strand DNA by electrotransformation has shown a promise in few *lactobacilli* strains [96].

10.4.2.4 PROTOPLAST FUSION

Protoplast fusion method was first attempted successfully in LAB for genetic exchange between a few strains of *Lc. lactis*. The method of protoplast fusion includes:(i) Enzymatic degradation of cell wall without affecting the viability of the cell; (ii) Intercellular membrane fusion by Polyethylene glycol (PEG);(iii) New cell-wall formation around fusants, yielding hybrid cell. However, the method could not show any encouraging promise for genetic improvement of LAB mainly because of lack of stringent protoplast formation and regeneration condition protocol for individual strain. Recently, researchers have reinitiated the proptoplast fusion method between strains of *Lb. reuteri* and have reported its efficacy [74].

10.5   IMPROVEMENT OF ACTIVITY OF STARTER CULTURES: RECENT ADVANCEMENTS IN GENETIC ENGINEERING

Knowledge of global gene regulation, in-depth understanding of genetic components of starter bacteria, regulation of HGT, and genome sequencing of LAB are the basis of employing techniques for genetic improvement of starter cultures. It is prerequisite to consider safety and sustainability of genetically modified organisms in milk industry. Profound understanding of genetic regulation of dairy starter cultures (in many cases advantageous use of natural gene transfer technique for strain development, and selection of food-grade vectors like LAB plasmids) may certainly ensure GRAS status of genetically modified starter organisms. Recombination mediated genetic engineering (Recombineering) is a new approach for safe, food-grade genetic manipulation [97].

The success of industrial production of fermented milks especially variety of cheeses, which are kept for curing for development of typical flavor and aroma by starter and non-starter LAB, hugely depends on the appropriate choice of organisms. Genomic analysis can help to select starter strains based on metabolic characteristics, safety analysis, etc. The volume and extent of genetic interventions are vast to improve starter strains worldwide, and therefore, it is beyond the scope of this chapter to be discussed. Few important genetic strategies for advancements of dairy starter cultures and LAB, bacteria are highlighted.

10.5.1   COMBINED 'OMIC' APPROACH TO REVEAL STRESS RESPONSE: MECHANISMS OF LAB

In the post-genome sequencing era, the introduction of computational genomics has brought a revolutionary change in the approach of the application of microbial technology. Anthology and annotation of the whole genome sequence of a large number of LAB, many pathogenic organisms and bacteriophages can enable researchers to understand molecular mechanisms of microbial physiology, metabolism, and immunity. Comparative genomics also provides a powerful tool: (1) for studying evolutionary changes among organisms; (2) to identify genes that are conserved or common among species; an (3) genes that give each organism its unique characteristics.

Now-a-days, popular approach among biotechnologists include [102]: (1) using knowledge of comparative and functional genomics, metabolomics, transcriptomics, and proteomics; (2) in some cases combined approach to use combination of these techniques; (3) to explain stress tolerance mechanisms of starter LABs in milk environment as well as in gut.

Adaptation of *S. thermophilus*LMD9 in the dairy environment has been studied by comparative genomics approach by Goh and his co-workers, who reported the stress tolerance mechanism of this strain under pH and oxidative stress [42]. *Lb. helveticus* 5463 and DPC4571 have been studied to establish mechanisms of their adaptation to gut and dairy niche, respectively by the help of comparative genomics [79]. Combined knowledge of metabolic and proteomics have been employed to explore acid stress response of *Lc. Lactis* [17]. More such studies are being carried out across globe to reveal molecular mechanisms of stress response by different starter cultures. In future, this information will be the yardstick to design protocols by engineering the dairy starter cultures with better industrial adaptability.

10.5.2 ENHANCED FLAVOR PRODUCTION

Metabolic engineering for increased diacetyl production by lactococcal strains involves a genetic manipulation strategy. Diacetyl and its derivatives, important flavoring components of fermented milk products, are being produced from pyruvate via α-acetolactic acid (main precursor of diacetyl). Enhanced production of diacetyl can be ensured in an *ldh* (lactate dehydrogenase) encoding gene; and deficiency of mutant strain of *Lc. Lactisssplactisbiovar. diacetylactis*. *Ldh* results into overproduction of α-acetolactic acid, which in turn enhances diacetyl production [38]. Inactivation of *aldB* gene, responsible to produce acetoin from α-acetolactic acid, is another popular approach for enhancing production of diacetyl flavor [48].

Optimum ratio of $NADH: NAD^+$ co-factor influences the direction of red-ox reaction in the cell. Overproduction of *nox* gene product (NADH oxidase) can diminish NADH but will increase NAD^+ within the cell of lactococcal organisms, thus redirecting pyruvate from NADH dependent homo-lactic pathway to NADH independent α-acetolactic acid production pathway, and ensure more diacetyl/acetoin production [85].

Metabolism of amino acids plays a vital role to develop the characteristic flavor of different cheeses and fermented dairy products, because amino acids are important precursors of flavoring components [34]. Often amino acid metabolism genes are incomplete in facultative anaerobic LAB that lacks a complete TCA cycle cascade. The mechanism of flavor producing pathways from amino acids has been identified in many LAB, using the genetic and biochemical tool [43]. Flavor development mechanism by LAB involves glycolysis, proteolysis, and lipolysis, of which proteolytic system contributes significantly [80].

Certain LAB can utilize citrate for production of diacetyl and acetoin flavor [63]. Casein (the milk protein) is an important substrate of proteolytic action of LAB. The proteolysis cascade starts with casein degradation by two types of cell-envelope proteinases: CEP and Prt [88]. Wide diversity in CEP characteristics have been found among 213 Lactobacillus and associated genera [90]. Total of 60 genes for CEPs were reported in the study and its high correlation with phylogenetic clades was observed. Mostly LAB possesses one CEP that can act on different substrates, producing various products of proteolysis, which contribute to the delicate flavor of the cheese. However, the presence of four different genes *in Lb. helveticus* may explain the higher proteolytic activity of this organism [14].

Another manipulation for enhancement of flavor production by LAB involves reverse pathway engineering (RPE). A small molecule is targeted and looked for enzymatic or chemical reactions that can track its precursor. Unknown metabolic pathways can be revealed with the help of retrosynthesis and genomic information [54].

10.5.3   GENOME-SCALE METABOLIC MODEL (GSMM)

The scope of metabolic engineering has broadened due to steady and dynamic advancement of genome sequencing of LAB and related information management in the form of publicized databases. A holistic approach has been adapted to propose dynamic design strategy of metabolic engineering through integrating various biological data, including genomic scale metabolic model (GSMM), transcriptomic, and metabolomics data [94]. GSMM has potential to study microbial metabolism and strain improvement [75, 104]. The first GSMM was developed for *Lc. lactis* ssp. *lactis* IL1403 with 621 reactions and 509 metabolites [68]. This model enabled to design reactions to achieve enhanced production of metabolites with minimal nutrients. However, GSMM has been constructed for a few strains of LAB [102].

10.5.4   ENHANCED PHAGE RESISTANCE

Bacteriophage attack to LAB poses a serious threat to the technological as well as economic viability of fermentation industry. Destruction of phage genome by host restriction-modification system and abortive infection mechanisms are well adapted, in order to control phage different strategies like prevention of phage adsorption, blocking the entry of phage DNA [39]. Bacteriophage resistance of Lactococcal species is mainly plasmid linked. Conjugation of such plasmids can ensure phage resistance in a population. Continuous conjugation may enable a bacterial strain to possess more than one plasmid containing complementary phage defense mechanism. Rotation of different phage resistance and abortive infection mechanisms within a single host cell can efficiently remove phages from the system [49].

Efficiency of this strategy may be limited as it depends on the natural mobility of plasmids. In order to overcome this uncertainty, plasmids containing complementary phage defense mechanisms are introduced within a host cell by electroporation. Another novel mechanism has been isolated in almost all bacteria, including LAB that may offer improved immunity against foreign DNA like plasmid or phage.

Clustered regularly interspaced short palindromic repeats (CRISPR) along with CRISPR associated gene (*cas*) have been identified in the genome of bacteria [5, 6]. CRISPR, a series of short palindromic sequences and intermittently separated by spacer region, is located adjacent to *cas* gene. The spacer region bears significant homology with foreign DNA. This finding helps to explain that CRISPR accumulates homologous sequence of invading foreign genes between two repeats followed by its transcription into small interfering RNA called CRISPR-RNA. The later molecule may help cellular protein complex to identify and degrade foreign DNA [5, 28].

Development of strategy to improve CRISPR activity in a wide range of industrially important starter strains may ensure remarkable improvement in Phage resistance management of industries. Hyper variable spacer region of CRISPR may be used for strain typing studies [5].

10.5.5 FOOD GRADE VECTOR SYSTEM

Suitable vector selection for gene cloning or gene delivery is a great challenge for food biotechnologists. Main criteria for selecting vector must include the following considerations:

- Presence of a phenotype that may act as selective markers in the transformed cell;
- Presence of 'multiple cloning site' on hosts DNA to allow easy insertion of foreign DNA without causing any damage to functional elements;
- Smaller size to ensure easy transformation into host cells.

The extensive use of replicative extra-chromosomal vectors and integrative vectors were common practices for genetic improvements of industrial strains. However, those traditional gene delivery systems cannot guarantee the safety aspect of recombinant DNA due to use of antibiotic resistance gene as selective markers.

Food grade vectors must contain some food-grade selection systems without any adverse health interference, and foreign DNA from the homologous host or GRAS organisms can only be used. Auxotrophic complementation, resistance to LAB bacteriocins, resistance to cadmium or copper, property of thermos-stability and ability to ferment new carbohydrates may be the choice of new generation selective markers [91, 100]; while presence of inducible gene expression system in food-grade vector may be another additional safety measure. Inducible gene expression may be advantageous when the product is harmful for the expression host [71].

In case of constructing food-grade gene expression systems, the inducer (that is used and added for overexpression) obviously has to be non-toxic, safe or food-approved. NICE (Nisin-Controlled Expression) system has been used popularly as an inducible gene expression system in Gram-positive bacteria [60]. Use of quorum sensing approach for induced gene expression has also been introduced in lactobacilli such as *Lb. sakei*, *Lb. plantarum* [86, 87]. LAB cryptic plasmids may have potential to be used as food-grade cloning vectors. Vectors based on these plasmids have been developed and used in the cloning and expression of several heterologous genes [81]. Remarkable advancements in food-grade cloning systems may come up in the future.

10.5.6   SAFETY ASSESSMENT BY GENOMIC ANALYSIS

Safety assessment by the genomic analysis of starter organisms provides stronger ground for the selection of a suitable starter. *S. thermophilus*, a viridian streptococci, is used extensively for yogurt production. Comparative genomic analysis of two *S. thermophiles* strains revealed absence or inactivation of virulence-related genes, commonly present in other viridian's streptococci [11], thus substantiating their safe use in the food industry. Presence of 51 antibiotic resistance-associated genes, 126 virulence-associated genes, and 23 adverse metabolism-associated genes were found in *Lb. plantarum* JDM1 in a comprehensive genetic study [106].

Potentially hazardous biogenic amines (like tyramine, histamine, etc.) are produced mainly by amino acid decarboxylase activity of microorganisms and pose a serious threat to the fermented food consumers [4]. A real-time quantitative PCR assay has been designed targeting histidine decarboxylase gene sequence to quantify histamine producing strains in milk and cheese [33]. Safety assessment of industrial strains of starter cultures must be examined periodically by genomic analysis to check whether any antibiotic resistance gene or virulence-related genes are acquired by HGT.

10.6 SUMMARY

The chapter outlines the present trend of the activities related to genetic improvement of dairy starter cultures; scope of genetic study to improve dairy starter strains for industrial application; basis of genetic characterization of starter cultures with few examples; various methods employed for genetic improvement of dairy starter organisms; and different strategies and recent approaches for accomplishing genetic improvement of dairy starter cultures. The information in this chapter will be useful to have a glimpse of modern approaches that are used for exploring food-grade dairy starter and LAB cultures by genetic manipulation.

KEYWORDS

- **bacteriophage**
- **conjugation**
- **CRISPR**
- **electrotransformation**
- **genome-scale metabolic model**
- **horizontal gene transfer**
- **mobile genetic element**
- **protoplast fusion**
- **recombineering**
- **reverse pathway engineering**
- **transposon**

REFERENCES

1. Ainsworth, S., Stockdale, S., Bottacini, F., Mahony, J., & Van Sinderen, D., (2014). The *Lactococcuslactis* plasmidome: Much learnt, yet still lots to discover. *FEMS Microbiology Review, 38*, 1066–1088.
2. Alduina, R., & Pisciotta, A., (2015). Pulsed field gel electrophoresis and genome size estimates. *Methods in Molecular Biology, 123*(1), 1–14.
3. Arber, W., (2000). Genetic variation: Molecular mechanisms and impact on microbial evolution. *FEMS Microbiology Review, 24*(1), 1–7.
4. Arena, M. P., Russo, P., Capozzi, V., Beneduce, L., & Spano, G., (2010). Effect of abiotic stress conditions on expression of the *Lactobacillus brevis* IOEB 9809 tyrosine decarboxylase and agmatinedeiminase genes. *Annals of Microbiology, 61*, 179–183.
5. Barrangou, R., & Horvath, P., (2012). CRISPR: New horizons in phage resistance and strain identification. *Annual Review of Food Science and Technology, 3*, 143–162.
6. Barrangou, R., Fremaux, C., & Deveau, H., (2007). CRISPR provides acquired resistance against viruses in prokaryotes. *Science, 315*, 1709–1712.
7. Baugher, J. L., Durmaz, E., & Klaenhammer, T. R., (2014). Spontaneously induced prophages in *Lactobacillus gasseri* contribute to horizontal gene transfer. *Applied Environmental Microbiology, 80*(1), 3508–3517.
8. Berg, O. G., & Kurland, C. G., (2002). Evolution of microbial genomes: Sequence acquisition and loss. *Molecular Biology and Evolution, 19*(12), 2265–2276.
9. Biet, F., Cenatiempo, Y., & Fremaux, C., (1999). Characterization of pFR18, a small cryptic plasmid from *Leuconostocmesenteroides* ssp. mesenteroides FR52, and its use as a food grade vector. *FEMS Microbiology Letters, 179*(2), 375–383.
10. Bolotin, A., Mauger, S., Malarme, K., Ehrlich, S. D., & Sorokin, A., (1999). Low-redundancy sequencing of the entire *Lactococcuslactis* IL1403 genome. *Antonie Van Leeuwenhoek, 76*, 27–76.
11. Bolotin, A., Quinquis, B., Renault, P., Sorokin, A., & Ehrlich, S. D., (2004). Complete sequence and comparative genome analysis of the dairy bacterium *Streptococcus thermophilus*. *Nature Biotechnology, 22*(12), 1554–1558.
12. Bolotin, A., Wincker, P., Mauger, S., & Jaillon, O., (2001). The complete genome sequence of the lactic acid bacterium *Lactococcuslactis* ssp. lactis IL1403. *Genome Research, 11*(5), 731–753.
13. Bourgoin, F., Pluvinet, A., Gintz, B., Decaris, B., & Guedon, G., (1999). Are horizontal transfers involved in the evolution of the *Streptococcus* thermos-sphilusexopolysaccharide synthesis loci? *Gene, 233*(1), 151–161.
14. Broadbent, J. R., Cai, H., Larsen, R. L., & Hughes, J. E., (2011). Genetic diversity in proteolytic enzymes and amino acid metabolism among *Lactobacillus helveticus* strains. *Journal of Dairy Science, 94*, 4313–4328.
15. Burrus, V., Roussel, Y., Decaris, B., & Guedon, G., (2000). Characterization of a novel integrative element, ICE-*St1*, in the lactic acid bacterium *Streptococcus thermophilus*. *Applied Environment and Microbiology, 66*, 1749.
16. Cancilla, M. R., Powell, I. B., Hillier, A. J., & Davidson, B. E., (1992). Rapid genomic fingerprinting of *Lactococcuslactis* strains by arbitrarily primed polymerase chain reaction with 2p and fluorescent labels. *Applied and Environmental Microbiology, 58*, 1772–1775.

17. Carvalho, A. L., Turner, D. L., Fonseca, L. L., & Solopova, A., (2013). Metabolic and transcriptional analysis of acid stress in *Lactococcuslactis*, with a focus on the kinetics of lactic acid pools. *PLoS One*, *8*(7), E-article ID68470.

18. Chaffanel, F., Bourgoin, F. C., Libante, V., Bourget, N. L., & Payot, S., (2015). Resistance genes and genetic elements associated with antibiotic resistance in clinical and commensal isolates of *Streptococcus salivarius*. *Applied and Environmental Microbiology*, *81*(12), 4155–4163.

19. Chen, C., Ai, L. Z., Zhou, F. F., Ren, J., Sun, K. J., Zhang, H., Chen, W., & Guo, B. H., (2012). Complete nucleotide sequence of plasmid pST-III from *Lactobacillus plantarum* ST-III. *Plasmid*, *67*, 236–244.

20. Cho, G. S., Huch, M., Mathara, J. M., Van Belkum, M. J., & Franz, C. M., (2013). Characterization of pMRI 5. 2, a rolling-circle-type plasmid from *Lactobacillus plantarum* BFE 5092 which harbors two different replication initiation genes. *Plasmid*, *69*, 160–171.

21. Chris, S. C., & Garcillán-Barcia, M. P., (2010). Mobility of plasmids. *Microbiology and Molecular Biology Review*, *74*(3), 434–452.

22. Claesson, M. J., Li, Y., Leahy, S., Canchaya, C., & Van Pijkeren, J. P., (2006). Multi-replicon genome architecture of *Lactobacillus salivarius*. *Proceedings of the National Academy of Sciences*, USA, *103*, 6718–6723.

23. Claesson, M. J., Van Sinderen, D., & O'Toole, P. W., (2007). The genus *Lactobacillus*–a genomic basis for understanding its diversity. *FEMS Microbiology Letters*, *269*, 22–28.

24. Claesson, M. J., Van Sinderen, D., & O'Toole, P. W., (2008). *Lactobacillus* phylogenomics–towards a reclassification of the genus. *International Journal of Systematic and Evolutionary Microbiology*, *58*, 2945–2954.

25. Cruz, F. D. L., (2012). The role of conjugation in the evolution of bacteria: Chapter 18. In: Kolter, R. R., & Maloy, S., (eds.), *Microbes and Evolutions: The World That Darwin Never Saw* (pp. 133–138). ASM Press, Washington DC.

26. Cui, Y., Hu, T., Qu, X., Zhang, L., Ding, Z., & Dong, Z., (2015). Plasmids from food lactic acid bacteria: Diversity, similarity, and new developments. *International Journal of Molecular Sciences*, *16*, 13172–13202.

27. De Vuyst, L., & Leroy, F., (2007). Bacteriocin from lactic acid bacteria: Production, purification and food application. *Journal of Molecular Microbiology and Biotechnology*, *13*, 194–199.

28. Deveau, H., Garneau, J. E., & Moineau, S., (2010). CRISPR/*Cas* system and its role in phage-bacteria interactions. *Annual Review of Microbiology*, *64*, 475–493.

29. Douillard, F. P., & De Vos, W. M., (2014). Functional genomics of lactic acid bacteria: From food to health. *Microbial Cell Factories*, *13*, S1–S8.

30. Dunny, G. M., (2007). The peptide pheromone-inducible conjugation system of *Enterococcus faecalis* plasmid pCF10: Cell-cell signaling, gene transfer, complexity and evolution. *Philosophical Transactions of the Royal Society B: Biological Sciences*, *362*, 1185–1193.

31. Dunny, G. M., & Leonard, B. A. B., (1997). Cell-cell communication in gram-positive bacteria. *Annual Review of Microbiology*, *51*, 527–564.

32. Durmaz, E., & Klaenhammer, T. R., (2000). Genetic analysis of chromosomal regions of *Lactococcuslactis* acquired by recombinant lytic phages. *Applied and Environmental Microbiology*, *66*, 895–903.

33. Fernandez, M., Del Rio, B., Linares, D. M., Martin, M. C., & Alvarez, M. A., (2006). Real-time polymerase chain reaction for quantitative detection of histamine-producing bacteria: Use in cheese production. *Journal of Dairy Science, 89,* 3763–3769.

34. Fernandez, M., & Zuniga, M., (2006). Amino acid catabolic pathways of lactic acid bacteria. *Critical Reviews in Microbiology, 32*(3), 155–183.

35. Foley, S., Bruttin, A., & Brussow, H., (2000). Widespread distribution of a group I intron and its three deletion derivatives in the lysin gene of *Streptococcus thermophiles* bacteriophages. *Virology, 74*(2), 611–618.

36. Foligne, B., Daniel, C., & Pot, B., (2013). Probiotics from research to market: The possibilities, risks and challenges. *Current Opinion in Microbiology, 16*(3), 284–292.

37. Frazier, C. L., Filippo, J. S., Lambowitz, A. M., & Mills, D. A., (2003). Genetic manipulation of *Lactococcuslactis* by using targeted group II Introns: Generation of stable insertions without selection. *Applied Environmental Microbiology, 69*(2), 1121–1128.

38. Garcia-Quintans, N., Repizo, G., Martin, M., Magni, C., & Lopez, P., (2008). Activation of the diacetyl/acetoin pathway in *Lactococcuslactis* subsp. lactisbv. diacetylactis CRL264 by acidic growth. *Applied Environmental Microbiology, 74,* 1988–1996.

39. Garneau, J. E., & Moineau, S., (2011). Bacteriophages of lactic acid bacteria and their impact on milk fermentations. *Microbial Cell Factory, 10,* S1–S20.

40. Gevers, D., Huys, G., & Swings, J., (2001). Applicability of rep-PCR fingerprinting for identification of Lactobacillus species. *FEMS Microbiology Letters, 205,* 31–36.

41. Goessweiner-Mohr, N., Arends, K., Keller, W., & Grohmann, E., (2013). Conjugation in gram-positive bacteria. *Microbiology Spectrum, 2,* E-article. http://www.asmscience. org/content/journal/microbiolspec/10.1128/microbiolspec.PLAS-0004–2013 (Accessed on 20 July 2019).

42. Goh, Y. J., Goin, C., O'Flaherty, S., Altermann, E., & Hutkins, R., (2011). Specialized adaptation of a lactic acid bacterium to the milk environment: the comparative genomics of *Streptococcus thermophilus* LMD-9. *Microbial Cell Factory, 10,* S15–S22.

43. Gómez de Cadiñanos, L. P., García-Cayuela, T., Yvon, M., Martínez-Cuesta, M. C., Peláez, C., & Requena, T., (2013). Inactivation of the pan-E gene in *Lactococcuslactis* enhances formation of cheese aroma compounds. *Applied and Environmental Microbiology, 79,* 3503–3506.

44. Goupil-Feuillerat, N., Cocaign-Bousquet, M., Godon, J. J., Ehrlich, S. D., & Renault, P., (1997). Dual role of α-acetolactate decarboxylase in *Lactococcuslactis* subsp. *lactis*. *Journal of Bacteriology, 179,* 6285–6293.

45. Grohmann, E., Muth, G., & Espinosa, M., (2003). Conjugative plasmid transfer in gram-positive bacteria. *Microbiology Molecular Biology Review, 67,* 277–301.

46. Haugen, P., Simon, D. M., & Bhattacharya, D., (2005). The natural history of group I introns. *Trends in Genetics, 21*(2), 111–119.

47. Holzapfel, W. H., & Wood, B. J. B., (2014). *Lactic Acid Bacteria: Biodiversity and Taxonomy* (p. 324). Wiley-Blackwell, New York, NY.

48. Hugenholtz, J., Kleerebezem, M., Starrenburg, M., Delcour, J., De Vos, W., & Hols, P., (2000). *Lactococcuslactis* as a cell factory for high-level diacetyl production. *Applied and Environmental Microbiology, 66*(9), 4112–4114.

49. Hyman, P., & Abedon, S. T., (2010). Bacteriophage host range and bacterial resistance. *Advances in Applied Microbiology, 70,* 217–248.

50. Johansen, E., (1999). Genetic engineering: Modification of bacteria. In: Robinson, R., Batt, C., & Patel, P., (eds.), *Encyclopedia of Food Microbiology* (pp. 917–921). Academic Press, London.

51. Kelly, W. J., Ward, L. J., & Leahy, S. C., (2010). Chromosomal diversity in *Lactococcuslactis* and the origin of dairy starter cultures. *Genome Biology and Evolution, 2,* 729–744.

52. Landete, J. M., Arqués, J. L., Peirotén, A., Langa, S., & Medina, M., (2014). An improved method for the electro-transformation of lactic acid bacteria: A comparative survey. *Journal of Microbiological Methods, 105,* 130–133.

53. Li, Y., Canchaya, C., Fang, F., & Raftis, E., (2007). Distribution of megaplasmids in *Lactobacillus salivarius* and other *Lactobacilli. Journal of Bacteriology, 189,* 6128–6139.

54. Liu, M., Bienfait, B., Sacher, O., Gasteiger, J., Siezen, R. J., Nauta, A., & Geurts, J. M., (2014). Combining chemo-informatics with bioinformatics: Silico prediction of bacterial flavor-forming pathways by a chemical systems biology approach "reverse pathway engineering." *Plos One, 9,* e84769.

55. Mahony, J., Ainsworth, S., Stockdale, S., & Sinderen, D., (2012). Phages of lactic acid bacteria: The role of genetics in understanding phage-host interactions and their co-evolutionary processes. *Virology, 434*(2), 143–150.

56. Makarova, K. S., & Koonin, E. V., (2007). Evolutionary genomics of lactic acid bacteria. *Journal of Bacteriology, 189*(4), 1199–1208.

57. Marco, M. B., Moineau, S., & Quiberoni, A., (2012). Bacteriophages and dairy fermentations. *Bacteriophage, 2,* 149–158.

58. McDonnell, B., Mahony, J., Neve, H., Hanemaaijer, L., Noben, J. P., Kouwen, T., & Van Sinderen, D., (2016). Identification and analysis of a novel group of bacteriophages infecting the lactic acid bacterium *Streptococcus thermophilus. Applied and Environmental Microbiology, 82,* 5153–5165.

59. Meinhardt, F., Schaffrath, R., & Larsen, M., (1997). Microbial linear plasmids. *Applied Microbiology and Biotechnology, 47*(4), 329–336.

60. Mierau, I., & Kleerebezem, M., (2005). Ten years of the nisin-controlled gene expression system (NICE) in *Lactococcuslactis. Applied Microbiology and Biotechnology, 68*(6), 705–717.

61. Mills, S., Griffin, C., O'Sullivan, O., Coffey, A., McAuliffe, O. E., Meijer, W. C., Serrano, L. M., & Ross, R. P., (2011). A new phage on the 'Mozzarella' block: Bacteriophage 5093 shares a low level of homology with other *Streptococcus thermophilus* phages. *International Dairy Journal, 21*(12), 963–969.

62. Mills, S., McAuliffe, O. E., Coffey, A., Fitzgerald, G. F., & Ross, R. P., (2006). Plasmids of lactococci-genetic accessories or genetic necessities? *FEMS Microbiol. Rev., 30,* 243–273.

63. Mortera, P., Pudlik, A., Magni, C., Alarcon, S., & Lolkema, J. S., (2013). Ca2þ-citrate uptake and metabolism in *Lactobacillus casei* ATCC 334. *Applied and Environmental Microbiology, 79,* 4603–4612.

64. Munoz-Lopez, M., & Garcia-Perez, J. L., (2010). DNA transposons: Nature and applications in genomics. *Current Genomics, 11*(2), 115–128.

65. Nielsen, H., (2012). Group-I intron ribozymes. *Method in Molecular Biology, 848,* 73–89.

66. O'Connor, E. B., O'Sullivan, O., & Stanton, C., (2007). The pEOC01, a plasmid from *Pediococcusacidilactici,* which encodes an identical streptomycin resistance (*aadE*) gene to that found in *Campylobacter jejuni. Plasmid, 58,* 115–126.

67. Ogier, J. C., & Serror, P., (2008). Safety assessment of dairy microorganisms: The enterococcus genus. *International Journal of Food Microbiology, 126*(3), 291–301.

68. Oliveira, A. P., Nielsen, J., & Forster, J., (2005). Modeling *Lactococcuslactis* using a genome-scale flux model. *BMC Microbiology*, *5*(1), 39.

69. O'Sullivan, O., O'Callaghan, J., Sangrador-Vegas, A., & McAuliffe, O., (2009). Comparative genomics of lactic acid bacteria reveals a niche-specific gene set. *BMC Microbiology*, *9*(1), 50–55.

70. Pedersen, M. B., Iversen, S. L., Sørensen, K. I., & Johansen, E., (2005). The long and winding road from the research laboratory to industrial applications of lactic acid bacteria. *FEMS Microbiology Review, 29*(3), 611–624.

71. Peterbauer, C., Maischberger, T., & Haltrich, D., (2011). Food-grade gene expression in lactic acid bacteria. *Biotechnology Journal*, *6*(9), 1147–1161.

72. Pilar, C. M., Samuel, A., Karola, B., Trinidad, M., Ananias, P., & Jorge, B. V., (2008). Current applications and future trends of lactic acid bacteria and their bacteriocins for the biopreservation of aquatic food products. *Food and Bioprocess Technology*, *1*(1), 43–63.

73. Raha, A. R., Hooi, W. Y., Mariana, N. S., Radu, S., Varma, N. R., & Yusoff, K., (2006). DNA sequence analysis of a small cryptic plasmid from *Lactococcuslactis* subsp. *Lactis* M14. *Plasmid*, *56*, 53–61.

74. Rosander, A., Connolly, E., & Roos, S., (2008). Removal of antibiotic resistance gene carrying plasmids from *Lactobacillus reuteri* ATCC 55730 and characterization of the resulting daughter strain, *L. reuteri* DSM 17938. *Applied Environment and Microbiology, 74*, 6032–6040.

75. Saha, R., Chowdhury, A., & Maranas, C. D., (2014). Recent advances in the reconstruction of metabolic models and integration of Omics data. *Current Opinion in Biotechnology*, *29*, 39–45.

76. Sanders, M. E., Leonhard, P. J., Sing, W. D., & Klaenhammer, T. R., (1986). Conjugal strategy for construction of fast acid- producing, bacteriophage resistant *lactic streptococci* for use in dairy fermentations. *Applied Environmental Microbiology*, *52*, 1001.

77. Sandine, W. E., Elliker, P. R., Allen, L. K., & Brown, W. C., (1962). Symposium on lactic starter cultures, II: Genetic exchange and variability in lactic *Streptococcus* starter organisms. *Journal of Dairy Science*, *45*, 1266.

78. Schroeter, J., & Klaenhammer, T., (2009). Genomics of lactic acid bacteria. *FEMS Microbiology Letters*, *292*, 1–6.

79. Senan, S., Prajapati, J. B., & Joshi, C. G., (2014). Comparative genome-scale analysis of niche-based stress-responsive genes in *Lactobacillus helveticus* strains. *Genome*, *57*, 185–192.

80. Settanni, L., & Moschetti, G., (2010). Non-starter lactic acid bacteria used to improve cheese quality and provide health benefits. *Food Microbiology*, *27*(6), 691–697.

81. Shareck, J., Choi, Y., Lee, B., & Miguez, C. B., (2004). Cloning vectors based on cryptic plasmids isolated from lactic acid bacteria: Their characteristics and potential applications in biotechnology. *Critical Reviews in Biotechnology*, *24*(4), 155–208.

82. Shimizu-Kadota, M., & Sakurai, T., (1982). Prophage curing in *Lactobacillus casei* by isolation of a thermos-inducible mutant. *Applied and Environmental Microbiology*, *43*(6), 1284–1287.

83. Siezen, R. J., Van Enckevort, F. H., Kleerebezem, M., & Teusink, B., (2004). Genome data mining of lactic acid bacteria: The impact of bioinformatics. *Current Opinion Biotechnology*, *15*, 105–115.

84. Siguier, P., Gourbeyre, E., & Chandler, M., (2014). Bacterial insertion sequences: Their genomic impact and diversity. *FEMS Microbiology Review*, *38*, 865–891.

85. Song, A. A., In, L. L., Lim, S. H., & Rahim, R. A., (2017). Review on *Lactococcuslactis*: From food to factory. *Microbial Cell Factory*, *16*(1), 55–60.

86. Sorvig, E., Gronqvist, S., Naterstad, K., Mathiesen, G., Eijsink, V. G., & Axelsson, L., (2003). Construction of vectors for inducible gene expression in *Lactobacillus sakei* and *L. plantarum*. *FEMS Microbiology Letters, 229*(1), 119–126.

87. Sorvig, E., Mathiesen, G., Naterstad, K., Eijsink, V. G., & Axelsson, L., (2005). High level, inducible gene expression in *Lactobacillus sakei* and *Lactobacillus plantarum* using versatile expression vectors. *Microbiology, 151*, 2439–2449.

88. Stefanovic, E., Fitzgerald, G., & McAuliffe, O., (2017). Advances in the genomics and metabolomics of dairy *lactobacilli*: A review. *Food Microbiology*, *61*, 33–49.

89. Stentz, R., Jury, K., Eaton, T., Parker, M., Narbad, A., Gasson, M., & Shearman, C., (2004). Controlled expression of CluA in *Lactococcuslactis* and its role in conjugation. *Microbiology, 150*, 2503–2512.

90. Sun, Z., Harris, H. M., McCann, A., Guo, C., & Argimon, S., (2015). Expanding the biotechnology potential of *lactobacilli* through comparative genomics of 213 strains and associated genera. *Nature Communications*, *6*, 8322.

91. Takala, T., & Saris, P., (2002). A food-grade cloning vector for lactic acid bacteria based on the nisin immunity gene nisI. *Applied Microbiology and Biotechnology*, *59*(4/5), 467–471.

92. Temmerman, R., Huys, G., & Swings, J., (2004). Identification of lactic acid bacteria: Culture dependent and culture-independent methods. *Trends in Food Science and Technology*, *15*, 348–349.

93. Thompson, J. K., McConville, K. J., McReynolds, C., & Collins, M. A., (1999). Potential of conjugal transfer as a strategy for the introduction of recombinant genetic material into strains of *Lactobacillus helveticus*. *Applied Environmental Microbiology*, *65*, 1910–1915.

94. Tomar, N., & De, R. K., (2013). Comparing methods for metabolic network analysis and an application to metabolic engineering. *Gene*, *521*, 1–14.

95. Toro, N., Jiménez-Zurdo, J. I., & García-Rodríguez, F. M., (2007). Bacterial group II introns: Not just splicing. *FEMS Microbiology Reviews, 31*(3), 342–358.

96. VanPijkeren, J. P., & Britton, R. A., (2012). High efficiency recombineering in lactic acid bacteria. *Nucleic Acids Research*, *40*, E76–E81.

97. Van Pijkeren, J. P., & Britton, R. A., (2014). Precision genome engineering in lactic acid bacteria. *Microbial Cell Factories*, *13*, 1–10.

98. Villena, J., Oliveira, M. L., Ferreira, P. C., Salva, S., & Alvarez, S., (2011). Lactic acid bacteria in the prevention of pneumococcal respiratory infection: Future opportunities and challenges *International Immunopharmacology*, *11*, 1633–1645.

99. Villion, M., & Moineau, S., (2009). Bacteriophages of *lactobacillus*. *Frontiers in Bioscience*, *14*, 1661–1683.

100. Wong, W. Y., Su, P., Allison, G. E., Liu, C. Q., & Dunn, N. W., (2003). A potential food grade cloning vector for *Streptococcus thermophilus* that uses cadmium resistance as the selectable marker. *Applied and Environmental Microbiology*, *69,* 5767–5771.

101. Wright, L. D., & Grossman, A. D., (2016). Autonomous replication of the conjugative transposon Tn916. *Journal of Bacteriology*, *198*(24), 3355–3366.

102. Wu, C., Huang, J., & Zhou, R., (2017). Genomics of lactic acid bacteria: Current status and potential applications. *Critical Reviews in Microbiology*, *43*(4), 393–404.

103. Wyszyńska, A., Kobierecka, P., & Bardowski, J., (2015). Lactic acid bacteria: 20 years exploring their potential as live vectors for mucosal vaccination. *Applied Microbiology and Biotechnology*, *99*(7), 2967–2977.

104. Xu, C., Liu, L., Zhang, Z., Jin, D., Qiu, J., & Chen, M., (2013). Genome-scale metabolic model in guiding metabolic engineering of microbial improvement. *Applied Microbiology and Biotechnology, 97*, 519–539.

105. Zhang, W. Y., & Zhang, H. P., (2014). Genomics of lactic acid bacteria. In: Zhang, H. P., & Cai, Y. M., (eds.), *Lactic Acid Bacteria-Fundamentals and Practice* (1st edn., pp. 235–238). Springer Publishing Inc., New York, NY, USA.

106. Zhang, W. Y., Liu, C., Zhu, Y. Z., & Wei, Y. X., (2012). Safety assessment of *Lactobacillus plantarum* JDM1 based on the complete genome. *International Journal of Food Microbiology, 153*(1/2), 166–170.

107. Zhou, H., Hao, Y., Xie, Y., Yin, S., Zhai, Z. Y., & Han, B. Z., (2010). Characterization of a rolling-circle replication plasmid pXY3 from *Lactobacillus plantarum* XY3. *Plasmid, 64*, 36–40.

CHAPTER 11

METABOLIC ENGINEERING OF LACTIC ACID BACTERIA (LAB)

MITAL R. KATHIRIYA, J. B. PRAJAPATI, and YOGESH V. VEKARIYA

ABSTRACT

Metabolic engineering of lactic acid bacteria (LAB) involves in altering the metabolic pathways to yield the desired final product. The genetically modified LAB may be used to develop novel fermented foods with functional properties like cholesterol assimilation, rich in antioxidants, antidiabetic compounds, other bioactive peptides, anticarcinogenic activity, improved sweetness naturally, enhanced body and texture, desired flavor, etc. However, still there is no commercial application of metabolically engineered LAB in fermented food products that have been documented opportunities available to make robust GM LAB with the capacity to provide consumer acceptance. There is a huge scope to explore GM LAB as factories for the production of metabolites of probiotics.

11.1 INTRODUCTION

Lactic acid bacteria (LAB) are nonsporulating, facultative anaerobic rods or cocci, Gram-positive existing naturally in foods [23]. They show wide differences among the bacteria in a group with G+C contents in the range of 34 to 53%, which comprises of *lactobacilli*, *streptococci*, *lactococci*, *enterococci*, *pedicocci*, and *leuconostoc* [31]. LAB are having known health beneficial activities so that they are frequently used to prepare functional and probiotic dairy products [37]. They are used as starter culture too, which are related to acid production, flavor development, desired body and texture and the nutritional benefits of the fermented foods. However, the natural availability of all these nutritionally and functionally important metabolites is limited. To avoid this constraint, metabolic engineering comes into the picture, which involves deletion of unwanted genes or overexpression of available or new ones to make the desired route of the metabolic fluxes that will provide required new metabolic products.

Lactobacilli and *Lactococci* are most commonly utilized for lactic acid production; however, it was observed that *Enterococcus* isolate capable of producing a high amount of lactic acid may be a potential candidate for metabolic engineering pathway due to simple nutritional requirements unlike other LAB [32].

LAB show simple and well-described metabolism, where the lactose is metabolized usually to lactic acid. LAB finds their application in industry, biotechnology, and medicine fields. The available large number of genome sequences of LAB is helpful in learning of genetic potential of this group. As they are having small genome size (easy for genetic manipulation) and capable of producing the end product in larger quantity, therefore a number of trials have been conducted on rearranging metabolic pathway of LAB. Advancements in the application of these bacteria require analyzing the gene sequence responsible for the expression of promising characteristics, such as flavoring compound production, secretion of antimicrobial compounds, resistance to the phage attack, production of organic acid, etc. Based on the demand, the genetic makeup of Lab can be altered by applying metabolic engineering.

This aim of this chapter is to review and discuss the capability of LAB for producing several industrial important products through the manipulation of metabolic pathways of LAB.

11.2   WHY METABOLIC ENGINEERING?

Metabolic engineering can be exploited to achieve many goals, such as:

- To create robust strain;
- To eliminate the unwanted byproduct production;
- To increase the amount of metabolite production;
- To increase the rate the process;
- To optimize the energy consumption.

11.3   WHY SHOULD WE USE LAB IN METABOLIC ENGINEERING?

- For altering the metabolic pathways of LAB, many methods have been reported [37].

- LAB is safely used since a long time (Generally Regarded as Safe Status, GRAS) and they multiply at a faster rate on milk-based growth media, where lactose is utilized as a source of energy and produces a variety of useful metabolites.
- Metabolic pathways are not usually linked with synthesis pathways. Therefore, these pathways can be altered without interfering with the biosynthesis process of bacteria [15].
- The genomic size of LAB is comparatively small (~2–3 Mb), and they do not have complex metabolism [26].
- They utilized worldwide as probiotics and starters in the fermentation process, while metabolically engineered LAB is exploit for the preparation of food with improved physiochemical, sensorial, and functional properties.

11.4 PATHWAYS OF METABOLIC ENGINEERING IN LAB

11.4.1 LACTIC ACID

It consists of D and L as optical stereoisomers. As L-form is natural intermediate metabolite in a human metabolic pathway, it is more preferably used for the food. While D-form is not found naturally in human metabolism, it could cause a detrimental effect on consumers suffering from gastrointestinal (GI) related problems [19]. In one of the study, to modify a host cell for the production of only L-form of lactic acid, two *Lactobacillus helveticus* CNRZ32 strains were engineered by replacing the genes responsible for the secretion of D-lactic acid.

Out of two strains, one was modified by removing the promoter region genes while the second strain was modified by removing structural genes for D-lactic acid and adding structural genes (from same species) for L-lactic acid. Both the promoter and structural genes are involved in the expression of the enzyme (lactate dehydrogenase) responsible for the production of lactic acid. The modified strains were named as GRL86 and GRL89, respectively. Both these strains were capable of producing a higher amount of L-form of lactic acid (i.e., GRL 86–53% and GRL 89–93%) compared to the original (without engineered) *Lactobacillus helveticus* [16, 17]. The strain of *Lactobacillus helveticus* was the first modified LAB, which produced pure L-form of lactic acid [4].

11.4.2 FLAVOR PRODUCTION

11.4.2.1 DIACETYL

It is the main flavor compound in cultured butter, which is produced by the oxidative decarboxylation of α- acetolactate. *Lactococcus lactis* subsp. *Lactis* biovar *diacetylactis* in co-metabolic fermentation of citrate and lactose produces diacetyl [14]. In milk and milk products, citrate is available in minute amount; therefore, lactose is chosen most commonly to boost the diacetyl production by reconstructing *Lactococcus lactis* [10, 14]. In a study amount of diacetyl metabolite can be increased by increasing the concentration of α-acetolactate synthase enzyme, which is involved in the decomposition of pyruvate into α-acetolactate [14]. However, the conversion of pyruvate into acetoin can only be achieved by the expression of this enzyme at a higher level in *Lactococcus lactis* in anaerobic condition [14, 27].

11.4.2.2 ACETALDEHYDE

Acetaldehyde is another flavor compound in yogurt. Yogurt is prepared using *Lactobacillus delbrueckii* subsp. *Bulgaricus* and *Streptococcus thermophilus*, and this LAB give typical flavor to the product by the production of acetaldehyde [2]. For the production of acetaldehyde, the LAB has many metabolic pathways, which includes amino acids, nucleotide, and pyruvate metabolism [3]. Chaves et al., found that expression of *glyA* gene at high level, which codes for an enzyme serine hydroxymethyltransferase produced acetaldehyde in the large amount.

The serine hydroxymethyltransferase enzyme converts threonine into acetaldehyde and glycine. On the expression of *glyA* gene at a high level, the acetaldehyde production was found to be increased in the range of 80–90% [3]. Bongers et al., found that the transfer of *Zymomonasmobilis* gene that codes for pyruvate decarboxylase into *Lactococcus lactis* increased the concentration of acetaldehyde compared to wild type strain. In this study, expression of for pyruvate decarboxylase at a high level was directed by the Nisin Controlled Gene Expression (NICE) system. In addition to this, when Nicotinamide adenine dinucleotide (NADH) oxidase (*nox*) gene was overexpressed, it raises the availability of pyruvate for acetaldehyde production. They observed that the mutant cell containing NADH oxidase and pyruvate decarboxylase activity transformed about 50% of the pyruvate into acetaldehyde under anaerobically [2].

11.4.3 LOW-CALORIE SWEETENERS

11.4.3.1 MANNITOL

Mannitol is sweet in taste, and conversion of less sweet lactose into sweeter mannitol leads to improve the sweetness of fermented dairy products. In our intestine, mannitol is found to be decomposed into short-chain fatty acids that are capable of inhibiting the colon cancer development [25]. In addition to this, it also reduces calories because mannitol is found to be poorly decomposed compared to lactose while poor digestion of mannitol causes GI discomfort [25]. LAB, which does not have lactose dehydrogenase (LDH), were observed to produce mannitol as a metabolite in the absence of oxygen. It was observed that when glucose was completely depleted from the growth medium of *Lactococcus lactis*, mannitol was utilized as a source of energy. Therefore, upon removal of *mtlA* or *mtlF* genes that codes for the mannitol transport system in *Lactococcus lactis*, the mannitol cannot be transported inside the cell to prevent utilization of mannitol from the medium. Such modified *Lactococcus lactis* strain was found to decompose around 33% available glucose into mannitol [5].

In another study, about half of the glucose was rerouted towards mannitol production using *Lactococcus lactis* strain from, which genes responsible for LDH were deleted; mannitol 1-phosphate dehydrogenase gene were expressed at a high level (obtained from *Lactobacillus plantarum*), and mannitol 1-phosphate phosphatase genes (obtained from *Eimeriatenella* a protozoan parasite) were inserted [36].

11.4.3.2 SORBITOL

Sorbitol is 20 times more soluble in water compared to mannitol. It is also about 60% relatively sweeter in comparison to sucrose. Therefore, sorbitol is mostly proffered polyol in a wide variety of food products, where sweetness and solubility are required to obtain the desired body, texture, taste, appearance, *etc.* [28]. *Lactobacillus plantarum* (predominantly found in the human gastrointestinal tract (GIT)) was genetically bioengineered by overexpression of *srlD1* and *srlD2* genes codes for two sorbitol-6-phosphate dehydrogenase (Stl6PDH) enzyme, which converts glucose into sorbitol and removal of genes codes for L- and D- lactate dehydrogenase that converts glucose into lactic acid. This genetically engineered *Lactobacillus plantarum* strain was found to produce sorbitol from about 65% of available glucose under pH-controlled system [18].

11.4.3.3   L-ALANINE

L-Alanine is also one of the sweeteners used in the preparation of fermented milk products (i.e., sweet dahi, yogurt, buttermilk, etc.). It is most commonly produced by LAB [14]. Hols et al., observed that an increased amount of L-alanine was obtained by expression of L-alanine dehydrogenase (ALAD) through *ala-D* gene of *Bacillus sphearicus* at a high level [8]. Generally in lactococci around 30–40% pyruvate was found to be converted into alanine when ammonia was present in larger amounts, and glucose was used as a source of energy for resting cells of lactococci. Under such conditions, pyruvate was converted to alanine with the help of ALAD enzyme. In addition to this, complete conversion of pyruvate can be achieved by using genetically modified lactococci with no LDH activity, insertion of *ala-D* gene, and incubating the cells at optimum conditions. The pure L-isomer of alanine was obtained by deleting the gene, which codes for D-isomer in the lactococcal cell [8].

11.4.4   VITAMINS

11.4.4.1   RIBOFLAVIN

In *Lactococcus lactis* subsp. *cremoris* NZ9000, riboflavin biosynthetic operon was found. Riboflavin is utilized by *Lactococcus lactis* subsp. *cremoris* NZ9000 as growth factor. This riboflavin utilizing cells were transformed into riboflavin producing cells by metabolic engineering [11]. It was achieved by expression of ribG, ribH, ribB, and ribA genes at high level in *Lactococcus lactis*

11.4.4.2   FOLIC ACID

The LAB *Lactococcus lactis* is capable of producing folate and is stored in the form of polyglutamyl. From total synthesized folate, only little amount is secreted out in the medium. In *Lactococcus lactis* strain MG1363, five genes *folA, folB, folKE, folP*, and *folC* were found to be involved in the synthesis of folate. Upon expression of *folKE* at a high level, the total amount of folate production was observed to increase by three times while extracellular folate by ten times compared to wild strain; while expression of *folA* gene at a high level reduced the folate production in the medium [33]. In the production of folate, Para-aminobenzoic acid (pABA) acts as precursor; and expression of pABA producing gene at a high level in *Lactococcus lactis* was found to synthesize folate in milligram per liter [35].

11.4.5   EXOPOLYSACCHARIDES (EPS)

Many LAB are capable of producing EPs during the fermentation of dairy products. EPs play an important role in providing consistency, mouth feel, viscosity to the final fermented product. In addition to this, it also acts as prebiotic [7], cholesterol-lowering nutraceuticals [24] and helps in the proper functioning of the immune system in the host [9, 13]. In one of the study, it was found that combined expression of *galU* codes for UDP-glucose pyrophosphorylase and *pgmA* codes for phosphoglucomutase enzyme at high level leads to an increase in the concentration of EPs metabolite. A metabolically engineered *Streptococcus thermophilus* strain LY03 with galactose fermentation ability was observed to possess improved activities of enzymes involved in the Leloir pathway and increased the EPs production by 1.4 folds compared to wild strain.

The EPs production was further increased by the insertion of *galU* in *Streptococcus thermophilus* strain LY03. The study showed that EPs concentration as metabolite could be increased by deleting or inserting suitable genes coding for the enzymes involved in the fermentation of carbohydrate [20]. Looijesteijn and Hugenholtz [22] found that *Lactococcus lactis* cells, when grown in medium containing fructose, the formation of EPs, were less compared to glucose or lactose as a source of carbon. This may be due to the comparatively low activity of fructose biphosphatase, which is involved in secretion of EPs when cells were grown on fructose. Upon expression of *fbp* gene encodes fructose biphosphatase enzyme at a high level in *Lactococcus lactis* with the NICE system, resulting into increased production of EPs from the degradation of fructose [21].

11.5   GENOME-SCALE METABOLIC MODELS (GSMM) VERSUS METABOLIC ENGINEERING

By using genetic information and mathematical models, one can predict the behavior of the cell as well as required alteration can be carried out in the cell to improve the cell's activity [12]. For example, the genome-scale metabolic model (GSMM) helps in analyzing the behavior of a particular organism theoretically [34]. These models correlate the genotypical and phenotypical information [29]. In order to obtain desired characteristics in fermented milk products, the developing such type of mathematical models will be beneficial [6, 30].

11.6 SUMMARY

Metabolic engineering is a modern scientific approach to produce high purity compounds with a reasonable cost. It can eliminate the use of chemical process and reduce the risk to industrial operations, which is always a national interest.

KEYWORDS

- **acetaldehyde**
- **exopolysaccharide**
- **lactic acid bacteria (LAB)**
- **L-alanine**
- **mannitol**
- **metabolic engineering**
- **para-amino-benzoic acid**
- **polyols**
- **prebiotics**
- **probiotics**
- **pyruvate**
- **riboflavin**
- **sorbitol**

REFERENCES

1. Bailey, J. E., (1991). Toward a science of metabolic engineering. *Science, 252,* 1668–1675.

2. Bongers, R. S., Hoefnagel, M. H., & Kleerebezem, M., (2005). High-level acetaldehyde production in Lactococcus lactis by metabolic engineering. *Applied Environmental Microbiology*, *71*, 1109–1113.

3. Chaves, A. C., Fernandez, M., & Lerayer, A. L., (2002). Metabolic engineering of acetaldehyde production by *Streptococcus thermophilus*. *Applied Environmental Microbiology*, *68*, 5656–5662.

4. Flahaut, N. L. A., & De Vos W. M., (2015). System biology and metabolic engineering of lactic acid bacteria for improved fermented foods: Chapter 8, In: Holzapfel, W., (ed.), *Advances in Fermented Foods and Beverages* (pp. 177–196). Woodhead Publishing Series in Food Science, Technology and Nutrition, Elsevier, New York.

5. Gaspar, P., Neves, A. R., & Ramos, A., (2004). Engineering *Lactococcus lactis* for production of mannitol: High yields from food-grade strains deficient in lactate dehydrogenase and the mannitol transport system. *Applied Environmental Microbiology*, *70*, 1466–1474.

6. Gaspar, P., Carvalho, A. L., Vinga, S., Santos, H., & Neves, A. R., (2013). From physiology to systems metabolic engineering for the production of biochemicals by lactic acid bacteria. *Biotechnology Advances*, *31*, 764–788.

7. Gibson, G. R., & Roberfroid, M. B., (1995). Dietary modulation of the human colonic microbiota: Introducing the concept of prebiotics. *The Journal of Nutrition*, *124*, 1401–1412.

8. Hols, P., Kleerebezem, M., & Schranck, A. N., (1999). Conversion of *Lactococcuslactis* from homolactic to homoalanine fermentation through metabolic engineering. *Nature Biotechnology*, *17*, 588–592.

9. Hosono, A., Lee, J., & Ametani, A., (1997). Characterization of a water soluble poly-saccharide fraction with immunopotential activity from *Bifidobacterium adolescentis* M101–4. *Bioscience, Biotechnology and Biochemistry*, *61*, 312–316.

10. Hugenholtz, J., Kleerebezem, M., & Starrenburg, M., (2000). *Lactococcus lactis* as a cell factory for high level diacetyl production. *Applied Environmental Microbiology*, *66*, 4112–4114.

11. Hugenholtz, J., & Smid, E. J., (2002). Nutraceutical production with food-grade microorganisms. *Current Opinion in Biotechnology*, *13*, 497–507.

12. King, Z. A., Lloyd, C. J., Feist, A. M., & Palsson, B. O., (2015). Next-generation genome-scale models for metabolic engineering. *Current Opinion in Biotechnology*, *35*, 23–29.

13. Kitazawa, H., Yamaguchi, T., & Miura, M., (1993). B-cell mitogen produced by slime forming, encapsulated *Lactococcuslactis* subspecies cremoris isolated from sour milk. *Journal of Dairy Science*, *76*, 1514–1519.

14. Kleerebezem, M., Hols, P., & Hugenholtz, J., (2000). Lactic acid bacteria as a cell factory: Rerouting of carbon metabolism in *Lactococcus lactis* by metabolic engineering. *Enzyme and Microbial Technology*, *26*, 840–848.

15. Kleerebezem, M., & Hugenholtz, J., (2003). Metabolic pathway engineering in lactic acid bacteria. *Current Opinion in Biotechnology*, *14*, 232–237.

16. Kylä-Nikkilä, K., Hujanen, M., & Leisola, M., (2000). Metabolic engineering of *Lactobacillus helveticus* CNRZ32 for production of pure L-(+)-lactic acid. *Applied Environmental Microbiology*, *66*, 3835–3841.

17. Kylä-Nikkilä, K., (2016). *Genetic Engineering of Lactic Acid Bacteria to Produce Optically Pure Lactic Acid and to Develop a Novel Cell Immobilization Method Suitable for Industrial Fermentations* (p. 117). PhD Thesis, University of Helsinki, Helsinki–Finland.

18. Ladero, V., Ana, R., & Anne, W., (2007). High level production of the low calorie sugar sorbitol by *Lactobacillus plantarum* through metabolic engineering. *Applied and Environmental Microbiology, 73*, 1864–1872.

19. Lapierre, L., Germond, J. E., & Ott, A., (1999). D-Lactate dehydrogenase gene (ldhD) inactivation and resulting metabolic effects in the *Lactobacillus johnsonii* strains La1 and N312. *Applied Environmental Microbiology, 65*, 4002–4007.

20. Levander, F., Svensson, M., & Rådström, P., (2001). Enhanced exopolysaccharide production by metabolic engineering of *Streptococcus thermophilus*. *Applied and Environmental Microbiology, 68*, 784–790.

21. Looijesteijn, P. J., Boels, I. C., & Hugenholtz, J., (1999). Regulation of exopolysaccharide production by *Lactococcus lactis* subsp. Cremoris by the source of sugar. *Applied Environmental Microbiology, 65*, 5003–5008.

22. Looijesteijn, P. J., & Hugenholtz, J., (1999). Uncoupling of growth and exopolysaccharide production by *Lactococcus lactis* subsp. cremoris NIZO B40 and optimization of its biosynthesis. *Journal of Bioscience and Bioengineering, 88*, 178–182.

23. Mckay, L. L., & Baldwin, K, A., (1990). Applications for biotechnology present and future improvements in lactic acid bacteria. *FEMS Microbiology Reviews, 87*, 3–14.

24. Nakajima, H., Suzuki, Y., & Kaizu, H., (1992). Cholesterol lowering activity of ropy fermented milk. *Journal of Food Science, 57*, 1327–1329.

25. Neves, A. R., Pool, W. A., & Kok, J., (2005). Overview on sugar metabolism and its control in *Lactococcus lactis*–the input from *in vivo* NMR. *FEMS Microbiology Review, 29*, 531–554.

26. Papagianni, M., (2012). Recent advances in engineering the central carbon metabolism of industrially important bacteria. *Microbial Cell Factories, 11*, 50–63.

27. Platteeuw, C., Hugenholtz, J., & Starrenburg, M., (1995). Metabolic engineering of *Lactococcus lactis*: Influence of the overproduction of alpha-acetolactate synthase in strains deficient in lactate dehydrogenase as a function of culture conditions. *Applied Environmental Microbiology, 61*, 3967–3971.

28. Silveira, M. M., & Jonas, R., (2002). The biotechnological production of sorbitol. *Applied Microbiology and Biotechnology, 59*, 400–408.

29. Steele, J., Broadbent, J., & Kok, J., (2013). Perspectives on the contribution of lactic acid bacteria to cheese flavor development. *Current Opinion in Biotechnology, 24*, 135–141.

30. Stefanovic, E., Fitzgerald, G., & McAuliffe, O., (2017). Advances in the genomics and metabolomics of dairy lactobacilli: A review. *Food Microbiology, 61*, 33–49.

31. Stiles, M. E., & Holzapfel, W. H., (1997). Lactic acid bacteria of foods and their current taxonomy. *International Journal of Food Microbiology, 36*, 1–29.

32. Subramanian, M. R., (2014). Production of lactic acid using a new homofermentative *Enterococcus fecalis* isolate. *Microbial Biotechnology, 8*, 221–229.

33. Sybesma, W., Starrenburg, M., & Kleerebezem, M., (2003). Increased production of folate by metabolic engineering of *Lactococcus lactis*. *Applied and Environmental Microbiology, 69*, 3069–3076.

34. Teusink, B., Bachmann, H., & Molenaar, D., (2011). Systems biology of lactic acid bacteria: A critical review. *Microbial Cell Factories, 10*, 1–11.

35. Wegkamp, A., Van Oorschot, W., & De Vos, W. M., (2007). Characterization of the role of para-amino benzoic acid biosynthesis in folate production in *Lactococcus lactis*. *Applied and Environmental Microbiology, 73*, 2673–2681.

36. Wisselink, H. W., Moers, A. P., & Mars, A. E., (2005). Overproduction of heterologous mannitol 1-phosphatase: A key factor for engineering mannitol production by *Lactococcus lactis. Applied Environmental Microbiology, 71,* 1507–1514.
37. Yebra, M. J., Monedero, V., & Martínez, G. P., (2012). Genetically engineered Lactobacilli for technological and functional food application: Chapter 9, In: Valdez, B., (ed.), *Food Industrial Processes–Methods and Equipment* (pp. 905–909). Intech Open Source Publisher, Madrid, Spain.

CHAPTER 12

METABOLITES OF LACTIC ACID BACTERIA (LAB) FOR FOOD BIOPRESERVATION

SANTOSH KUMAR MISHRA, REKHA CHAWLA,
VEENA NAGARAJAPPA, and KRISHAN KUMAR MISHRA

ABSTRACT

Application of bacteriocins in milk sectors can make way for milk products without any chemical preservatives, which can further help in reducing the heat intensity during milk processing and manufacturing of milk products that are naturally processed and preserved. There is significant scope for research in the area of bacteriocins for traditional foods. The knowledge of the proper selection of bioprotective cultures along with proper formulation of bacteriocin treatment (single or along with other antimicrobials) can pave the way for safe products especially for developing nation, where milk safety is the most neglected area.

12.1 INTRODUCTION

There is a quest for safe food in the current scenario. The public is willing to accept levels of risk in other aspects of their life but not in foods. This is a consequence of the special role of foods in society. For many persons, it is the only component of their environment over which they believe they have the control. Under these circumstances, the public insists on absolutely safe food products.

In the developed world, consumers often question the safety of the thousands of non-food preservatives and other additives that are incorporated in foods. It has also encouraged them to voice their feelings against the use of these chemicals in foods and also to look for foods that are "natural," "healthy" and not treated with harsh processing steps. Although there is a lot of advancement in modern technology, yet there is a large scope for the preservation of foods for both developing as well as modern industrialized countries.

There are many challenges available for the food industry, which involves lowering of processing cost, lower economical losses due to microbial spoilage, reducing microbial hazards due to pathogens transmission through food. In addition to this, there is a continuous increase in demand for quality and safe food in the world. From ancient times, microorganisms are being used for the preservation of traditional food products [34]. Bacteriocin produced by LAB has gained popularity because of its specific action against specific bacteria and green nature. Bacteriocins produced by LAB have been studied widely by many researchers [9, 11, 12].

This chapter focuses on food safety issues and the importance of alternatives for chemical preservatives. Application of metabolites of Lactic acid bacteria (LAB) including bacteriocins in the food sector by means of *ex-situ* and *in-situ* production and their role in biopreservation of foods is also included.

12.2 SAFETY CONCERNS ON CHEMICAL PRESERVATIVES

Consumers generally prefer foods that do not have any chemical preservatives as it gives the impression that these foods are healthy and natural. This is why; the safety of some preservatives, as well as some other additives, has been questionable, such as NO_2, sulfides, sodium diacetate, beta propiolactones, and therapeutic antibiotics. Reports are available on possible health hazards from consumption of some preservatives and additives such as nitrite and saccharine, and additives previously used but currently not been permitted for use such as cyclamate. Some dyes have shaken the faith of consumers, thus prompting them to question the safety and wholesomeness of foods that are consumed.

Antimicrobial chemical preservatives act by hampering or retarding the microbial spoilage and are used to enhance food safety. They produce their antimicrobial effect by interfering with the structural and active functional components of microorganisms. Moreover, in recent years, the chemicals used in food items have increased exponentially to several thousands. The effect of actual consumption of these chemicals in multiple products over a substantial length of time should be an important consideration in judging their safety.

Therefore, the basis of selection of antimicrobial biopreservatives for use in foods should not only be their effectiveness against all pathogens and spoilage organisms, but also their proven safety standards and their acceptability by health-conscious consumers and health regulatory agencies. Several antimicrobial metabolites of lactic acid bacteria (LAB) associated with fermented foods have provided huge interest in recent times [2].

12.3   END-PRODUCTS OF LACTIC ACID BACTERIA (LAB)

LAB are known to produce several antimicrobial substances, which are inhibitory to many organisms. These antimicrobials include hydrogen peroxide, lactic acid, acetoin, free fatty acids (FFAs), 2,3-butan-diol, acetaldehyde, diacetyl, and bacteriocins. They can be used as biopreservatives by modifying some known food properties so that there is a least negative impact of spoilage and foodborne pathogenic microorganisms including psychotropic spoilage bacteria, aerobic, and anaerobic spore-forming bacteria and some foodborne pathogenic bacteria, etc. (Figure 12.1).

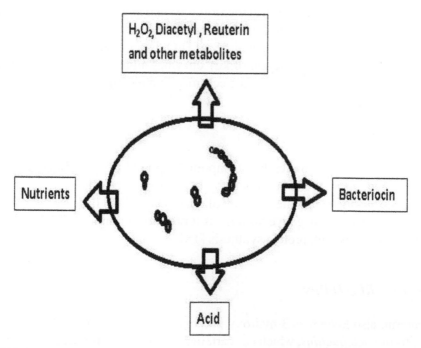

FIGURE 12.1   Production of various metabolites by a lactic culture including acid, H_2O_2, diacetyl, reuterin, and bacteriocin.

12.3.1   ORGANIC ACID

Lactic acid possesses antibacterial activity by reducing the pH of the medium. The inhibitory action of lactic acid is mostly because of its ability to disintegrate the cytoplasmic membrane of microorganisms in the un-dissociated form, by which intracellular pH gets reduced [33]. Several organic acids are known to be produced by LAB [39, 23]. Propionic and sorbic acids are reported to increase the shelf-life of many foods. Acetic acid is majorly applied in the food industry.

12.3.2   HYDROGEN PEROXIDE

Hydrogen peroxide is an unstable compound and terminates bacterial enzymatic activity [5]. H_2O_2 released by one bacterium suppresses other species: is a classic example of antagonistic interaction. LAB has not been able to utilize the cytochrome system as it does not have heme-protein; thus, they cannot convert oxygen to water. Instead of the cytochrome system, lactobacilli possess flavor-proteins, by which they can reduce oxygen to H_2O_2. Therefore, the absence of both cytochrome system and catalase (CAT) results in accumulation of abundance of H_2O_2. LAB themselves cannot degrade this excess H_2O_2. The antimicrobial activity of H_2O_2 is because of oxidation of SH groups resulting in denaturation of key enzymes along with peroxidation of membrane lipid, thus increasing the cell membrane permeability.

12.3.3   DIACETYL

Diacetyl is another antimicrobial compound reported from LAB. Diacetyl is an intermediate compound produced by LAB by decarboxylation of α- acetolactate during the metabolic pathway of citric acid [17]. It has a bacteriostatic effect against Gram-negative bacteria as it prevents arginine assimilation and finally interferes with protein synthesis [18].

12.3.4   REUTERIN

Reuterin, also known as 3-hydroxy propenaldehyde, is produced by certain strains of *Lactobacillus*, which has certain inhibitory effective against a broad range of microorganisms [4, 29, 30]. This is an aldehyde compound produced from glycerol as a sole carbon source by *Lactobacillus reuteri*. Antimicrobial activity of reuterin maybe because of its interference of activity of RNA reductase enzyme, which is required for DNA synthesis [21].

12.3.5 BACTERIOCINS OF LAB

Bacteriocins of LAB have been frequently overlooked. Recently, bacteriocins have gained momentum, and the role of bacteriocin for biopreservation of foods has been the subject of novel research.

Bacteriocins are basically ribosomally synthesized proteinaceous compounds, which can display an inhibitory action exclusively against closely related bacterial species [2]. The main difference between antibiotics and bacteriocins is that bacteriocins have narrow inhibitory spectrum against closely related bacteria, while antibiotics may have wide inhibitory spectrum against a large group of bacteria (Table 12.1).

TABLE 12.1 Difference Between Bacteriocins and Conventional Antibiotics

Bacteriocins	Antibiotics
Bacteriocins are generally applied for shelf-life increment and biopreservation of foods	Antibiotics are generally used for clinical purpose
Bacteriocins are synthesized by ribosomes of bacteria	Antibiotics are the secondary metabolites
Bacteriocins generally act through pore formation in the cytoplasmic membrane of bacteria	Antibiotics acts on specific targets
No residual toxic effect of bacteriocins on eukaryotic cells	Toxic effect of antibiotics on eukaryotic cells after longer use
They are generally effective against closely related bacteria (Short spectrum of antimicrobial activity)	Antibiotics are generally having a broad spectrum of antimicrobial activity

LAB has GRAS status. Use of bacteriocins for shelf-life increment of foodstuff will ultimately result in decreasing the usage of chemical preservatives. Spoilage bacteria may enter and proliferate in food through various sources, which include raw material, processing parameters, storage, and transportation. All these factors increase the chances of foodborne illness. Therefore, there is a need to check the actual action of bacteriocins in the food system so that it can be used successfully against specific organisms in food. Preservation of food by natural methods including bacteriocins approach can reduce spoilage of raw materials and food products, foodborne illness, and so many other current food-related issues.

12.4   BACTERIOCIN-BASED FORMULATIONS FOR FOOD PRESERVATION

Benefits of bacteriocins for biopreservation of foods [38] are (Table 12.2):

- Development of novel foods with high water activity and low heat treatment during processing;
- Provide additional protection for temperature change during transportation, storage, and distribution;
- Reduction in economic losses due to deterioration of food during storage and transport;
- Reduction in foodborne disease outbreak;
- Reduction in use of chemical preservatives which is generally not liked by the consumers;
- Shelf life enhancements of several foods, including dairy foods.

In the current situation, there is a need to utilize these benefits to confirm the proper level of food safety and to fulfill the consumer's demand of hygienic, safe, and minimally processed food with an enhanced shelf life.

TABLE 12.2   Examples of Bacteriocins

LAB Strain	Bacteriocin
Enterococcus faecium	Enterocin
Lactobacillus gasseri	Gassericin
Lactobacillus helveticus	Helveticin
Lactobacillus plantarum	Plantaricin
Lactobacillus reuteri	Reutericin
Lactobacillus sakei	Sakacins
Lactococcus lactis ssp. *cremoris*	Lactococcin
Lactococcus lactis ssp. *lactis*	Nisin
Pediococcus acidilactici	Pediocin

12.5   BACTERIOCINS FOR FOOD SAFETY

Food safety can be enhanced by two methods, either addition of already produced bacteriocin in food or by spiking of bioprotective culture in food, which can produce antimicrobial compounds inside the gut after ingestion under favorable conditions [36, 37]. Bacteriocins combinations can also be prepared from the bioprotective culture in a fermentor for large-scale production as purified or partially purified concentrates. Till now, Nisin is an only antimicrobial compound that got permission by FDA to be used as a food preservative. In the market, several different antimicrobial concentrates like ALTA™ 2341 or Microgard™ are available, but recently some different milk-based antimicrobial preparations have also been reported like Lacticin 3147 [15] and Pediocin 34 [27].

LAB, which can produce bacteriocins, can also be used as a starter or bioprotective culture [9]. This is an excellent way to preserve foods with least effect on processing cost; but prior to use of bioprotective culture, strains should be selected cautiously with its effect on food environment during refrigerated storage [12]. Bacteriocin production inside the gut by bioprotective culture provides several advantages over bacteriocin added outside in the food itself both in terms of final product cost and legal requirements. Cost reduction during processing may be very helpful for developing nations, where the safety aspect is often neglected. Several studies have been carried out for isolation and characterization of bioprotective cultures for food biopreservation [10, 25].

12.5.1   DAIRY FOODS

Common foodborne pathogens, which can reside and proliferate in raw materials, are *Salmonella*, *Shigella*, *Listeriamonocytogenes*, *Staphylococcus aureus*, and coliforms especially *E. coli* [7]. Nisin has been used in many dairy foods to prevent late blowing of certain cheeses by *Clostridium* spp. [8], and in other dairy foods like processed cheese to reduce the growth of spore-forming bacteria and certain post-processing contaminant bacteria like *Listeriamonocytogenes* [6, 38]. Certain starter cultures, which can produce Lacticin 3147, have been utilized to improve cheese quality by reducing the growth of certain non-lactic LAB during the maturing stage of cheese [35].

Pediocin produced by certain *Pediococcuspentosaceous* strains has been widely used in many dairy foods [27]. In one such study, a hurdle was created through different antimicrobials (pediocin, nisin along with NaCl, K-sorbate, and EDTA) to enhance the shelf-life of paneer for more than two months [27]. Some common applications of bacteriocins for milk and dairy products are as below:

- Control of aerobic spore-forming bacteria like *Bacillus cereus* in butter, cream.

- Inhibition of psychotropic bacteria in milk and cream.
- Prevention of entry of *L. monocytogenes* in milk products after processing.
- Prevention of over-acidification in *dahi*, *lassi*, and other fermented foods.
- Prevention of proliferation of spoilage and pathogenic foodborne organisms in food.
- Shelf life enhancement of paneer.
- To make starter culture dominance over other contaminants if any during fermentation.

12.5.2 MEAT PRODUCTS

These products are a good source for the survival and growth of a number of microorganisms based on the processing and storage conditions. Under low-temperature storage, presence of oxygen generally enhances the aerobic Gram-negative spoilage bacteria, while LAB generally grows better in the anaerobic environment [3, 14]. Enterocin has been applied on meat sausage, which resulted in a decrease of *L. monocytogenes* [1].

Another study reported that bacteriocins added to meat resulted in fewer requirements of high pressure (HPs) treatments [16]. It was reported that application of hurdle technology by means of nisin, pediocin, and high-pressure processing (HPP) reduced the spore-forming *Clostridium* spp. in beef [20]. Meat and meat products contain LAB as their natural microbiota; hence, their use as a bioprotective culture to preserve meat products for a longer time can be utilized by the meat industry. In one such study, it was revealed that *L. sakei* in fermented meat sausage could reduce the *L. monocytogenes* counts [24].

12.6 APPLICATION OF BACTERIOCINS FOR DEVELOPMENT OF ANTIMICROBIAL PACKAGING (AMP) SYSTEM FOR ENHANCED FOOD SAFETY

There is a possibility that bacteriocins, along with other antimicrobial compounds, can be incorporated in bioactive packaging films by immobilization techniques. The biofilms prepared by use of proteins or certain polysaccharides along with various concentrations of different antimicrobials was able to decrease *Salmonella* counts in poultry skin [31, 32]. Similarly, a gel containing gelatin protein [13] having various concentrations of different antimicrobials on cooked ham showed a reduction in numbers of various spoilage organisms including *L. monocytogenes*. Application of zein protein having various concentrations of sodium lactate, and nisin completely killed the *L. monocytogenes* [26].

HPP [19] has also been used along with antimicrobial packaging (AMP) system. Similarly, enterocins in alginate film along with HPP were used to control the growth of *L. monocytogenes* [28].

12.7   SUMMARY

The chapter highlights food safety issues, need of alternatives for chemical preservatives, metabolites of LAB including bacteriocins, application of bacteriocins in milk sector and their role in biopreservation of food, and hurdle technology in AMP. The topics in this chapter will be of fundamental importance for future applications in the milk industry.

KEYWORDS

- **antibacterial**
- **bacteriocins**
- **diacetyl**
- **lactic acid bacteria**
- **lacticin**
- **lantibiotics**
- **microbial metabolites**
- **non-lantibiotics**
- **pediocin**
- **reuterin**

REFERENCES

1. Ananou, S., Maqueda, M., Martınez-Bueno, M., Galvez, A., & Valdivia, E., (2005). Control of *Staphylococcus aureus* in sausages by enterocin AS-48. *Meat Science, 71,* 549–576.

2. Balciunas, E. M., Martinez, F. A. C., Todorov, S. D., De Melo Franco, B. D. G., Converti, A., & De Souza, O. R. P., (2013). Novel biotechnological applications of bacteriocins: A review. *Food Control, 32*, 134–142.

3. Borch, E., Kant-Muermans, M. L., & Blixt, Y., (1996). Bacterial spoilage of meat and cured meat products. *International Journal of Food Microbiology, 33*, 103–120.

4. Cleusix, V., Lacroix, C., Vollenweider, S., & Le Blay, M. D. G., (2007). Inhibitory activity spectrum of reuterin produced by *Lactobacillus reuteri* against intestinal bacteria. *BMC Microbiology, 7*, 101–109.

5. Collins, E. B., & Aramaki, K., (1980). Production of hydrogen peroxide by *Lactobacillus acidophilus*. *Journal of Dairy Science, 63*, 353–357.

6. Davies, E. A., & Delves-Broughton, J., (1999). Nisin. In: Robinson, R., Batt, C., & Patel, P., (eds.), *Encyclopedia of Food Microbiology* (pp. 191–198). Academic Press, London.

7. De Buyser, M. L., Dufour, B., Maire, M., & Lafarge, V., (2001). Implication of milk and milk products in food-borne diseases in France and in different industrialized countries. *International Journal of Food Microbiology, 67*, 1–17.

8. De Vuyst, L., & Vandamme, E. J., (1994). Nisin, a lantibiotic produced by *Lactococcus lactis* subsp. *lactis*: Properties, biosynthesis, fermentation and applications. In: De Vuyst, L., & Vandamme, E. J., (eds.), *Bacteriocins of Lactic Acid Bacteria* (pp. 151–222). Blackie Academic and Professional, London.

9. Deegan, L. H., Cotterm, P. D., Hill, C., & Ross, P., (2006). Bacteriocins: Biological tools for bio-preservation and shelf-life extension. *International Dairy Journal, 16*, 1058–1071.

10. Foulquie, M. M. R., Sarantinopoulos, P., Tsakalidou, E., & De Vuyst, L., (2006). The role and application of enterococci in food and health. *International Journal of Food Microbiology, 106*, 1–24.

11. Franz, C. M. A. P., Van Belkum, M. J., Holzapfel, W. H., Abriouel, H., & Galvez, A., (2007). Diversity of enterococcal bacteriocins and their grouping into a new classification scheme. *FEMS Microbiology Reviews, 31*, 293–310.

12. Galvez, A., Abriouel, H., Lucas, L. R., & Ben Omar, N., (2007). Bacteriocin-based strategies for food biopreservation. *International Journal of Food Microbiology, 120*, 51–70.

13. Gill, A. O., & Holley, R. A., (2000). Inhibition of bacterial growth on ham and bologna by lysozyme, nisin and EDTA. *Food Research International, 33*, 83–90.

14. Gram, L., Ravn, L., Rasch, M., Bruhn, J. B., Christensen, A. B., & Givskov, M., (2002). Food spoilage-interactions between food spoilage bacteria. *International Journal of Food Microbiology, 78*, 79–97.

15. Guinane, C. M., Cotter, P. D., Hill, C., & Ross, R. P., (2005). Microbial solutions to microbial problems: *Lactococcal bacteriocins* for the control of undesirable biota in food. *Journal of Applied Microbiology, 98*, 1316–1325.

16. Hugas, M., Garriga, M., & Monfort, J. M., (2002). New mild technologies in meat processing: High pressure as a model technology. *Meat Science, 62*, 359–371.

17. Hugenholtz, J., Kleerebezem, M., Starrenburg, M., Delcour, J., De Vos, W., & Hols, P., (2000). *Lactococcuslactis* as a cell factory for high-level diacetyl production. *Applied and Environmental Microbiology, 66*, 4112–4114.

18. Jay, J. M., (1982). Antimicrobial properties of diacetyl. *Applied and Environmental Microbiology, 44*, 525–532.

19. Jofre, A., Garriga, M., & Aymerich, T., (2007). Inhibition of *Listeriamonocytogenes* in cooked ham through active packaging with natural antimicrobials and high-pressure processing. *Journal of Food Protection, 70*, 2498–2502.

20. Kalchayanand, N., Dunne, C. P., Sikes, A., & Ray, B., (2003). Inactivation of bacterial spores by combined action of hydrostatic pressure and bacteriocins in roast beef. *Journal of Food Safety, 23*, 219–231.

21. Karlson, M., (1989). *Lactobacillus reuteri* and the enteric microbiota. In: *The Regulatory and Protective Role of the Normal Microflora* (pp. 283–292). Wenner-Gren International Symposium Series, London.

22. Klaenhammer, T. R., (1993). Genetics of bacteriocins produced by lactic acid bacteria. *FEMS Microbiology Reviews, 12*, 39–86.

23. Lavermicocca, P., Valerio, F., Evidente, A., Lazzaroni, S., Corsetti, A., & Gobbetti, M., (2000). Purification and characterization of novel antifungal compounds from the sourdough *Lactobacillus plantarum* strain 21B. *Applied and Environmental Microbiology, 66*, 4084–4090.

24. Leroy, F., Lievens, K., & De Vuyst, L., (2005). Modeling bacteriocin resistance and inactivation of *Listeria innocua* LMG 13568 by *Lactobacillus sakei* CTC 494 under sausage fermentation conditions. *Applied Environmental Microbiology, 71*, 7567–7570.

25. Leroy, F., Verluyten, J., & De Vuyst, L., (2006). Functional meat starter cultures for improved sausage fermentation. *International Journal Food Microbiology, 106*, 270–285.

26. Lungu, B., & Johnson, M. G., (2005). Fate of *Listeria monocytogenes* inoculated onto the surface of model Turkey frankfurter pieces treated with zein coatings containing nisin, sodium diacetate, and sodium lactate at 4°C. *Journal of Food Protection, 68*, 855–859.

27. Malik, R. K., Rao, K. N., Bandhopadhyay, P., & Kumar, N., (2005). Bacteriocins: Natural and safe anti-microbial peptides for food preservation. *Indian Food Industry, 24*(1), 69–70.

28. Marcos, B., Aymerich, T., Monfort, J. M., & Garriga, M., (2008). High pressure processing and antimicrobial biodegradable packaging to control *Listeria monocytogenes* during storage of cooked ham. *Food Microbiology, 25*, 177–182.

29. Mishra, S. K., Malik, R. K., Kaur, G., Manju, G., Pandey, N., & Singroha, G., (2011). Potential bioprotective effect of reuterin produced by *L. reuteri* BPL-36 alone and in combination with nisin against foodborne pathogens. *Indian Journal of Dairy Science, 64*(5), 406–411.

30. Mishra, S. K., Malik, R. K., Manju, G., Pandey, N., Singroha, G., Bahare, P., & Kaushik, J. K., (2012). Characterization of a reuterin-producing *Lactobacillus reuteri* BPL-36 strain isolated from human infant fecal sample. *Probiotics and Antimicrobial Proteins, 4*(3), 154–161.

31. Natrajan, N., & Sheldon, B. W., (2000a). Efficacy of nisin-coated polymer films to inactivate *Salmonella Typhimurium* on fresh broiler skin. *Journal of Food Protection, 63*, 1189–1196.

32. Natrajan, N., & Sheldon, B. W., (2000b). Inhibition of *Salmonella* on poultry skin using protein- and polysaccharide-based films containing a nisin formulation. *Journal of Food Protection, 63*, 1268–1272.

33. Ray, B., & Sandine, W. E., (1992). Acetic, propionic, and lactic acids of starter culture bacteria as biopreservatives. In: *Food Preservatives of Microbial Origin* (pp. 103–136). CRC Press, Boca Raton, Florida.

34. Ross, R. P., Morgan, S., & Hill, C., (2002). Preservation and fermentation: Past, present and future. *International Journal of Food Microbiology, 79*, 3–16.

35. Ryan, M. P., Ross, R. P., & Hill, C., (2001). Strategy for manipulation of cheese flora using combinations of lacticin 3147 producing and resistant cultures. *Applied Environmental Microbiology, 67*, 2699–2704.

36. Schillinger, U., Geisen, R., & Holzapfel, W. H., (1996). Potential of antagonistic microorganisms and bacteriocins for the biological preservation of foods. *Trends in Food Science and Technology, 71*, 58–64.

37. Stiles, M. E., (1996). Biopreservation by lactic acid bacteria. *Antonie van Leeuwenhoek, 70*, 331–345.

38. Thomas, L. V., & Delves-Broughton, J., (2001). New advances in the application of the food preservative nisin. *Advanced Food Science, 2*, 11–22.

39. Wang, H., Yan, Y., Wang, J., Zhang, H., & Qi, W., (2012). Production and characterization of antifungal compounds produced by *Lactobacillus plantarum* IMAU10014. *PLoS ONE, 7*, e29452.

APPENDIX A   FERMENTED DAIRY PRODUCTS

Indian Dairy products	Origin	Description	English name
Lassi	Haryana, Rajasthan, Punjab	Lassi is a blend of yogurt, water, spices and sometimes fruit.	Fermented milk beverage
Bajra Lassi	Haryana, Rajasthan	Lassi is a blend of yogurt, water, spices including Bajra (Pearl Millet)	
Chaach	Gujarat, Haryana, Rajasthan, Punjab.	A buttermilk preparation from India. It is consumed all year round where it is usually taken along with meals. It contains raw milk, cream (malai) or yogurt which is blended manually in a pot with an instrument called madhani (whipper).	Cultured buttermilk
Dahi	Indian subcontinent	*Dahi* is obtained by curdling (coagulating) milk with rennet or an edible acidic substance such as lemon juice or vinegar, and then draining off the liquid portion. The increased acidity causes the milk proteins (casein) to tangle into solid masses, or curds.	Fermented milk curd
Raabadi	Haryana, Rajasthan.	Raabadi is prepared by fermenting pearl millet (*Pennisetum glaucum* (L.)) flour with butter milk, which is traditional popular beverage of North-Western states of India.	Cereal based fermented beverages
Probiotic Lassi	Haryana, Punjab, Rajasthan.	Lassi is a blend of probiotic yogurt, water and spices.	Fermented milk beverage
Probiotic Dahi	Indian subcontinent	Dahi is obtained by fermenting milk using combination of starter culture (Majorly *Lactococcus*) and probiotic culture.	Probiotic curd

Additional References:

Aneja, R. P., Mathur, B. N., Chandan, R. C., & Banerjee, A. K. (2002). Technology of Indian milk products. In: *Handbook on Process Technology Modernization for Professionals, Entrepreneurs and Scientists;* Dairy India Yearbook, New Delhi, India, pp. 1–462.

Hussain, S. A., Patil, G. R., Reddi, S., Yadav, V., Pothuraju, R., Singh, R. R. B., & Kapila, S. (2017). Aloe vera *(Aloe barbadensis Miller)* supplemented probiotic lassi prevents Shigella infiltration from epithelial barrier into systemic blood flow in mice model. *Microbial Pathogenesis, 102,* 143–147.

Modha, H., & Pal, D. (2011). Optimization of Raabadi-like fermented milk beverage using pearl millet. *Journal of Food Science Technology, 48* (2), 190–196.

Prabhakar, V. C. (2014). *Preparation of Low Fat, Mango Fortified Bajra Lassi.* PhD Dissertation; Mahatma Phule Krishi Vidyapeeth, Rahuri, pages 298.

Yadav, H., Jain, S., & Sinha, P. R. (2007). Antidiabetic effect of probiotic dahi containing *Lactobacillus acidophilus* and *Lactobacillus casei* in high fructose fed rats. *Nutrition, 23* (1), 62–68.

CHAPTER 13

SHELF-LIFE EXTENSION OF FERMENTED MILK PRODUCTS

V. SREEJA and KUNAL M. GAWAI

ABSTRACT

Microbes are major causative agents of spoilage in fermented dairy foods. Hence, aseptic manufacturing techniques are the first step towards ensuring a higher shelf-life. The shelf-life extending techniques in this chapter indicate promising applications for certain types of fermented dairy products. Most of the shelf-life extending studies have focused on a few products such as yogurt and cheeses. Hence, the applicability of such techniques in other fermented dairy products, especially the traditional products; need to be studied in detail. Application of bacteriocinogenic LAB (Lactic Acid Bacteria) strains has the edge over other techniques in fermented dairy products. Advances in packaging techniques, such as active packaging, can ensure a reduction in the risk of microbial contamination. However, combining two or more techniques seems to be more promising, which also should take into account the physical, chemical, and microbial characteristics of the product, the feasibility of application and economic aspects. However, application of such techniques should not be a substitute for high-quality raw materials and standard manufacturing practices, and rather these should be used as complementary techniques along with existing practices.

13.1 INTRODUCTION

In this era of increased food demand and rising food safety and quality concerns by the consumers, keeping the foods in its wholesome condition for a longer duration has become a necessity. This, in addition to preventing food loss, shelf-life can increase income, ensure availability of foods all year round, and improve food safety and security. The processes of fermentation, sun drying, salting, storage at low temperatures are some oldest means for extending shelf-life of foods in general. Milk and its products are well-known for their perishable nature, and fermentation is considered one of the oldest ways to preserve the vital nutrients of milk.

The process of fermentation of milk and the resulted fermented milk products have become an integral part of daily human nutrition. The process of fermentation of milk, in addition to extending the shelf-life of milk, has contributed to improve the taste, enhancing the digestibility and variety and value addition to fermented milk products. It also provides the consumer live and active beneficial cultures in significant numbers, which provide specific health benefits beyond conventional nutrition. Approximately 400 diverse products derived from the fermentation of milk are consumed around the world [20].

Lactic Acid Bacteria (LAB) are major groups responsible for such fermentation. LAB produces antimicrobial substances such as organic acids, H_2O_2, bacteriocins, diacetyl, ethanol, CO_2, etc., which exerts antimicrobial effect towards the growth of harmful bacteria, thereby prolonging the shelf-life of fermented dairy foods [45, 76]. Even though fermented products have extended shelf-life compared to its raw material, a limited shelf-life and the need to be kept under refrigeration are still major limitations with fermented dairy products. Like any other food, the shelf-life of fermented dairy products is affected by number of factors such as the microbial and physical quality of the raw material, processing parameters such as type and extent of heating and cooling, water activity of the food, pH/acidity of the food, presence, and extent of antimicrobial substances produced during fermentation, addition, and proper distribution of salt, sugar, etc.; cleaning and sanitization of equipment, environmental hygiene, packaging, storage temperature, humidity, etc.

During their storage, fermented dairy products undergo deterioration to a certain extent depending on packaging, post-production contamination, and storage conditions, which may include loss of nutritional value, changes in organoleptic properties and compromise on safety. Hence for the food industry, it becomes a challenge to control such deterioration or spoilage and to maintain the safety of the fermented foods.

This chapter encompasses microbial groups in the spoilage of fermented dairy products, different techniques for extending shelf-life of fermented milk products (such as aseptic manufacturing techniques, post-production heat treatment, use of preservatives, use of bacteriocins, physical methods, use of protective cultures and novel packaging techniques.

13.2   SHELF–LIFE OF FERMENTED DAIRY PRODUCTS

According to IFST Guidelines [70], shelf-life is defined as the time during, which the food product will [109]: (1) Remain safe; (2) Be certain to retain desired sensory, chemical, physical, and microbiological characteristics; and (3) Comply with any label declaration of nutritional data, when stored under the recommended conditions. Many factors influence the shelf-life of fermented dairy foods, which can be categorized into intrinsic and extrinsic factors [70]:

• Intrinsic factors such as water activity (a_w), pH, total acidity, redox potential (Eh), available oxygen, nutrients, inherent and acquired microbial load and preservatives used product formulation are important and contribute in acceptance of the final product by consumers. Intrinsic factors are influenced by variables such as raw material type, quality, and product formulation and structure.

• Final product comes across with extrinsic factors as it passes on through the food chain. These factors are (1) processing-related parameters like time and temperature, pressure over headspace, environment within packaging; (2) post-production treatment including cooking or warming before consumption and the manner by package handled by the consumers. Similarly, temperature variation, relative humidity (RH), exposure to light and microbial counts during processing, storage, and transportation also govern shelf-life.

All these factors, intrinsic and extrinsic factors can operate singly or interactively and many times unpredictable controlling the shelf-life of products. Fermented milk products vary in their shelf-life based on their method of preparation, storage conditions, composition, and the ability of the food matrix to support the growth of spoilage organisms. Some fermented products such as Kishk could be preserved for up to 2 years, while yogurt or dahi may get spoiled within 15 days even if stored under refrigerated conditions. On the basis of duration of their shelf-life, fermented milk products can be classified into three categories:

• **Short Shelf-Life:** Three weeks under refrigeration;
• **Medium Shelf-Life Products:** Several weeks under refrigeration; and
• **Long Shelf-Life Products:** Several months at room temperature.

Products, which have a normal shelf-life of less than three weeks, are termed as short shelf-life products such as cottage cheese, yogurt, dahi, cultured buttermilk [21, 128]. The shelf stability of such short shelf-life dairy product is governed by growth rate microorganisms and consequently degradation by their metabolic end products. And the shelf-life of intermediate and long-life dairy products is generally decided by enzymatic and/or by chemical degradation. Tables 13.1 and 13.2 depict average shelf-life of some fermented dairy foods.

13.3 MICROBIAL GROUPS ASSOCIATED WITH SPOILAGE OF FERMENTED DAIRY PRODUCTS

Spoilage of fermented milks is represented by flavor, texture, and appearance defects. The most defects may be caused by microorganisms [69] (including the starter cultures) or maybe due to manufacturing issues. The defect of whey syneresis is an example of a manufacturing issue. Even though rare, some chemically-derived flavor defects are also seen in fermented milks, which may be caused by using poor-quality milk. An example is rancid or oxidized flavors.

Various processing steps contribute and decide the dominance of types of spoilage microorganisms, during fermented dairy foods preparations. Such as formulation, processing, packaging, storage, distribution, and handling. These spoilage organisms mostly belong to psychrotrophic, Gram-negative bacteria group, yeasts, and molds, lactobacilli, and sometimes aerobic spore-forming bacteria. Coliforms, lactose fermenting yeasts, few heterofermentative LAB, and anaerobic clostridia spore have produced gas in cheeses. The extracellular hydrolytic enzymes produced by the psychrotrophic bacteria bring proteolytic and lipolytic changes in the cheese, causing flavor, body, and textural defects and decrease in yield. Psychrotrophs are a major cause of limited shelf-life of fresh cheese varieties, such as cottage cheese. Table 13.3 shows some commonly occurring defects in fermented dairy products.

13.3.1 YEASTS AND MOLDS

Yeasts and molds are primary causative agents of spoilage in fermented dairy products owing to their ability to grow at low pH, low temperature, and high sugar concentration, and low water activity, ability to ferment lactose, ability

TABLE 13.1 Examples of Some Fermented Dairy Foods with Average Shelf-Life When Stored at 4°C; Modified from [13, 15, 118]

Food	Description	Origin	Average Shelf-Life
Acidophilus milk	Low-fat heat-treated milk inoculated with *Lactobacillus acidophilus* or *Bifidobacterium bifidum*	USA, Russia	2–3 weeks
Cultured Buttermilk	Cow skim milk heated, homogenized, cooled, and inoculated with *Streptococcus cremoris*, *Streptococcus lactis*, *Streptococcus lactis ssp. diacetylactis*, *Leuconostoc cremoris*	USA	10 days
Filmjolk	Product made using pasteurized, homogenized, whole cow's milk fermented with ropy strains of *Streptococcus cremoris* and starter cultures of cultured buttermilk. Ropy strain is responsible for the slimy texture of the product	Sweden	10–14 days
Kefir	Acid-alcoholic effervescent milk made using Kefir grains or starter comprising *Lactobacillus casei, Streptococcus lactis, Lactobacillus bulgaricus, Leuconostoccremoris, Candida kefir, Kluyveromyces fragilis*, etc.	Russia	10–14 days
Kishk	Fermented milk mixed with parboiled wheat and dried	Egypt, the Arab world	1 year at room temperature
Kumiss	Similar to Kefir, from horse milk and frequently served with cereal	Russia	10–14 days
Quark	Low-fat acidic soft cheese made using rennet and starter culture of lactic acid bacteria	Germany	Varies
Ricotta	Whey cheese, perhaps with added skimmed, whole milk or cream, salt. Culture used is *Streptococcus thermophilus* and *Lactobacillus bulgaricus*. It is a hard variety	Europe	Varies
Skyr	Fermented milk made from ewe milk using rennet and starter	Iceland	–
Viili	Viscous milk fermented with lactic acid bacteria and mold	Finland	14 days
Yakult	High heat-treated milk fermented by *L. casei* Shirota strain	Japan	10–14 days
Yogurt	Custard-like sour fermented milk	Turkey	30–40 days

TABLE 13.2 Some Indian Fermented Dairy Foods with Their Average Shelf-Life when Stored at 4°C

Fermented Milk Product	Description	Average Shelf-Life
Dahi/probiotic dahi	Milk soured by using starter cultures. Probiotic dahi additionally contains probiotic culture(s)	15 days
Chhash/Buttermilk	Mesophilic fermented pasteurized milk	10 days
Lassi/probiotic lassi	Sour drink consumed sweetened with sugar/honey	15 days
Misti Doi	Misti doi is a fermented milk product with a cream to light brown color which has around 17–19% sugar	15 days
Shrikhand (thermized)	Concentrated fermented milk product prepared by partial removal of whey from the curd/dahi, followed by mixing the concentrated mass with sugar, flavoring, and spices. The thermized product is given a post-production treatment of thermization	120 days

TABLE 13.3 Some Commonly Occurring Defects in Fermented Dairy Products; Modified from [20, 119]

Defect	Causes	Products
Bitter	Contamination with Proteolytic strains of starters and/or contaminating microflora	Yogurt, dahi, buttermilk, sour cream, cheese
Colored spots	Growth of Pigment producing contaminants on the surface, mold growth	Yogurt, dahi, buttermilk, cheese
Crumbly, mealy, and short body	Excessive acid production and low moisture retention in cheese	Cheddar Cheese.
Early blowing/late blowing	Coliform contamination/ contamination with anaerobic spore formers	Cheese
Flat/sweet	Too low incubation temperature, slow cultures, improper acid development, contamination	Yogurt, dahi, buttermilk, sour cream, cheese
Gassiness/Bloat	Yeast or coliform contamination, Overgrowth of gas producers,	Yogurt, dahi, buttermilk, sour cream
Moldy/Musty	Mold growth	Yogurt, dahi, buttermilk, cheese
Rancid/soapy	Microbial lipases from contaminating microflora	Yogurt, dahi, buttermilk, sour cream, cheese
Ropiness	Contamination, too low temperature of incubation, ropi strain of cultures dominating	Yogurt, dahi, buttermilk, sour cream, cheese
Sour/High acid	Too high incubation temperature, improper cooling, an excessive amount of inoculum	Yogurt, dahi, buttermilk, sour cream, cheese
Whey separation	Over or under acidification, mechanical damage, low solids content, high incubation temperature, insufficient heat treatment of milk, etc.	Yogurt, *dahi*
Yeasty, fermented, fruity flavors	Contamination with yeasts	Yogurt, dahi, buttermilk, sour cream, cheese

to release extracellular proteolytic and lipolytic enzymes. Their growth is selectively favored at low pH of cultured dairy products such as in dahi, yogurt, and buttermilk leading to the production of fermented or yeasty off-flavors. *Geotrichumcandidum* contamination of cottage cheese many times decreases the diacetyl content. Diacetyl concentrations are observed to reduce by 52–56% in low-fat cottage cheese after 15–19 days of storage at 4–7C [8]. When yeast populations are increased up to 10^5–10^6 CFU/g that resulted in the development of yeasty fermented off-flavors with gassy appearance. Galactose utilizing strains of yeasts such as *Saccharomyces cerevisiae* and *Hansenulaanomala* has resulted in the spoilage of yogurt [53].

High acid and abundant nutrients in most of the cheeses are enhancing grow of spoilage causing yeasts. Surface moisture, often containing lactic acid, peptides, and amino acids favors rapid growth. The production of alcohol and carbon dioxide (CO_2) by yeasts results in yeasty flavor defect in cheeses [66].

Also, a large amount of CO_2 production by yeast can cause bulging of packages of cheese [154]. The lipolytic action of yeasts subsequently produced short-chain fatty acids and form fruity esters by combining with ethanol. Some proteolytic yeast strains developed an egg-like odor as a result of the production of sulfides. Routinely identifying yeasts, which contaminate cheeses, belong to *Candida spp.*, *Kluyveromycesmarxianus*, *Geotrichum candidum*, *Debaryomyces hansenii*, and *Pichia spp.* [75, 90].

Molds are ubiquitous in the dairy environment. Surface contamination of products such as dahi, yogurt, cheese by molds often leads to visible surface growth and product rejection. Presence of oxygen favors the growth of molds on the cheese surfaces. In packaged cheeses, the unavailability of oxygen restricts the growth of molds; however, certain molds are capable of growth under low oxygen level. *Penicillium spp.* and *Cladosporium spp.* are most commonly growing molds in vacuum-packaged cheeses [6, 64]. Certain mold species are known to produce mycotoxins. Some strains of mold species *Aspergillus flavus*, *Aspergillus paraciticus*, and *Aspergillus nominus* produce aflatoxins. Thus mold growth on cheese and fermented dairy products may pose potential hazards to food safety and human health [6, 157].

13.3.2 COLIFORMS

The coliforms include all aerobic and facultative anaerobic Gram-negative, non-sporulating bacilli, capable of producing acids and gas from fermentation of lactose. Their isolation from foods indicates evidence of poor hygiene or inadequate processing, especially heat-treatment, process failure, and post-process contamination of foods. Among the coliforms members, *E. coli* is commonly used as an index organism for potential fecal contamination in certain foods and the presence of enteric pathogens, such as Salmonella [142]. During cheese production, lactic acid production by slow strains of starter cultures favors the growth of coliforms, which have short generation times under such conditions, leading to the production of gas in the product.

In soft mold-ripened cheeses, the pH increases during ripening, thus increasing the growth potential of coliform bacteria [46]. In cheeses, coliforms grow during the early stages of cheese making and bring defect such as 'early blowing' due to the production of CO_2 and in particular, hydrogen, which has very low solubility in cheese [106].

13.3.3 SPORE-FORMING BACTERIA

Spore-forming bacteria can come in contact with fermented dairy products through the raw milk. Most important spore-forming bacteria in the dairy industry belong to genus *Bacillus* and genus *Clostridium*. Owing to its aerobic oxygen requirement, members of the genus *Bacillus* such as *Bacillus licheniformis*, *B. cereus*, *B. subtilis*, *B. mycoides* and *B. megaterium* are the most frequently found spore-formers in dairy products. However, subsequently with the depletion of oxygen in the packages, growth of *Clostridia* occurs that may cause the increase in pH due to extensive proteolysis during ripening of cheeses with the release of amino acids. *Clostridium tyrobutyricum* leads to the production of gas and butyric acid particularly in this favorable condition [86]. Spore-formers are also responsible for gassy defects, such as 'late blowing' in cheese [90].

13.3.4 PSYCHROTROPHS

The psychrotrophs are capable of growing at commercial refrigeration temperature irrespective of the optimum growth temperature. Psychrotrophs mainly include members of the genera *Pseudomonas*, *Bacillus*, *Micrococcus*, *Aerococcus*, *Lactococcus*, and members of the Enterobacteriaceae family [90]. Proteolytic psychrotrophs are capable of releasing proteolytic enzymes, which can attack particular fractions of casein resulting in a lower yield, body, and texture defects, bitter taste and whey separation in cheese and other cultured dairy products. Psychrotrophs (such as, *Pseudomonas fragi*, *Pseudomonas fluorescens*, *Achromobacter lipolyticum*, *Flavobacterium sp.*, *Alcaligenes ssp.*, and *Acinetobacter sp.*) produce lipolytic enzymes, which are heat stable and active at ambient or even refrigeration temperature. Such lipases can bring about hydrolysis of fat resulting in the production of short-chain fatty acids and corresponding flavor defects, such as hydrolytic rancidity in the products [63].

13.3.5 LACTIC ACID BACTERIA (LAB)

Even though used as starter cultures for fermented product manufacturing, the post-process activity of starter cultures and their unwanted presence in other cultured dairy products can bring defects in the products. Heterofermentative LAB (such as lactobacilli and leuconostoc) can develop off-flavors and gas in ripened cheeses. Non-starter culture, naturally occurring lactobacilli, propionibacteria, and *Lactococcus lactis* subsp. *lactis* produce small amounts of gas in cheeses through catabolism of amino acids [96]. Decarboxylation of glutamic acid by certain strains of *Streptococcus thermophiles* and *Lactobacillus helveticus* results in the formation of CO_2 and 4-aminobutyric acid causing excessive gas production and subsequent formation of cracks in cheese [161]. Few lactobacilli produced tyrosine lead to a pink to brown discoloration in ripened cheeses [90].

13.4 TECHNIQUES FOR ENHANCING SHELF-LIFE OF FERMENTED MILK PRODUCTS

The limited shelf-life of fermented dairy products both at room and refrigeration temperature is a major constraint in their effective marketing. Normally the shelf-life of fermented milks such as yogurt, dahi, and buttermilk is 2–3 days at ambient temperature and 2–3 weeks at refrigerated temperature. This limited shelf-life of these products is ascribed mainly to two factors: (i) excessive souring due to the continuous growth of starter; and (ii) spoilage owing to the presence of contaminants. These limitations are being tackled through methods of suppression of starter culture and prevention and control of contaminants, especially yeast and molds after the product are ready. Researchers have tried various techniques for extending the shelf-life of cultured milks, *viz.*, heat treatments, the addition of preservatives, use of gas, effective packaging, microwave treatment, irradiation, hurdle technology, bio-stabilization, etc. [119]. Some of the methods can be applied to extend shelf-life of fermented dairy products are discussed here in this section.

13.4.1 ASEPTIC MANUFACTURING TECHNIQUES

Combining aseptic manufacturing techniques along with the use of cultures having the mild post-acidification ability can enhance the shelf-life of fermented dairy products to a certain extent. The following requirements should be taken into consideration in the aseptic manufacturing of fermented dairy products:

- Closed production lines must be used.
- Filling and capping to be carried out by using aseptic or semi-aseptic machines.
- Milk meant for product manufacturing should be of the very low microbial count.
- Packaging materials/containers should be sterile.
- The production line must be designed for sterilization and aseptic operation.
- The starter/probiotic culture must be free of contaminants.

Aseptic processing can be done by inoculating the sterile milk under aseptic conditions with a special culture, followed by acidification, homogenization, and aseptic packaging. Another option is inoculation of the sterile milk (UHT sterilized) followed by aseptic packaging (Tetra-pack or Brick-pack) and acidification in packages. Such methods of aseptic acidification are used in the large-scale production of long-life yogurt [16].

13.4.2 POST PRODUCTION HEAT TREATMENT

Post-production heat treatment, such as mild pasteurization or thermization, can be used to extend the shelf-life of the fermented dairy products. This is mainly done for fluid products with low viscosity. The most popular example is thermization of buttermilk usually being carried out to get extended market life of buttermilk. A post-production heat treatment helps to improve the shelf-life since heating kills most of the starter bacteria and yeast and mold; and if post-processing contamination is avoided, then the shelf-life of the product can be extended up to eight weeks [140]. There is no post acidification in heated yogurt. Heated yogurt or pasteurized yogurt can be prepared by heating in the package at about 55°C for 30 minutes, followed by cooling. However, the limitations associated with such treatments are increased by whey syneresis, flavor loss, and a drastic reduction in total viable starter culture count [140, 147].

Thermization of *Misti Doi,* a sweetened curd type product at 60°C/5 min caused inhibition of 99.99% acid-producing bacteria [136] and was able to extend the product shelf-life from 48 hours to 3 weeks at 28–32°C [135]. Thermization of *shrikhand* at 70°C for 5 min enhanced its shelf-life from 45 days to 70 days without adversely affecting its composition, consistency, and organoleptic qualities [120]. Thermization in activates yeasts and molds in yogurt and increases shelf-life to 6–8 weeks at 12°C. Thermal treatment at 65–70°C retarded the lipolytic changes in *lassi,* a stirred curd and observed extended shelf-life up to 30–35 days at 7 ± 1°C [87, 135].

13.4.3   USE OF PRESERVATIVES

The addition of specific legally permissible chemicals to foods to inhibit microbial growth and chemical reactions is a major method of food preservation. However, in India, only very few preservatives are legally permitted in many cases, and there are specific limits on how much can be used and particularly in foods. Also, the use of some preservatives is limited to just a few types of food. Some of the preservatives used for fermented dairy products are mentioned discussed in this section.

13.4.3.1   NITRATES AND NITRITES

Nitrate and its nitrite salts are permitted in specific cheeses only in case of fermented dairy foods [151]. As per *Codex draft General Standard for Food Additives*, the maximum permitted level for nitrate and nitrite in cheese and cheese products is 37 mg/kg and 17 mg/kg, respectively [156].

13.4.3.2   SORBIC AND BENZOIC ACIDS

Sorbic and benzoic acids, and their salts, and sulfur dioxide and its derivatives, are the most widely used chemical preservatives [159]. Present Standards of FSSAI in 2011 have strictly lower down maximum permissible limits for the use of chemical preservatives [7]. This poses a significant problem for the food industry. Reduction in preservative level necessitates need to look for another preservation technique, which could better work with available systems [151].

Sorbates and natamycin are most popular and widely used mold inhibitors for surface application on cheeses. The limitation with sorbate is that it tends to diffuse into the cheese affecting its flavor and decreasing its concentration, whereas very little diffusion happens in case of natamycin [36]. Sorbates are more effective against yeasts and molds than bacteria. Their methods of application include dipping, spraying, or dusting of the product or incorporation into the wrapper. Environmental factors (such as pH, length, and temperature of storage, water activity, atmosphere, type of microflora, and certain food components) can affect the activity of sorbates [91].

13.4.3.3   CHITOSAN

This is an environment-friendly and relatively inexpensive antimicrobial agent, which can prolong the shelf-life of cheese. Chitosan is effective in inhibiting the growth of spoilage microorganisms (such as coliforms and Pseudomonas spp.), whereas it does not affect the growth of LAB, thus saving functional dairy microbiota [3].

13.4.3.4   BACTERIOCINS AS BIOPRESERVATIVE

Bacteriocins are increasingly finding application in the food industry as biopreservatives [49, 158]. The safe history to use of LAB in traditional fermented food products make LAB bacteriocins very attractive to be used as biopreservatives [19]. Bacteriocinogenic strains of LAB can be used for *in situ* production of bacteriocins in dairy products [83]. Bacteriocins produced by LAB possess antibacterial activity against foodborne pathogens [5, 144]. The antimicrobial effects of bacteriocins against sensitive microorganisms depend to a great extent on environmental factors, such as pH, temperature, composition, and constitution of food [45].

Bacteriocins can endure harsh treatments, such as boiling, without losing much of their activity. Genera from LAB (such as *Lactobacillus, Lactococcus, Leuconostoc, Pediococcus,* and *Enterococcus*) have reported to be produced>300 different bacteriocins. Bacteriocins are generally low molecular weight proteins that enter into target cells by attaching on the receptors of the cell surface. They acted on cells with specific bactericidal mechanisms, which include pore formation on the cell wall, decomposition of cellular DNA, interruption during specific cleavage of 16S rRNA, and prevent peptidoglycan synthesis [4].

There are some of the important aspects that need careful consideration for application of bacteriocin in foods, such as gathering of detailed on antibacterial spectrum of bacteriocin, biochemical, and genetic characteristics, effectiveness in food systems, regulatory implications, safety aspects, virulence or antibiotics resistance, economical aspects or cost of using a bacteriocin in foods [4].

Uses of bacteriocins have some drawbacks such as difficulty in obtaining large-scale preparations suitable for assays and application at pilot or industrial scale. The use of bacteriocins has been widely studied in different cheese production systems. Use of bacteriocins in cheese by combination with high-pressure (HP) treatments combined produced *in situ* on the survival of *Escherichia coli* O157:H7 was investigated. The application of reduced pressures combined with the bacteriocin-producing LAB is a feasible procedure to improve cheese safety [129].

Among the dairy products, most of the bacteriocins applications are for cheese. However, Benkerroum et al., [15] reported *in situ* bacteriocin production by bacteriocinogenic *Streptococcus salivarius* ssp. And *S. thermophilus* B during yogurt storage refrigeration or temperature abuse and found that it was effective in controlling *L. monocytogenes* and *Staphylococcus aureus*, resulting in a 5-day extension of shelf-life. Bacteriocins of *S. Thermophilus* have a broad inhibitory spectrum against several pathogens (such as *Listeria monocytogenes*, *Salmonella Typhimurium*, *Escherichia coli*, *Yersinia pseudotuberculosis*, and *Yersinia enterocolitica*, *Clostridium tyrobutyricum*); hence it may be of useful for yogurt biopreservation [11].

13.4.3.4.1 Natamycin

Natamycin produced by the bacterium *Streptomyces natalensis* [42] is found to inhibit microbial activity by binding to and altering fungal cell membrane sterols, causing leakage of the vital cell structures [33, 60]. Natamycin at a concentration of 5–10 ppm has been suggested for yogurt preservation [149]. Yogurt treated with natamycin showed no yeast and mold growth after 30 days of storage, whereas normal yogurt showed yeast and mold growth after 7–15 days [152].

Tortorello et al., [150] reported the use of 0.02% natamycin along with 10 IU/g nisin to extend shelf-life of cottage cheese from 21 to 35 days. Use of natamycin as a natural preservative in dairy products and other foods has been approved in more than 60 countries [36]. This preservative is particularly effective against yeasts and molds [43, 73, 77, 98]. Among the dairy products, natamycin is commonly used in cottage cheese, sour cream, and yogurt [23, 42]. Eissa et al., [44] found that yogurt treated with natamycin @10 ppm can extend the shelf-life up to 40 days with good sensory characteristics during the refrigerated storage.

13.4.3.4.2 Nisin

Nisin is a bacteriocin produced by *Lactococcus lactis* subsp. *Lactis*. It is used as a biopreservative. It is considered safe by the World Health Organization (WHO) and has generally recognized as safe (GRAS) status. U.S. Food and Drug Administration (US-FDA) has approved its use as a preservative in many foods, and it has been licensed as a food additive in over 45 countries [138]. Nisin is usually used in processed cheese products to prevent outgrowth of spores of *Clostridium tyrobutyricum* that may survive high heat treatments such as 85–105°C. Use of nisin helps in the formulation of cheese products with high moisture levels and low NaCl and phosphate contents. It also allows products to be stored outside chill cabinets without risk of spoilage. Nisin is reported to have antibacterial activity against Gram-positive bacteria, LAB, and *Listeria, Staphylococcus, Bacillus*, and *Clostridium* [47].

Nisaplin™ has been used to control *L. monocytogenes* in fresh cheese with a high water activity (a_w), such as ricotta-type cheese, to extend its shelf-life [37]. Nisin-producing lactococci were found to be effective against Clostridial [126] and Listerial [130] spoilage; however, interferences with starter performance and cheese ripening and high sensitivity to bacteriophages were demonstrated [114]. Nisin-Z, produced by *Lactococcuslactis* IPLA729, inhibited *Staphylococcus aureus*, as well as enterotoxigenic *Bacillus cereus* in different cheeses [107, 108, 127]. Nisin added @ 100 or 500 mg/kg in processed cheese suppressed total plate and anaerobic spore counts during 3 months of storage at 5 or 21 °C. Nisin @ 5 mg/kg prevented the growth of *B. stearothermophilus, B. cereus,* and *Bacillus subtilis*.

In Ricotta-type cheese, nisin @2.5 mg/l inhibited the growth of *L. monocytogenes* for more than 8 weeks, while cheese made without nisin contained unsafe levels of the bacteria within 1–2 weeks. Inactivation of endospores and mesophilic bacteria in cheese could be achieved using nisin in combination with HHP [9, 50]. Encapsulation of nisin in liposomes (nisin-Z) was done to solve the problem of non-specific inhibitory activity of nisin against Gram-positive bacteria, such as the growth of starter cultures in cheese fermentation. Use of liposome preparations inhibited *Listeria* in Cheddar cheese.

In Minas frescal cheese, encapsulated nisin gave best results compared to adding free nisin in the control of *L. monocytogenes* when stored at 7°C for at least 21 days [94]. Lipolysis in dahi during storage at 25°C was effectively controlled by using either nisin or nisin producing organisms capable of producing 1000 IU/g nisin [123] and nisin at a level of 15 RU/g in dahi retained all its desirable characteristics up to 35 days at 15±1°C [88]. Use of 25 RU/g nisin in dahi meant for *shrikhand* manufacturing reduced the spoilage of shrikhand (sweetened concentrated curd) due to LAB, proteolytic, and lipolytic organisms [136].

13.4.3.4.3 Lacticin 3147

Lacticin, produced by *Lactococcus lactis*, is an effective inhibitor of many Gram-positive food pathogens and spoilage microorganisms. It is used to control the growth of undesirable microorganisms during Cheddar cheese manufacturing. The lacticin strain *Lactococcus lactis* was applied on cheese surface to control the growth of *L. monocytogenes* [112]. In Cheddar cheese manufacturing, *Lactococcus lactis* was used as an adjunct to control non-starter LAB and to increase lysis of starter cells to speed up enzyme release for contributing to quality, texture, and flavor during cheese making [111, 113, 114]. O'Sullivan et al. [115] constructed food-grade Lactococcal strains to produce lacticin 3147 and lacticin 481 for food applications. Lacticin 3147 was found to be highly effective against *L. monocytogenes* in natural yogurt and in Cottage cheese.

13.4.4 PROTECTIVE CULTURES

The bacteriocin producing cultures are increasingly finding application as protective cultures to extend the shelf-life of fermented dairy products. Protective cultures, consisting of live pure cultures or culture concentrates, are added to foods to reduce the risks by spoilage microorganisms. They develop their protective effects via metabolic pathways in food. The protective cultures available in practice usually contain the same microorganisms as found in the starter cultures, but they do not usually determine the typical nature of a fermented food characterized by the starter culture. Protective cultures act through the principles of competitive exclusion and the formation of antagonistically active substances [155].

Protective cultures must fulfill certain criteria to be effective, which include: the ability to grow in the food, ability to inhibit or limit the growth of spoilage flora, not create unwanted changes to the product, not to produce substances, which may be harmful to humans [85]. Some commercial preparations, which have been developed as protective cultures or shelf-life extenders, include Microgard™, BioProfit, ALTA™ 2341, ALC01, and FARGO™ 23 [82].

13.4.4.1   MICROGARD™

Microgard™ is produced by fermentation of grade-A skim milk using *Propionibacterium freudenreichii* ssp. *shermanii*. Its antimicrobial spectrum includes some Gram-negative bacteria, yeasts, and fungi. It is added to a variety of dairy products such as cottage cheese and yogurt. In cottage cheese, it inhibits psychrotrophic spoilage bacteria. The inhibitory activity of microgard is mainly due to propionic acid, but it is reported that bacteriocin-like proteins are produced during the fermentation [65, 50, 104]. Microgard @ 10% extended the shelf-life of yogurt by more than 82 days at 5°C due to inhibition of yeasts and molds [132].

13.4.4.2   BIOPROFIT™

BioProfit™ (Valio, Helsinki, Finland) is a commercial product containing *Lactobacillus rhamnosus* LC 705 and *Propionibacterium freudenreichii* JS. In yogurt and cheese, BioProfit is used as a protective culture to inhibit yeasts [82, 85].

13.4.4.3   ALTA™ 2341, ALC01, AND FARGO™ 23

ALTA™2341is produced by fermentation using *Pediococcusacidilactici*. It is found effective against *L. monocytogenes* and is used as a barrier against this organism in Queso-Blanco type cheese. Its inhibitory effect is due to organic acids and bacteriocinpediocin. ALC01 (Niebull, Germany) is a patented anti-Listerial culture developed for soft cheese. Pediocin produced by *Lactobacillus plantarum* is responsible for its protective activity. ALC01 inhibited the growth of Listeria on the surface of Munster cheese. FARGO™ 23 (Quest International, USA) is usually added to raw milk meant that is used for raw milk cheese production [82].

13.4.5   ROLE OF BACTERIOCINS IN BIOACTIVE PACKAGING

Another potential application of bacteriocins is in bioactive packaging, where bacteriocins produced by LAB are incorporated into packaging materials in contact with food. In bioactive packaging either the bacteriocin is directly immobilized to the food packaging, or a sachet containing the bacteriocin is put into the packaged food. This helps the bacteriocin to be released during product storage [95, 97]. The level of *L. Innocua* and *S. aureus* in sliced cheese was reduced by the application of nisin-adsorbed bioactive inserts [117]. The efficacy of bacteriocins can be improved further by applying them in combination with other antimicrobials or barriers such as acidification, preservatives, modified atmospheres, etc.

Bacteriocins have been tested in different food systems as part of hurdle technology [48, 51] to enhance the efficiency of treatments, to reduce the required concentration of bacteriocin, to broaden the antimicrobial spectrum of treatments, to enhance bacterial endospore inactivation and as an additional barrier to growth of sub-lethally injured microbial cells, other microbes and endospores surviving treatments.

13.4.6   PHYSICAL METHODS

13.4.6.1   IRRADIATION

Irradiation using electron beam at an average dose of 0.21 and 0.42 kGy has been found to be effective in reducing mold populations in Cheddar cheese [41, 58]. Gamma-ray irradiation of plain yogurt extended its shelf-life up to 3 weeks, and the treatment did not adversely affect the chemical and sensory attributes of plain yogurt [59].

13.4.6.2   HIGH PRESSURE (HP) PROCESSING

High-pressure processing (HPP: 100–1000 MPa) is a promising method for preservation of foods at room temperature [22]. Use of HPP in case of yogurt and related products has given an extended refrigerated shelf-life of 60 days [89]. The process is suitable for almost all packages such as bottles, cups, pouches, and trays [122]. Application of HPP in combination with moderately elevated temperature can be used for enhancing the shelf-life of cultured milk products. HP treatment at 400 MPa for 30 min increased the shelf-life of yogurt due to complete inactivation of LAB [125]. De Ancos et al., [35] reported that a pressure treatment of >200 MPa at 20°C for 15 min caused a significant reduction in the viable counts of *S. thermophilus* and *Lb. delbrueckii subsp. bulgaricus* and prevented post-processing acidification in stirred low-fat yogurt.

Jankowska et al., [72] used HPP at 550 MPa for the preservation of yogurt up to 4 weeks. Kefir when treatment with HPP at 400 MPa for 30 min caused the deactivation of bacteria and yeasts present in it without affecting the structure of kefir [93]. High hydrostatic pressure treatment of 400–500 MPa in combination with thermal processing (85°C/30 min) showed reduced syneresis in low fat set yogurt [62]. Swiss cheese slurries treated at 345 and 550 MPa for 10 and 30 min at 25°C significantly reduced the population of starter bacteria, coliforms, *Staphylococcus*, and yeasts and molds during ripening at 30°C [74].

13.4.6.3   OHMIC HEATING

In Ohmic heating, an electric current is applied directly to the food. The heat generated due to the electrical resistance of the food causes the destruction of microbes. Ohmic heating causes minimal changes in nutritional and flavor characteristics of foods. In the case of cultured products, additives such as fruit pieces are sterilized/pasteurized by this method [12]. Fruit preparations meant for use in yogurt are pasteurized by commercial Ohmic heater [151].

13.4.6.4   ULTRA-VIOLET IRRADIATION

UV irradiation having wavelengths 220–300 nanometers can destroy molds, yeasts, and viruses. The highest bactericidal efficiency is seen between 250 to 260 nm [102]. Yogurt cups when exposedto100–200 MW/cm^2 for 2 seconds at a distance of 40 mm improved the shelf-life of stirred fruit yogurt up to 42 days [135].

13.4.6.5   CARBONATION

Carbonation of yogurt through post-production homogenization pressures of 175 kg/cm^2 enhanced its shelf-life. In another study, the keeping quality of yogurt was increased for more than 4 weeks at 15°C using post-incubation heat-treatment and gas flushing with carbon dioxide [135, 148]. In the preparation of cream-style cottage cheese, direct incorporation of CO_2 into the cream dressing prior to mixing with the curd improved its shelf-life [24–26]. Moir and others [101] reported that the shelf-life of cottage cheese could be significantly increased by the injection of 10 mM CO_2 into cottage cheese cream dressing and package headspace without adversely affecting pH or flavor of the product. Menon et al., [99] reported that shelf-life of *lassi* could be extended by up to 12 weeks at refrigerated conditions (7°C) by carbonation at 50 psi for 30s.

13.4.7   NOVEL PACKAGING TECHNIQUES

The packaging is usually done to protect food from spoilage due to physical damage, physicochemical degradation, and microbial spoilage. Packaging of fermented foods requires special care as the process of fermentation continues throughout the storage even though at a very slow pace. Additionally, the packaging has to be sturdy enough to protect the fragile coagulum of set type of fermented milks such as set yogurt and set dahi to maintain its proper body and textural characteristics. It should also prevent migration of chemical and microbial contaminants into the product from external environment and packaging materials. In lactic acid fermented products, the acid produced can penetrate into the packaging and form bacterial films on polymer surfaces [145]. In the case of probiotic fermented products, the viability of probiotics should be taken care of by maintaining suitable environmental conditions [79].

Polystyrene cups are the most convenient packaging for indigenous fermented products like dahi, shrikhand, and chakka. Polypropylene (PP), polyvinylchloride, and polyvinylidene chloride are other commonly used plastics for packaging of fermented milks [56, 57, 116]. Cup thermo-fill and sealing machine does thermoforming of cups from the plastic film, filling of product and sealing of lid in a continuous sequence of operation. This can prevent contamination during packaging [119]. Rigid containers such as metal tins and glass packages and flexible plastics films are also been used for packaging of yogurt and other fermented milks [92], whereas cheeses of convenient sizes (8–10 g) are individually wrapped with aluminum foil and placed in an outer plastic or paperboard container, in plastic pouches or tubs with a press on plastic lids, or in squeezable tubes or bottles [124]. The recent decade has seen the emergence of innovative packaging approaches to provide shelf-life extension and safety to the fermented milk products. Some of these are discussed in this section.

13.4.7.1   MODIFIED ATMOSPHERIC PACKAGING (MAP)

Packaging foods in the right atmosphere is one of the best ways to ensure its shelf-life. The gas atmosphere comprising of oxygen, nitrogen, or carbon dioxide is carefully controlled for achieving maximum shelf-life. Oxygen causes food spoilage reactions such as oxidation, browning, and color changes. It also favors the growth of aerobic bacteria. But the total exclusion of oxygen favors the growth of anaerobic bacteria such as *Clostridium botulinum*. Hence, in most cases, food is preserved by decreasing the oxygen concentration in the package and replacing it with CO_2 and/or nitrogen. CO_2 discourages the growth of aerobic bacteria and fungi at low temperatures and the packages filled with 100%; CO_2 collapse at such temperatures due to the solubility of a gas in water and food. Such property is an advantage in the packaging of cheese and related products. Nitrogen is a non-reactive gas with no smell or taste, usually used as a filler gas to replace oxygen and CO_2. It is not absorbed in food or water [27, 52].

Modified atmospheres can be created either passively by the commodity or intentionally via active packaging [160]. MAP has the potential to increase the shelf-life of cultured milk products especially cheeses. Growth of molds in cheese can be prevented by the use of MAP, and for different types of cheeses, the optimum MAP conditions can vary [110]. Low O_2 and high CO_2 atmospheres were found to be optimum for processed cheeses containing no-starter bacteria, but for cheeses containing active starter cultures, MAP of low O_2 and controlled CO_2 using a permeable film gave best results. Use of gas mixtures containing 40% CO_2 and 60% N_2 increased the shelf-life of cottage cheese significantly [105]. MAP in combination with oregano and thyme caused a significant reduction of *E. coli* O157:H7 and *L. monocytogenes* in Feta cheese [55].

Conte et al., [29] reported that MAP(95:5 CO_2:N_2) in combination with lysozyme and EDTA in Burrata cheese prolonged its shelf-life. Nitrogen and CO_2 atmosphere in the headspace have been used to retard the growth of many types of yeast or molds present in yogurt [17]. However, the industrial use of modified atmospheres in package headspaces of cultured products seems limited. Instead, the manufacturers prefer to concentrate more on cleanliness and sanitation to minimize contamination.

13.4.7.2 ACTIVE PACKAGING

Active packaging comprises of incorporation of additives into packaging film or within packaging containers by which the safety and shelf-life of a product are enhanced, and the quality is maintained [121]. Active packaging techniques include absorbers (scavengers), releasing systems and other systems. Absorbing systems remove undesirable compounds such as oxygen, carbon dioxide, ethylene, excessive water, taints, and other specific compounds. Addition or emission of compounds to the packaged food or into the head-space of the packages takes place in releasing systems. Examples are carbon dioxide emitters, ethanol emitters, antioxidant releasers, and anti-microbial releasers.

Active packaging systems may be in the form of sachet, label or film type, and its use depends on the physical form of commodity. Sachets containing absorbers and releasers are placed freely in the headspace of the package, but care should be taken to prevent its direct contact with the product and migration problem [2, 38].

13.4.7.3 USE OF OXYGEN SCAVENGERS

Oxygen scavengers control the interior atmosphere of the package by removing oxygen from inside and thus help in extending shelf-life of food products [100]. Iron powder, ascorbic acid, and enzymes in sachets are used as active oxygen scavengers [143]. Oxygen absorber depends upon activation mechanism (auto-activated, water-activated, and UV activated), scavenger form (sachet, label, and extrudable component) and reaction speed (fast, medium, and slow effect). Oxygen scavengers can be fixed into the closures and/or in crown caps or intermingled with the packaging materials [33]. Such materials can also be used as a layer in a multilayer package like Zero2 and AmoSorb or as a liner in a closure like Darex, Smartcap, and Pure Seal [81, 100, 131, 146, 153]. Oxygen scavenging adhesive labels such as Fresh Max, Fresh Card, and ATCO1 are applied into the lidding film or inner wall of the package [153].

The viability of bifidobacteria is affected by oxygen content and redox potential during storage of cultured products. Ascorbic acid can be used as an oxygen scavenger in such cases. The counts of microaerophilic LAB such as *S. thermophilus* and *Lb. delbrueckii subsp. Bulgaricus* is found to improve with increasing concentrations of ascorbic acid [31]. Miller et al., [100] reported that for set yogurt addition of O_2 scavenger $ZerO_2$ to the packaging material further reduced the O_2 content during the first few weeks of a 6-weeks storage trial. In case of In probiotic yogurt added with glucose oxidase, the levels of dissolved O_2 was maintained low and thus preserved the cell viability of *B. longum* and *L. acidophilus* up to 21st day of storage at a refrigerated temperature [30].

13.4.7.4   ANTIBACTERIAL PACKAGING

Spoilage of foods could be prevented by coating antimicrobial compounds on food packaging material, usually referred to as antimicrobial packaging (AMP), and it is usually regarded as a component of active packaging [117]. Such a system can kill or inhibit the growth of microbes and thus, can provide enhanced safety and extended shelf-life to perishable products [61]. Antimicrobial agents such as organic acids, acid salts, bacteriocins, antioxidants, phenolics, and plant/spice extracts have been tried to incorporate into conventional food packaging systems [61, 71, 137].

Cao-Hoang et al., [18] incorporated nisin in sodium caseinate based films for use in mini red Babybel cheese; and they reported that the films affected *Listeriainnocua* inoculated on the cheese surface. Antimicrobials may be either spread onto the packaging materials meant for the direct contact of cheese or incorporated into the plastic films used for packaging [29]. Nisin, natamycin, and their combination were incorporated into a cellulose polymer matrix meant for cheese in a study by dos Santos Pires et al. [40].

13.4.8   OTHER TECHNIQUES

13.4.8.1   MICROENCAPSULATION

Microencapsulation has applications in fermented dairy products, especially in probiotic products, where the viability of probiotics is of paramount importance. Encapsulated probiotics get enhanced protection from harsh food environments, thereby maintaining their characteristics, viability, and stability during storage for a longer period [39, 78, 103]; and thus increasing shelf-life for probiotic fermented dairy products [84, 140]. Kailasapathy [80] studied the survival and effect of free and encapsulated probiotic bacteria (*Lactobacillus acidophilus* and *Bifidobacteriumlactis*) on pH, exopolysaccharide production and the sensory attributes of yogurt over a storage period of 7 weeks, using calcium-induced alginate starch for encapsulation of probiotic bacteria. Encapsulated probiotic bacteria caused a slower post acidification in yogurt compared to yogurt with free probiotic bacteria [140].

In another study, microencapsulated *B. longum* showed a significant difference in the viable counts compared to free cells in set yogurt at 4.4°C for 30 days [1]. Microencapsulation of *L. acidophilus* and *Bifidobacterium spp.* enhanced their survival in yogurt [54]. In the set yogurt, encapsulated cultures of *L. acidophilus* and bifidobacteria showed better survival compared to free cells overstorage period of 8 weeks at 4°C [84].

13.5   FUTURE PERSPECTIVES

The traditional fermented dairy products are prone to spoilage if the proper care is not taken during its manufacture, storage, and distribution. The increasing demand for such products requires these to be transported to longer distances and different countries. The presence of live starter/probiotic cultures in cultured products makes them more sensitive towards temperature abuse. In such a scenario, the shelf-life extension of fermented dairy products has become a necessity. Developing better processing technologies for enhanced probiotic stability and functionality in probiotic products is a major thrust area.

The successful application of bacteriocins as bio-preservatives for cultured dairy foods requires a better understanding on the efficacy of bacteriocins, optimization of bacteriocins production, and their activity in dairy products. To decide the optimal conditions for application of bacteriocins in foods certain important aspects (such as the most effective conditions for application of each particular bacteriocin, antimicrobial effects of bacteriocins and bacteriocinogenic cultures in food ecosystems especially, in terms of microbial interactions, and knowledge of the characteristics of bacteriocin resistant variants and the conditions to prevent their emergence) should be studied in detail. Additionally, food legislation for their approval and acceptance as food preservatives is a must. Packaging concepts (such as intelligent packaging that can sense the presence of microorganisms in the food) should be studied thoroughly for practical applications in fermented dairy foods. Industry–based research is a must for realizing the application of shelf-life extension techniques at a commercial level.

13.6 SUMMARY

The process of fermentation used in the preparation of dairy products affects the extension of shelf-life to a certain extent compared to the raw materials. However, a limited shelf-life and the need for refrigeration conditions are still major limitations with most of the fermented dairy products. Number of parameters affect the shelf-life of such products and may include the quality of raw materials, processing parameters, packaging, storage conditions, handling, and storage during marketing, etc.

Most common defects in fermented dairy products are caused by microbial groups (such as yeasts and molds, coliforms, spore formers, psychrotrophs, and LAB). Research studies are being taken to restrict or prevent the growth of such spoilage organisms and to increase the shelf-life of fermented dairy products. This chapter encompasses microbial groups in the spoilage of fermented dairy products, different techniques for extending shelf-life of fermented milk products (such as aseptic manufacturing techniques, post-production heat treatment, use of preservatives, use of bacteriocins, physical methods, use of protective cultures and novel packaging techniques.

KEYWORDS

- **ALC01**
- **aseptic manufacturing techniques**
- **bacteriocins**
- **chitosan**
- **fermented dairy products**
- **lacticin**
- **microencapsulation**
- **natamycin**
- **nisin**
- **ohmic heating**
- **sorbic acid**
- **ultra-violet irradiation**

REFERENCES

1. Adhikari, K., Mustapha, A., Grun, I. V., & Fernando, L., (2000). Viability of micro-encapsulated bifidobacteria in set yogurt during refrigerated storage. *Journal Dairy Science, 83*, 1946–1951.
2. Ahvenainen, R., (2003). *Novel Food Packaging Techniques* (pp. 5–21). CRC Press Pub, Boca Raton–FL.
3. Altieri, C., Scrocco, C., Sinigaglia, M., & Del Nobile, M. A., (2005). Use of chitosan to prolong mozzarella cheese shelf-life. *Journal of Dairy Science, 88*, 2683–2688.
4. Altuntas, E. G., (2013). *Bacteriocins: A Natural Way to Combat With Pathogens* (pp. 1005–1015). http://www.formatex.info/microbiology4/vol2/1005–1015.pdf (Accessed on 20 July 2019).
5. Ananou, S., Maqueda, M., Martínez-Bueno, M., & Valdivia, E., (2007). *Communicating Current Research and Educational Topics and Trends in Applied Microbiology* (pp. 475–486). Formtex Ltd. Pub., Madrid–Spain, < http://www.formatex.info> (Accessed on 20 July 2019).
6. Angelidis, A. S., (2015). *Dairy Microbiology: A Practical Approach* (pp. 22–68). CRC Press, Boca Raton–FL.
7. Anonymous, (2012). *Food Safety and Standards Act 2006, Rules 2011, Regulation-2011* (9th edn., p. 497). Pub. ILBCO International Law Book Company, New York.
8. Antinone, M. J., & Ledford, R. A., (1993). Reduction of diacetyl in cottage cheese by *Geotrichumcandidum. Cultured Dairy Products Journal, 28*, 26–30.
9. Arqués, J. L., Rodríguez, E., Gaya, P., Medina, M., & Nuñez, M., (2005). Effect of combinations of high pressure treatment and bacteriocin producing lactic acid bacteria on the survival of *Listeria monocytogenes* in raw milk cheese. *International Dairy Journal, 15*, 893–900.
10. Arqués, J. L., Rodríguez, E., Langa, S., Landete, J. M., & Medina, M., (2015). Antimicrobial activity of lactic acid bacteria in dairy products and gut: Effect on pathogens. *Bio-Medical Research International* (pp. 1–9). E-article ID: 584183, http://dx.doi.org/10.1155/2015/584183 (Accessed on 20 July 2019).
11. Aslam, M., Shahid, M., Rehman, F. U., Naveed, N. H., Batool, A. I., Sharif, S., & Asia, A., (2011). Purification and characterization of bacteriocin isolated from *Streptococcus thermophilus. African Journal of Microbiology Research, 5*, 2642–2648.
12. Bakalis, S., Cox, P. W., & Fryer, P. J., (2001). *Thermal Technologies for Food Processing* (pp. 113–137). Woodead Publishing Pvt. Ltd. Pub., London–England.
13. Bamforth, C. W., (2005). *Food, Fermentation and Micro-Organisms* (pp. 143–153). Blackwell Publishing. Pub., London–UK.
14. Benkerroum, N., (2013). Traditional fermented foods of North African countries: Technology and food safety challenges with regard to microbiological risks. *Comprehensive Reviews in Food Science and Food Safety, 12*, 54–89.
15. Benkerroum, N., Oubel, H., & Ben, M. L., (2002). Behavior of *Listeria monocytogenes* and *Staphylococcus aureus* in yogurt fermented with a bactriocin producing thermophilic starter. *Journal Food Protection, 65*, 799–805.
16. Von Bockelmann, B., & Von Bockelmann, I., (1998). *Long Life Acidified Products* (pp. 193–196). Åkarp. Pub, Linnévägen-4, SE- 232–52, Sweden.
17. Calvo, M. M., Montilla, A., & Cobos, A., (1999). Lactic acid production and rheological properties of yogurt made from milk acidified with carbon dioxide. *Journal of the Science of Food and Agriculture, 79*, 1208–1212.

18. Cao-Hoang, L., Chaine, A., Grégoire, L., & Waché, Y., (2010). Potential of nisin incorporated sodium caseinate films to control Listeria in artificially contaminated cheese. *Food Microbiology, 27*, 940–944.
19. Caplice, E., & Fitzgerald, G. F., (1999). Food fermentations: Role of microorganisms in food production and preservation. *International Journal of Food Microbiology, 50*, 131–149.
20. Chandan, R. C., (2014). *Food Processing: Principles and Applications* (pp. 405–436). John Wiley & Sons, Ltd. Pub., New York.
21. Chandan, R. C., (2006). *Manufacturing Yogurt and Fermented Milks* (pp. 3–16). Blackwell Publishing. Pub., London–UK.
22. Cheftel, J. C., (1992). *High Pressure and Biotechnology* (pp. 195–209). John Libbey Eurotex, Ltd. Pub., Paris–France.
23. Chen, G. Q., Lu, F. P., & Du, L. X., (2008). Natamycin production by *Streptomyces gilvosporeus* based on statistical optimization. *Journal of Agricultural and Food Chemistry, 56*, 5057–5061.
24. Chen, H. H., & Hotchkiss, J. H., (1993). Growth of *Listeria monocytogenes* and *Clostridium sporogenes* in cottage cheese in modified atmosphere packaging. *Journal of Dairy Science, 76*, 972–977.
25. Chen, J. H., & Hotchkiss, J. H., (1991a). Long shelf-life cottage cheese through dissolved carbon dioxide and high-barrier packaging. *Journal Dairy Science, 74*, 125.
26. Chen, J. H., & Hotckiss, J. H., (1991b). Effect of dissolved carbon dioxide on the growth of psychro-trophic organisms in cottage cheese. *Journal Dairy Science, 74*, 2941–2945.
27. Church, N., (1994). Developments in modified atmosphere packaging and related technologies. *Trends in Food Science and Technology, 5*, 345–352.
28. Conte, A., Brescia, I., & Del Nobile, M. A., (2011). Lysozyme/EDTA disodium salt and modified-atmosphere packaging to prolong the shelf-life of *Burrata* cheese. *Journal Dairy Science, 94*, 5289–5297, doi: 10.3168/jds.2010–3961.
29. Conte, A., Scrocco, C., Sinigaglia, M., & Del Nobile, M. A., (2007). Innovative active packaging systems to prolong the shelf-life of Mozzarella cheese. *Journal Dairy Science, 90*, 2126–2131.
30. Cruz, A. G., Castro, W. F., Faria, J. A. F., & Bolini, H. M. A., (2013). Stability of probiotic yogurt added with glucose oxidase in plastic materials with different permeability oxygen rates during the refrigerated storage. *Food Research International, 51*, 723–728.
31. Dave, R. I., & Shah, N. P., (1997). Effectiveness of ascorbic acid as an oxygen scavenger in improving viability of probiotic bacteria in yogurts made with commercial starter cultures. *International Dairy Journal, 7*, 435–443.
32. Davies, E. A., Bevis, H. E., & Delves-Broughton, J., (1997). The use of the bacteriocin, nisin, as a preservative in ricittatyoe cheeses to control the foodborne pathogen *Listeria monocytogenes. Letters in Applied Microbiology, 24*, 343–346.
33. Dawson, P. L., (2003). *Beverage Quality and Safety.* CRC Press, Boca Raton–FL, eBook ISBN: 9780203491201.
34. Deacon, J. W., (1997). *Modern Mycology* (pp. 289–290). Blackwell Science. Pub., Oxford UK.
35. De Ancos, B., Cano, M. P., & Gómez, R., (2000). Characteristics of stirred low fat yogurt as affected by high pressure. *International Dairy Journal, 10*, 105–111.
36. Delves-Broughton, J., (2013). *Protective Cultures, Antimicrobial Metabolites and Bacteriophages for Food and Beverage Biopreservation* (pp. 63–99). Woodhead Publishing Ltd. Pub., Cambridge, UK.

37. De Ruig, W. G., & Van Den Berg, G., (1985). Influence of the fungicides sorbate and natamycin in cheese coatings on the quality of the cheese. *Netherlands Milk Dairy Journal, 39*, 165–172.
38. Dobrucka, R., (2013). Application of active packaging systems in probiotic foods. *Scientific Journal of Logistics, 9*, 167–175.
39. Donkor, O. N., Nilmini, S. L. I., Stolic, P., Vasiljevic, T., & Shah, N. P., (2007). Survival and activity of selected probiotic organisms in set type yogurt during cold storage. *International Dairy Journal, 17*, 657–665.
40. Dos Santos Pires, A. C., Ferreira Soares, N., & De Andrade, N. J., (2008). Development and evaluation of active packaging for sliced mozzarella preservation. *Packaging Technology Science, 21*, 375–383.
41. Doyle, M. P., & Marth, E. H., (1975). Thermal inactivation of conidia from *Aspergillus flavus* and *Aspergillus parasiticus*. Part I: Effects of moist heat, age of conidia and sporulation medium. *Journal of Milk Food Technology, 38*, 678–682.
42. Dzigbordi, B., Adubofuor, J., & Faustina, D. W. M., (2013). The effects of different concentrations of natamycin and the point of addition on some physicochemical and microbial properties of vanilla flavored yogurt under refrigerated condition. *International Food Research Journal, 20*, 3287–3292.
43. EFSA (European Food Safety Authority), (2009). Scientific opinion on the use of natamycin (E 235) as a food additive. *EFSA Journal, 7*, 1412.
44. Eissa, S. A., Ibrahim, E. M. A., Abdou, A. M., & Mohammed, H. A., (2014). The role of natamycin fortification to extend shelf-life of plain yogurt. *Benha Veterinary Medical Journal, 27*, 140–149.
45. Erginkaya, Z., Ünal, E., & Kalkan, S., (2011). *Science Against Microbial Pathogens: Communicating Current Research and Technological Advances* (pp. 1342–1348). Formatex Ltd. Pub., Madrid–Spain.
46. Frank, J. F., (2001). *Food Microbiology: Fundamentals and Frontiers* (pp. 111–126). ASM Press. Pub., Washington–D. C.
47. Fujita, K., Ichimasa, S., Zendo, S. T., & Koga, S., (2007). Structural analysis and characterization of Lacticin Q: Novel bacteriocin belonging to a new family of unmodified bacteriocins of Gram-positive bacteria. *Applied Environmental Microbiology, 73*, 2871–2877.
48. Gálvez, A., Abriouel, H., López, R. L., & Omar, N. B., (2007). Bacteriocin based strategies for food biopreservation. *International Journal of Food Microbiology, 120*, 51–70.
49. Gálvez, A., Abriouel, H., Lucas, R., & Grande, B. M. J., (2005). *Natural Antimicrobials in Food Safety and Quality* (pp. 39–61). CAB International, Cambridge, USA.
50. Gálvez, A., Burgos, M. J. G., Lopez, R. L., & Pulido, R. P., (2014). *Food Biopreservation* (pp. 3–22). Springer briefs in food, health and nutrition, New York.
51. Gálvez, A., Lopez, R. L., Abriouel, H., Valdivia, E., & Omar, N. B., (2008). Application of bacteriocins in the control of food borne pathogenic and spoilage bacteria. *Critical Reviews in Biotechnology, 28*, 125–152.
52. Garcıá-Esteban, M., Ansorena, D., & Astiasarán, L., (2004). Comparison of modified atmosphere packaging and vacuum packaging for long period storage of dry cured ham: Effects on color, texture and microbiological quality. *Meat Science, 67*, 9–63.
53. Giudici, P., Masini, G., & Caggia, C., (**1996**). The role of galactose fermenting yeast in plain yogurt spoilage. *Annali di Microbiolia Ed Enzimologia* (Annals of Microbiolia and Enzymology), *46*, 11–19.

54. Godward, G., & Kalilasapathy, K., (2003). Viability and survival of free, encapsulated and co-encapsulated probiotic bacteria in yogurt. *Milchwissenschaft, 58*, 161–164.
55. Govaris, A., Botsoglou, E., Sergelidis, D., & Chatzopoulou, P. S., (2011). Antibacterial activity of oregano and thyme essential oils against *Listeria monocytogenes* and *Escherichia coli* O157:H7 in feta cheese packaged under modified atmosphere. *Food Science Technology, 44*, 1240–1244.
56. Goyal, G. K., (1986). Packaging of dairy products in United States of America. *Indian Dairyman, 38*, 287–290.
57. Goyal, G. K., & Rajorhia, G. S., (1991). Role of modern packaging in marketing of indigenous dairy products. *Indian Food Industry, 10*, 32–34.
58. Greg-Blank, G., Shamsuzzaman, K., & Sohal, S., (1992). Use of electron beam irradiation for mold decontamination on cheddar cheese. *Journal of Dairy Science, 75*, 13–18.
59. Ham, J. S., Jeong, S. G., Lee, S. G., Han, G. S., Jang, A., & Yoo, Y. M., (2009). Quality of irradiated plain yogurt during storage at different temperatures. *Asian-Australasian Journal of Animal Sciences, 22*, 289–295.
60. Hamilton-miller, J. M. T., (1974). *Advances in Applied Microbiology* (pp. 109–134). Academic Press. Pub., New York.
61. Han, J. H., (2005). *Innovations in Food Packaging* (pp. 80–107). Elsevier Ltd. Pub., New York.
62. Harte, F., Luedecke, L., Swanson, B., & Barbosa-Canovas, G. V., (2003). Low fat set yogurt made from milk subjected to combinations of high hydrostatic pressure and thermal processing, *Journal Dairy Science, 86*, 1074–82.
63. Havranek, J. L., & Hadžiosmanović, M., (1996). Shelf-life as requirement for quality of milk products. *Mljekarstvo, 46*, 197–206.
64. Hocking, S. L., & Faedo, M., (1992). Fungi causing thread mold spoilage of vacuum packaged Cheddar cheese during maturation. *International Journal of Food Microbiology, 16*, 123–130.
65. Holo, H., Faye, T., Brede, D. A., Nilsen, T., Ødegård, I., Langsrud, T., Brendehaug, J., & Nes, I. F., (2002). Bacteriocins of propionic acid bacteria. *Lait, 82*, 59–68.
66. Horwood, J. F., Stark, W., & Hull, H. H., (1987). Fermented, yeasty flavor defect in Cheddar cheese. *Australian Journal of Dairy Technology, 42*, 25–26.
67. Hotchkiss, J. H., Werner, B. G., & Lee Edmund, Y. C., (2006). Addition of carbon dioxide to dairy products to improve quality: Comprehensive review. *Comprehensive Reviews in Food Science and Safety, 5*, 158–168.
68. http://www.milkingredients.ca/index-eng.php?id=180 (Accessed on 20 July 2019).
69. Hutkins, R. W., (2006). *Microbiology and Technology of Fermented Foods* (pp. 107–144). Blackwell Publishing Ltd. Pub., New York–USA.
70. IFST, (1993). *Shelf-Life of Foods–Guidelines for its Determination and Prediction* (pp. 61–77). Institute of Food Science & Technology (IFST), London.
71. Iseppi, R., Pilati, F., Marini, M., Toselli, M., & De Niederhäusern, S., (2008). Anti-listerial activity of a polymeric film coated with hybrid coatings doped with enterocin 416K1 for use as bioactive food packaging. *International Journal of Food Microbiology, 123*, 281–287.
72. Jankowska, A., Reps, A., Proszek, A., & Krasowska, M., (2003). Applying high pressure to yogurt preservation. In: *Proceedings of the 17th Forum Applied Biotechnology, Parts I and II* (pp. 477–480). Universitet Gent, Gent, Belgium.
73. Jay, J. M., Loessner, M. J., & Golden, D. A., (2005). *Modern Food Microbiology* (p. 328). Springer. Pub., New York.

74. Jin, Z. T., & Harper, W. J., (2003). Effect of high pressure (HP) treatment on microflora and ripening development in Swiss cheese slurries. *Milchwissenschaft*, *58*, 134–137.

75. Johnson, M. E., (2001). *Applied Dairy Microbiology* (pp. 345–384). Marcel Dekker, New York.

76. Josephsen, J., & Jespersen, L., (2004). *Handbook of Food and Beverage Fermentation Technology* (pp. 23–49). Marcel Dekker, Inc. Pub., New York.

77. Juneja, V. K., Dwivedi, H. P., & Yan, X., (2012). Novel natural food antimicrobials. *Annual Review of Food Science and Technology*, *3*, 381–403.

78. Kailasapathy, K., (2002). Microencapsulation of probiotic bacteria: Technology and potential applications. *Current Issues in Intestinal Microbiology*, *3*, 39–48.

79. Kailasapathy, K., (2010). *Fermented Foods and Beverages of the World* (pp. 415–434). CRC Press, Boca Raton–FL.

80. Kailasapathy, K., (2006). Survival of free and encapsulated probiotic bacteria and their effect on the sensory properties of yogurt. *Food Science and Technology*, *39*, 1221–1227.

81. Kaufman, J., LaCoste, A., Schulok, J., Shehady, E., & Yam, K. K., (2000). *An Overview of Oxygen Scavenging Packaging and Applications*. www.packaingnetwork.com (Accessed on 20 July 2019).

82. Kesenkas, H., Gursoy, O., Kinik, O., & Akbulut, N., (2006). Extension of shelf-life of dairy products by biopreservation: Protective cultures. *GIDA*, *31*, 217–223.

83. Khalid, K., (2011). An overview of lactic acid bacteria. *International Journal of Biosciences*, *1*, 1–13.

84. Khalida, S., Godward, G., Reynolds, N., & Arumugaswamy, R., (2000). Encapsulation of probiotic bacteria with alginate starch and evaluation of survival in simulated gastro intestinal conditions and in yogurt. *International Food Microbiology*, *62*, 47–55.

85. Khurana, H. K., & Kanawjia, S. K., (2007). Recent trends in development of fermented milks. *Current Nutrition and Food Science*, *3*, 91–108.

86. Klijn, N., Nieuwendorf, F. F. J., & Hoolwerf, J. D., (1995). Identification of *Clostridium butyricum* as the causative agent of late blowing in cheese by species-species PCR amplification. *Applied and Environmental Microbiology*, *61*, 2919–2924.

87. Kumar, A., Solanky, M. J., & Chauhan, A. K., (2003). Storage related lipolysis changes in lassi. *Indian Journal of Dairy Science*, *56*, 20–22.

88. Kumar, R., Sarkar, S., & Misra, A. K., (1998). Effect of nisin on the quality of dahi. *Journal of Dairying Food and Home Science*, *17*, 13–24.

89. Leadley, C. E., & Williams, A., (2006). *Food Processing Handbook* (pp. 201–235). Wiley-VCH Verlag GmbH & Co. KGaA, Weinheim.

90. Ledenbach, L. H., & Marshall, R. T., (2009). *Compendium of the Microbiological Spoilage of Foods and Beverages, Food Microbiology and Food Safety* (pp. 41–67). Springer Science & Business Media, LLC, New Yrk.

91. Liewen, M. B., (1991). *Hand Book of Applied Mycology: Foods and Feeds* (pp. 541–552). Marcel Dekker, Inc. Pub., New York.

92. Macbean, R. D., (2010). Packaging and the shelf-life of yogurt: Chapter 8. In: Robertson, G. L., (ed.), *Foods Packaging and Shelf-life–A Practical Guide* (pp. 143–154). Taylor and Francis Group, LLC, New York.

93. Mainville, I., Montpetit, D., Durand, N., & Farnworth, E. R., (2001). Deactivating the bacteria and yeast in kefir using heat treatment, irradiation and high pressure. *International Dairy Journal*, *11*, 45–49.

94. Malheiros, P. S., Sant'Anna, V., Barbosa, M. S., Brandelli, A., & Franco, B. D. G. M., (2012). Effect of liposome encapsulated nisin and bacteriocin like substance P34 on

Listeria monocytogenes growth in Minasfrescal cheese. *International Journal of Food Microbiology*, *156*, 272–277.

95. Malhotra, B., Keshwani, A., & Kharkwal, H., (2015). Antimicrobial food packaging: Potential and pitfalls. *Frontiers Microbiology*, *6*, 1–9.

96. Martley, F. G., & Crow, V. L., (1993). Interactions between non-starter microorganisms during cheese manufacture and ripening. *International Dairy Journal*, *3*, 461–464.

97. Mauriello, G., De Luca, E., & La Storia, A., (2005). Anti-microbial activity of a nisin activated plastic film for food packaging. *Letters in Applied Microbiology*, *41*, 464–469.

98. Medina, A., Jimenez, M., Mateo, R., & Magan, N., (2007). Efficacy of natamycin for control of growth and ochratoxinA production by *Aspergillus carbonarius* strains under different environmental conditions. *Journal of Applied Microbiology*, *103*, 2234–2239.

99. Menon, R. R., Rao, K. J., Nath, B. S., & Ram, C., (2014). Carbonated fermented dairy drink–effect on quality and shelf-life. *Journal of Food Science and Technology*, *51*, 3397–3403.

100. Miller, C. W., Nguyen, M. H., & Rooney, M., (2003). The control of dissolved oxygen content in probiotic yogurts by alternative packaging materials. *Packaging Technology and Science*, *16*, 61–67.

101. Moir, C. J, Eyles, M. J., & Davey, J. A., (1993). Inhibition of pseudomonads in cottage cheese by pack aging in atmospheres containing carbon dioxide. *Food Microbiology*, *10*, 345–351.

102. Morgan, R., (1989). UV green light disinfections. *Dairy Industry International*, *54*, 33–40.

103. Mortazavian, A., Razavi, S. H., Ehsani, M. R., & Sohrabvandi, S., (2007). Principles and methods of microencapsulation of probiotic microorganisms. *Iranian Journal of Biotechnology*, *5*, 1–18.

104. Mudgal, S., (2015). *Dairy Product Technology: Recent Advances* (pp. 283–307). Daya Publishing House. Pub., New Delhi.

105. Mullan, M., & McDowell, D., (2003). *Food Packaging Technology* (pp. 304–339). Blackwell Publishing Ltd. Pub., New York.

106. Mullan, W. M. A., (2000). Causes and control of early gas production in Cheddar cheese. *International Journal of Dairy Technology*, *53*, 63–68.

107. Munoz, A., Ananou, S., & Galvez, A., (2007). Inhibition of *Staphyloccousaureus* in dairy products by entrocin AS-48 produced *in situ* and *ex situ* Bactericidal synergism with heat. *International Dairy Journal*, *17*, 760–769.

108. Munoz, A., Maqueda, M., Galvez, A., Martinez-Bueno, M., Rodriguez, A., & Valdivia, E., (2004). Bio control of psychrotropicentero-toxigenic *Bacillus cereus* in a non-fat hard cheese by an enterococcal strain producing enterocin AS-48. *Journal of Food Protection*, *67*, 1517–1521.

109. Nicoli, M. C., (2012). *Shelf-life Assessment of Food* (p. 302). CRC Press, Boca Raton.

110. Nielsen, P. V., & Haasum, I., (1997). Packaging conditions hindering fungal growth on cheese. *Scandinavian Dairy Information*, *11*, 22–25.

111. O'Sullivan, L., Morgan, S. M., Ross, R. P., & Hill, C., (2002a). Elevated enzyme release from Lactococcal starter cultures on exposure to the lantibioticlacticin 481, produced by *Lactococcuslactis* DPC5552. *Journal Dairy Science*, *85*, 2130–2140.

112. O'Sullivan, L., O'Connor, E. B., Ross, R. P., & Hill, C., (2006). Evaluation of live culture producing lacticin 3147 as a treatment for the control of *Listeria monocytogenes* on the surface of smear ripened cheese. *Journal of Applied Microbiology*, *100*, 135–143.

113. O'Sullivan, L., Ross, R. P., & Hill, C., (2003a). A lacticin 481-producing adjunct culture increases starter lysis while inhibiting nonstarter lactic acid bacteria proliferation during Cheddar cheese ripening. *Journal of Applied Microbiology, 95,* 1235–1241.

114. O'Sullivan, L., Ross, R. P., & Hill, C., (2002b). Potential of bacteriocin producing lactic acid bacteria for improvements in food safety and quality. *Biochimie, 84,* 593–604.

115. O'Sullivan, L., Ryan, M. P., Ross, R. P., & Hill, C., (2003b). Generation of food grade lactococcal starters which produce the lactibioticslacticin 3147 and lacticin 481, *Applied Environmental. Microbiology, 69,* 3681–3685.

116. Paltani, I. P., & Goyal, G. K., (2007). Packaging of dahi and yogurt- A review. *Indian Journal Dairy Science, 60,* 1–10.

117. Pérez-Pérez, C., & Regalado-González, C., (2006). Incorporation of antimicrobial agents in food packaging films and coatings. *Advances in Agricultural and Food Biotechnology: Research Signpost, 37*(2), 193–216.

118. Prajapati, J. B., & Nair, B. M., (2008). *Fermented Functional Foods* (pp. 1–25). CRC Press, Boca Raton–FL.

119. Prajapati, J. B., & Sreeja, V., (2013). *Dahi and Related Products* (pp. 30–32). ND Publishers and SASNET-FF. Pub., New Delhi.

120. Prajapati, J. P., Upadhyay, K. G., & Desai, H. K., (1993). Quality appraisal of heated shrikhand stored at refrigerated temperature. *Cultured Dairy Product Journal, 28,* 14–17.

121. Priyanka, P., & Anita, K., (2014). Active packaging in food industry: A review. *J. Environ. Sci. Toxicol. Food Technol., 8,* 1–7.

122. Puthal, K., (2015). Indian dairy industries–an over view, high pressure processing technology and it's applications in dairy industry. *Food and Beverage Processing, 1,* 28.

123. Rajmohan, S., & Prasad, V., (1995). Effect of nisin on the chemical changes in dahi. *Indian Journal of Dairy Science, 48,* 633.

124. Raju, P. N., & Singh, A. K., (2016). *Fermented Milks and Dairy Products* (pp. 637–672). CRC Press and Taylor and Francis Co. Ltd., Boca Raton–FL.

125. Reps, A., Warminska-Radyko, I., Krzyzewska, A., & Tomasik, J., (2001). Effect of high pressure on *Streptococcus salivarius subsp. Thermophilus. Milchwissenschaft, 56,* 131–33.

126. Rilla, N., Martinez, B., Delgado, T., & Rodriguez, A., (2003). Inhibition of *Clostridium tyrobyricum* in Vadiago cheese by *Lactococcus lactic sub splactis* IPLA 729, a nisin Z-producer. *International Journal of Food Microbiology, 85,* 23–33.

127. Rilla, N., Martinez, B., & Rodriguez, A., (2004). Inhibition of a methicillin resistant *Staphyloccusaureus* strain in Afuega'lpitu cheese by nisin Z-producing strain *Lactococcus lactic sub splactis* IPLA 729. *Journal of Food Protection, 67,* 928–933.

128. Robinson, R. K., Tamime, A. Y., & Wszolek, M., (2002). *Dairy Microbiology Handbook* (pp. 367–430). John Wiley. Pub., New York.

129. Rodriguez, E., Arques, J. L., Nuñez, M., Gaya, P., & Medina, M., (2005). Combined effect of high pressure treatments and bacteriocin producing lactic acid bacteria on inactivation of *Escherichia coli* O157:H7 in raw milk cheese. *Applied and Environmental Microbiology, 71,* 3399–3404.

130. Rodríguez, E., Gaya, P., Nuñez, M., & Medina, M., (1998). Inhibitory activity of a nisin producing starter culture on *Listeria innocua* in raw ewe's milk Manchego cheese. *International Journal of Food Microbiology, 39,* 129–132.

131. Rooney, M. L., (2005). Introduction to active food packaging technologies. In: *Innovations in Food Packaging* (pp. 63–77). Elsevier Academic Press, Oxford, UK.

132. Sahil, M. A., & Sandine, W. E., (1990). Inhibitory effect of Microgard™ on yogurt and cottage cheese spoilage organisms. *Journal of Dairy Science, 73*, 887–893.

133. Sarkar, S. P., Dave, J. M., & Sannabhadti, S. S., (1992a). A note on the effect of thermization of mistidahi on the acid producers count. *Indian Journal Dairy Science, 45*, 131–134.

134. Sarkar, S. P., Dave, J. M., & Sannabhadti, S. S., (1992b). Effect of thermization of mistidahi on shelf-life and Beta-D-gatactosidase activity. *Indian Journal of Dairy Science, 45*, 135–139.

135. Sarkar, S., (2006). Shelf-life extension of cultured milk products. *Nutrition and Food Science, 36*, 24–31.

136. Sarkar, S., Kuila, R. K., & Misra, A. K., (1996). Effect of incorporation of Gelodan™ SB 253 (stabilizer cum preservative) and nisin on the microbiological quality of shrikhand. *Indian Journal of Dairy Science, 49*, 176–184.

137. Scannell, A. G. M., Hill, C., Ross, R. P., & Marx, S., (2000). Development of bioactive food packaging materials using immobilized bacteriocinslacticin 3147 and nisaplin. *International Journal of Food Microbiology, 60*, 241–249.

138. Settanni, L., & Corsetti, A., (2008). Application of bacteriocins in vegetable food biopreservation. *International Journal of Food Microbiology, 121*, 123–138.

139. Shah, N. P., (2014). Other dairy products: yogurt, kefir and kumys: Chapter 9. In: Bemforth, C. W., & Ward, R. E., (eds.), *The Oxford Handbook of Food Fermentation* (pp. 385–407). Oxford University Press, Cambridge- UK.

140. Shah, N. P., & Ravula, R. R., (2000). Microencapsulation of probiotic bacteria and their survival in frozen fermented dairy desserts. *The Australian Journal of Dairy Technology, 55*, 139–144.

141. Shi, Y., Chen, Y., Li, Z., Yang, L., Chen, W., & Mu, Z., (2015). Complete genome sequence of *Streptococcus thermophilus* MN-BM-A02, a rare strain with a high acid producing rate and low post acidification ability. *Genome Announce, 3*, E-article ID: 00979-15, doi: 10.1128/genomeA.00979-15.

142. Singh, P., & Prakash, A., (2008). Isolation of *Escherichia coli, Staphylococcus aureus* and *Listeria monocytogenes* from milk products sold under market conditions at Agra region. *Actaagriculturae Slovenica, 92*, 83–88.

143. Singh, P., Wani, A. A., & Saengerlaub, S., (2011). Active packaging of food products: Recent trends. *Nutrition and Food Science, 41*, 249–260.

144. Soomro, A. H., Masud, T., & Anwaar, K., (2002). Role of lactic acid bacteria (LAB) in food preservation and human health–A Review. *Pakistan Journal of Nutrition, 1*, 20–24.

145. Steinka, I., & Morawska, M., (2006). The influence of biological factors on properties of some traditional and new polymers used for fermented food packaging. *Journal of Food Engineering, 77*, 771–775.

146. Suppakul, P., Miltz, J., Sonneveld, K., & Bigger, S. W., (2003). Active packaging technologies with an emphasis on antimicrobial packaging and its applications. *Journal of Food Science, 68*, 408–420.

147. Tamime, A. Y., & Robinson, R. K., (1999). *Yogurt Science and Technology* (pp. 261–366). Woodhead Publishing. Pub., Cambridge–UK.

148. Tamime, A. Y., & Deeth, H. C., (1980). Yogurt technology and bio chemistry. *Journal of Food Protection, 12*, 916–977.

149. Thomas, L. V., & Delves-Broughton, J., (2001). Applications of the natural food preservative natamycin. *Research Advances in Food Science, 2*, 1–10.

150. Tortorello, M. L., Best, S., Batt, C. H., Woolf, H. D., & Bender, J., (1991). Extending the shelf-life of cottage cheese: Identification of spoilage flora and their control using food grade preservatives. *Cultured Dairy Product Journal, 26*, 11–12.

151. Tucker, G. S., (2003). Food biodeterioration and methods of preservation: Chapter 2. In: Coles, R. McDowell, D., & Kirwan, M. J., (eds.), *Food Packaging Technology* (pp. 32–64). Blackwell Publishing Ltd., New York.

152. Var, I., Sahan, N., Kabak, B., & Golge, O., (2004). The effects of natamycin on the shelf-life of yogurt. *Archiv fur lebensmittelhygiene, 55*, 7–9.

153. Vermeiren, L., (1999). Development in active packaging of foods. *Trends in Food Science and Technology, 10*, 77–86.

154. Vivier, D., Rivemale, M., Reverbel, J. P., Ratomahenina, R., & Galzy, P., (1994). Study of the growth of yeasts from Feta cheese. *International Journal of Food Microbiology, 22*, 207–215.

155. Vogel, R. F., Hammes, W. P., Habermeyer, M., Engel, K. H., Knorr, D., & Eisenbrand, G., (2011). Microbial food cultures: Opinion of the Senate Commission on Food Safety (SKLM) of the German Research Foundation (DFT). *Molecular Nutrition & Food Research*. E-article: 10.1002/mnfr. 201100010.

156. WHO, (2003). Safety evaluation of food additives. *Nitrite and Nitrate Intake Assessment (WHO Food Additives Series No. 50, JECFA Monograph No 1059),* WHO, Geneva, http://monographs.iarc.fr/ENG/Monographs/vol94/mono94–6.pdf (Accessed on 20 July 2019).

157. Yadav, J. S., & Batish, V. K., (1982). Mycotoxins in fermented dairy products and their possible health hazards. *Indian Dairyman*, 533–538.

158. Yang, S. C., Lin, C. H., Sung, C. T., & Fang, J. Y., (2014). Antibacterial activities of bacteriocins: Application in foods and pharmaceuticals. *Frontiers Microbiology, 5*, 241.

159. Zacharis, C. K., & Tzanavaras, P. D., (2013). Preservatives: Chapter 15. In: Nollet, L. M. L., & Toldrà, F., (eds.), *Food Analysis by HPLC* (3rd edn., 529–550). CRC Press and Taylor and Francis Group, Boca Raton–FL.

160. Zagory, D., & Kader, A. A., (1988). Modified atmosphere packaging of fresh produce. *Food Technology, 42*, 70–77.

161. Zoon, P., & Allersma, D., (1996). Eye and crack formation in cheese by carbon dioxide from decarboxylation of glutamic acid. *Netherlands Milk and Dairy Journal, 50*, 309–318.

CHAPTER 14

PROBIOTICS: FROM SCIENCE TO TECHNOLOGY

VANDNA KUMARI and NARENDRA KUMAR

ABSTRACT

The concept of probiotics is now widely used in industrial and scientific fields due to its health benefits. There are several molecular techniques for identification of LAB isolates, but every technique cannot be used for all purposes. For safety assessments, the detailed fingerprint of individual isolates is important by using several techniques on the strain level. For species-level identification, biochemical technique or 16S rDNA sequencing or DNA–DNA hybridization is used. DGGE is used for a rapid microbial analysis of complex bacterial communities in fermented milk foods, but for routine analyses of LAB in food samples novel techniques are needed. Although clinical trials have been for probiotic efficiency, yet novel research on GI-tract diagnostics and immunology, biomarkers, and functionality is needed for human studies. While developing probiotic products in food, its viability and survivability is an important parameter, and this can be achieved by using microencapsulation and cell immobilization. Thus, probiotics can be used as a viable ingredient of our daily foods, and probiotics can be applied beyond the pharmaceutical and supplement uses.

14.1 INTRODUCTION

The word "probiotics" is a Greek word meaning "for life [46]" and it was first used by Kollath [71], who described it as various substances such as vitamins, enzymes, and aromatic compounds present in foods. These compounds are required in the form of different organic and inorganic supplements by malnourished patients for health restoration. Vergin [155] and Kolb [70] suggested that various antibiotic treatments in our body lead to gut microbiota imbalance, which may be restored by the intake of probiotics in our daily diet.

According to Lilly and Stillwell [78], probiotics are the microbial substances of one microorganism, which in turn promotes the growth of another microorganism. Sperti [140] and Fujii et al., [33] gave a similar definition of probiotics as compounds produced during microorganism growth, stimulated host-microbial growth improved the immunity. Build-up resistance against infection was given by the report on probiotics as organisms as well as substances that cause intestinal microbial balance according to Parker [114], which is similar to the modern concept. Most accepted theory of probiotics was proposed by Fuller [34], who defined them as *"a live microbial feed supplement, which beneficially affects the host by improving its intestinal microbial balance."* Further, Salminen [128] recommended incorporation of non-viable bacteria also in the definition. The FAO/WHO [25] defines probiotics as *"live microorganisms that when administered in adequate amounts confer health benefits to the host,"* which is widely accepted worldwide.

This chapter is a holistic approach to the emergence of concept on probiotics from the science of beneficial food ingredients, which grab the interest of scientists and food technologists globally due to its numerous health effects on humans to the most promising technology in the field of the food industry. The current status of probiotics and its future aspects have also been discussed.

14.2 IDENTIFICATION AND CHARACTERIZATION OF LACTIC ACID BACTERIA (LAB) STRAINS

14.2.1 LAB: PHYSIOLOGICAL PROPERTIES AND MODERN TAXONOMY

Taxonomic methods used for lactic acid bacteria (LAB) include both phenotypic characterization (analysis of cell wall composition, protein fingerprinting and electrophoretic mobility of certain enzymes) and genotypic analysis. Analysis of DNA-DNA homology is used as a reference method for identification of LAB on a molecular level. For several *Lactobacillus* species, nucleic acid probes either alone or in combination with priming methods have been developed [22, 149]. Classification using genotypic methods such as ribotyping, RAPD-PC, or similar techniques has also been used.

LAB is gram-positive, non-sporing, catalase (CAT)-negative organism, non-respiratory, aero-tolerant, fastidious, acid-tolerant, and ferment glucose to lactic acid as the major end product, or to lactic acid, ethanol, and CO_2. The major genera of LAB include *Lactobacillus, Streptococcus, Pediococcus, Lactococcus, Leuconostoc, Enterococcus, Carnobacterium, Bifidobacterium, Weissella,* and *Tetragenococcus.* LAB is originated from milk, and since then these have been used in many foods and fermented products. As these are ubiquitous in nature and having GRAS (Generally Regarded as Safe) status, they are considered as industrially important microorganisms. Commonly used microorganisms as probiotics comprise species from the genera *Lactobacillus, Bifidobacterium, and Enterococcus,* but strains of *Lactobacillus* spp. and *Bifidobacterium* sp. are important commercially (Table 14.1).

TABLE 14.1   Selected Microorganisms Commonly used as Probiotics

Genus	Species
Bacillus	*Cereus var. toyoi*
Bifidobacterium	*Longum, lactis, breve, bifidum, animalis, adolescentis, infantis*
Enterococcus	*Faecalis, faecium*
Escherichia	*Coli strain nissle*
Lactobacillus	*Rhamnosus, reuteri, plantarum, paracasei, johnsonii, gasseri, gallinarum, delbrueckii subsp. bulgaricus, crispatus, casei, acidophilus, amylovorus*
Lactococcus	*Lactis*
Leuconostoc	*Mesenteroides*
Pediococcus	*Acidilactici*
Propionibacterium	*Freudenreichii*
Saccharomyces	*Cerevisiae, boulardii*
Streptococcus	*Thermophilus*

Lactobacilli are Gram-positive, CAT negative, non-spore-forming microorganisms, rods, or coccobacilli appearance. They are micro-aerophilic, chemoorganotrophic, and have a GC content of <54 mol%. The genus *Lactobacillus* belongs to the phylum *Firmicutes*, class Bacilli, order *Lactobacillales,* and family *Lactobacillaceae.* This genus comprises of 106 described species commonly isolated from the human intestine, e.g., *Lactobacillus acidophilus, L. casei, L. salivarius, L. fermentum, L. plantarum, L. brevis,* and *L. reuteri* [96]. Among them, functional, and safety aspects of *L. casei, L. acidophilus, L. johnsonii,* and *L. Rhamnosus* strains have been widely studied and recognized.

Bifidobacteria are branched rod-shaped, anaerobic organisms firstly isolated from feces of breast-fed infants. They are Gram-positive, CAT-negative, non-spore forming and non-motile having a particular metabolic pathway called the bifid shunt, which produces acetic acid along with lactic acid in the molar ratio of 3:2 using the key enzyme fructose-6-phosphate phosphoketolase (F6PPK) and is the generally used as a diagnostic test for this *Bifidobacteria*. This family includes five genera, i.e., *Bifidobacterium*, *Mycobacterium*, *Propionibacterium*, *Brevibacterium*, and *Corynebacterium*. Genus *Bifidobacterium* has 32 species, 12 of which are of human origin (*B. adolescentis*, *B. breve*, *B. bifidum*, *B. infantis*, *B. catenulatum*, *B. dentium*, *B. angulatum*, *B. longum*, *B. pseudocatenulatum*, and *B. adolescentis*): 15 from intestinal tracts of animal or rumen, 3 from honeybees and 2 from fermented milk and sewage.

14.2.2   IDENTIFICATION AND TYPING OF LAB

Reliable identification of LAB is important for industrial application due to its extensive use in food and fermentation industry. Over the past decade, special attention has been given to correct identification of the bacteria for safe use of human consumption. Presently a wide range of methods and techniques are available with different levels of discrimination, reproducibility, and workload. Majority of the identification methods are culture-dependent, and they are reported to give excellent results, but due to the lower time requirement of analysis technique without any cultivation of bacteria have advantages over earlier one.

14.2.2.1   PHENOTYPIC METHODS

Identification of LAB uses various phenotypic methods such as morphological and physiological characterization, protein profiling, and carbohydrate fermentation profile on a regular basis. Although phenotypic methods are cheaper and useful for identification, yet the requirement of technical skill and standardized assay and time-consuming analysis are not desirable. Moreover, these methods have poor reproducibility, discriminatory power, and ambiguity of some techniques. Recently, morphological, and physiological characteristics based approach was used for identification of 113 LAB isolates from a Ugandan traditional fermented beverage, but the results were found uncertain for identification at the species level [100, 159]. The limitations of phenotypic methods have been shown by different studies due to their poor reproducibility and lower taxonomic resolution as compared to genotypic that allows differentiation at the genus level.

Sodium Dodecyl Sulphate–Polyacrylamide Gel Electrophoresis (SDS–PAGE) analysis of whole-cell proteins is more reliable identification of LAB [4, 120]. Recently, 355 bacteria isolated from different probiotic products were identified using protein SDS–PAGE [148]. The major drawback of this technique is heavy workload despite of its reproducibility and reliability; moreover, discriminatory power is low for sub-species level in the *L. acidophilus* group.

Another phenotypic technique is Fatty Acid Methyl Ester (FAME) analysis and Thin layer chromatography of organic acids but with limited success for identification of LAB. Due to phenotypic heterogeneity of classical methods, which makes it ambiguous and unreliable, thus there has been a consequent shift towards the use of genotypic methods for characterization of LAB to have more reliable classification and differentiation [92].

14.2.2.2   GENOTYPIC METHODS

In the last two decades, various genotypic techniques based on nucleic acid and other macromolecules have been developed having discriminatory power varying from species to strains (typing) level. These techniques are mainly PCR-based, which cause amplification of specifically targeted DNA fragments by using oligonucleotide primers under controlled reaction conditions.

14.2.2.2.1   16S-Ribosomal RNA (rRNA) Probing

The 16S-rRNA (rDNA) gene is a conserved region, which is considered as a universal marker and is relatively unaltered by environmental changes. That is why, it is the most common gene targeted in studying bacterial diversity. 16S-rRNA has nine variable regions (V1 to V9) separated by variable space regions, some of the genes are taxa-specific and give clear genetic diversity between closely related species, e.g., the pmoA gene for methanotrophs and dsr-gene for sulfate-reducing bacteria [14, 21]. 16S-rRNA probes have been used for the identification of LAB that discriminates between *Lactobacillus, Lactococcus, Leuconostoc, Pediococcus* and *Enterococcus spp.* from group and genus to species and subspecies level. These probes have also been used to distinguish and identify many *Bifidobacterium spp.* and species-specific probes [108].

The variability of 16S-23S rRNA intergenic transcribed sequence (ITS) is higher than the 16S-rRNA structural gene within closely related taxonomic groups; thus it is considered as a suitable target for identification as well as typing by using species-specific primers. Although 16S rRNA genes are a better quantitative representation of bacterial species, yet some shortcomings are also associated with its use, such as it has a limited resolving power, also while targeting different bacterial species lower representation of some species is produced due to variation in copy numbers of the gene among bacterial species [1].

Several molecular typing techniques are also available for the identification and classification of bacteria up to strain level. Genetic-based molecular methods are considered as the most powerful of these and are known as DNA fingerprinting techniques, such as amplified fragment length polymorphism (AFLP), randomly amplified polymorphic DNA (RAPD), ribotyping, and pulsed-field gel electrophoresis (PFGE). These methods have been widely used for the intra-specific identification, and genotyping of LAB and bifidobacteria isolates from various sources [92].

In RAPD analysis, an anonymous stretches of DNA are amplified by using a primer with an arbitrary sequence of 10-base-pair sequence primer and each particular primer will adhere to the template DNA randomly; therefore the nature of the obtained products is unknown. After synthesizing, these sequences PCR reactions are carried out with less stringency annealing conditions to enable the primer to anneal to the template DNA; thus, the primer will bind to regions with the nearest homology. This method is used for identification of bacteria with low discrimination sequence. Careful controlled conditions are required for this method as slight changes in reaction conditions will lead to change in the banding pattern. The low reproducibility of RAPD can be overcome by Triplet Arbitrary Primed (TAP) PCR, in which PCR reaction is carried out at three different annealing temperatures (38, 40 and 42°C) simultaneously and banding patterns are compared by using gel electrophoresis.

In Restriction fragment length polymorphism (RFLP), DNA is digested with rare-cutting restriction enzymes (8–6 bp) and relatively large fragments bacterial DNA are produced, which are then fractionated according to their size using PFGE. The pattern of DNA fragments created is referred to as an RFLP, which is highly characteristic of the particular species or strain of bacteria. This high discriminatory property of RFLP has been studied for the differentiation between some important probiotic bacterial strains, for example, *L. johnsonii* and *L. helveticus*, *B. animalis* and *B. longum*, *L. casei* and *L. rhamnosus*, *L. acidophilus* complex [123].

In Ribotyping, nucleic acid probes are used to recognize ribosomal genes. This technique is a combination of Southern hybridization of the DNA finger-printing with rDNA-targeted probing (16S, 23S, or 5S rRNA genes) [107]. Many *Bifidobacterium* and *Lactobacillus* strains of commercial products and human origin have been characterized [42]. Higher discrimination power is obtained while using multiple ribopatterns for each ribotype, especially when restriction enzymes are used with particular recognition sequences. AFLP is a more advanced fingerprinting technique, which combines RFLP and PCR-based methods by ligating the adaptors that is a primer-recognition sequence to the digested DNA.

Two types of restriction enzymes are used in this technique:(1) one having an average while another with a higher cutting frequency that digests the genomic DNA. In this method, double-stranded nucleotide adapters are ligated to the DNA fragments to provide binding sites for primer during PCR amplification yielding strain-specific amplification patterns. This method has been successfully reported for strain typing of the isolates of *Lb. johnsonii* and *Lb. acidophilus* group [35, 153]; and (2) Amplified rDNA restriction analysis (ARDRA) is another technique in which 16S rRNA gene is amplified using primers targeted at universally conserved regions within this gene. The amplification created is then restricted with an appropriate restriction enzyme followed by size separation of resulting restriction fragments using agarose gel electrophoresis, resulting in a characteristic RFLP. Various lactobacilli species, including *Lactobacillus delbrueckii* (sp. *lactis, delbrueckii, and bulgaricus*), *Lactobacillus helveticus* and *Lactobacillus acidophilus* have been discriminated by ARDRA [123]. DGGE or TGGE has been used for the rapid evaluation of diversity and simultaneous analysis of multiple bacterial samples by profiling the 16S-rRNA population [101, 102, 163].

The principle involves the separation of individual rRNA genes on the basis of their differences in melting temperature or chemical stability. In DGGE, formamide, and urea are used at high temperatures in a linearly increasing denaturing gradient to denature the double-stranded DNA, which results in a mixture of amplified PCR products and after staining a banding pattern will be formed depending upon the melting behavior of the different sequences. In TGGE, a linear temperature gradient is used. These techniques are useful for studying specific bacterial groups or species in a mixed bacte-rial populations and its subsequent identification.

Limitations of these techniques are: weak detection of the minor popula-tion (<1%) of bacterial community and detection of heteroduplex formed by heterogeneous rRNA operons, results in incorrect estimation of the bacterial diversity. In repetitive element sequence-based-PCR (Rep-PCR), primers complementary primers to interspersed repetitive consensus sequences are used to differentiate microbes by amplifying diverse-sized DNA fragments, which consist of repetitive elements sequences. For typing purposes, two main sets of repetitive elements are used [110], i.e., Repetitive extragenic palindromic (REP) elements (-38-bp seq) and Enterobacterial repetitive intergenic consensus (ERIC)-126 bp sequence.

In Multi-locus sequence typing (MLST), an automated DNA sequencing method is used to distinguish the alleles present at different housekeeping gene loci. This method has high discriminatory power and gives unambiguous results due to increased sensitivity of nucleotide sequencing, and that can be compared directly between different laboratories using a central MLST website [http://www.mlst.zoo.ox.ac.uk/]. Recently, DNA chips or DNA arrays have come in focus for assessing genetic diversity among microorganisms. These are ordered arrays of oligonucleotides derived from every region of the genome varying from several hundreds to almost a million. Oligonucleotides are immobilized on an organic substrate and depending upon the hybridization of isolated microbial DNA the array performs. Because of its high degree of specificity and sensitivity, it can be used for large-scale testing.

14.3   SELECTION OF PROBIOTIC BACTERIA

There are various selection criteria for probiotics has been suggested and followed. However; Figure 14.1 depicted the most accepted one for key selection [25]. In the first step of the selection process taxonomic classification of probiotics is determined, which have already been discussed; and which suggests the habitat, physiology, and origin of the bacterial strain.

14.3.1   SAFETY OF PROBIOTICS

The important safety aspects of probiotics include origin, infectivity, virulence, and pathogenicity factors. According to WHO [25], the specificity of probiotic action is more significant than its origin because of the ambiguity in origin of the human intestinal microflora. The antibiotic resistance of LAB (including the lactobacilli and bifidobacteria) has been a matter of great concern because they may be a source of antibiotic-resistance genes; many strains of *L. acidophilus, L. plantarum, L. salivarius, L. casei,* and

L. leishmannii have found to be resistance against vancomycin because of presence of D-alanine: D-alanine ligase-related enzymes [24]. Recently *Lactobacillus* of blood origin have shown low Minimum Inhibition Concentration (MICs) to erythromycin, clindamycin, piperacillin-tazobactam, and imipenem, but their susceptibility varies to cephalosporins and penicillin [126].

As antibiotic-resistant genes are mainly plasmid-encoded transferable between microorganisms, therefore antibiotic resistance of probiotics is considered as an important safety aspect. The transfer of these genes depends upon the type of the genetic material (transposons, plasmid) to be transferred, the nature of the donor and recipient strains and their interactions with the environment [90]. Therefore, natural antibiotic resistance of probiotic strains should be tested to prevent the transfer of resistant genes to other endogenous bacteria.

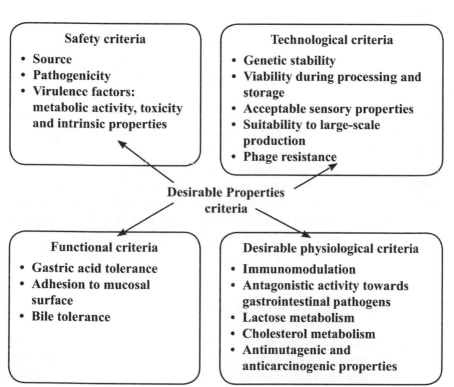

FIGURE 14.1   Desirable criteria for the selection of probiotics in commercial applications.

The probiotic bacteria can be considered safe for use when there is an absence of pathogenicity and infectivity. Despite immense use of *Lactobacillus* species in clinical and food industry, a very rare case of pathogenicity has been reported. In the case of immune-compromised patients or patients with underlying severe illnesses, the incidence of opportunistic infections has been found higher as they are more vulnerable to infections.

Intestinal mucosal injury, immunodeficiency, and an unbalanced intestinal microflora of the host induce bacterial translocations resulting in the weaker intestinal barrier that causes passage of bacteria across the epithelium and mucous membrane. When these translocated bacteria reach to mesenteric lymph nodes (MLN) and other organs, these cause bacteremia. In clinical trials, *Lactobacillus* strains have been administered to premature babies, children, adults, people suffering from Crohn's disease or diarrhea, and HIV-positive persons but no side-effects have been reported [47, 74, 86, 127, 129, 143, 157, 158, 160].

Several approaches have been proposed for the safety assessment of probiotics, which include studies of intrinsic properties of the probiotic strain, pharmacokinetics study, and interactions among the probiotic bacteria and the host. One of the prerequisites for the probiotic bacteria to be safe for use is that it should not produce any harmful substances during metabolic activities. Some intestinal bacteria act on proteins and produce ammonia, indol, phenols, and amines [23]. Secondary bile acids produced by intestinal bacteria have harmful effects on our body, which may act as carcinogenic substances (by effecting the mucous-secreting cells) or promoters for carcinogenesis [17]. Although many intestinal bacteria, including *Bifidobacterium* and *Lactobacillus* species, are able to deconjugate conjugated bile acids, yet some of Lactobacillus (5 species), *Bifidobacterium* (15 species), *S. thermophilus* and *Leuconostoc lactis subsp. lactis,* have been reported 7α -dehydroxylase deficient that is responsible for the secondary bile acids production [26, 144].

14.4 FUNCTIONAL PROPERTIES OF PROBIOTICS

For proper functionality of probiotic, strains must remain alive during the stressful conditions of manufacturing and storage of food product. The *in vitro* selection studies are tolerance to the harsh conditions of food manufacturing that constitute a critical issue for the inclusion of probiotics in most food product categories. It is determined by their ability to tolerate the hostile conditions of the stomach and the small intestine, and their adhere ability to intestinal surfaces.

14.4.1   ACID RESISTANT

Probiotic strains must be able to tolerate the acidic conditions in our stomach to reach the intestine. More than two liters of gastric juice having pH as low as 1.5 are secreted every day, which constitutes a primary defense mechanism against most ingested microorganisms. Therefore, the determination of acid resistance by various *Lactobacillus* and *Bifidobacterium* isolated strains were conducted as preliminary experiments. Enteric lactobacilli have been reported to tolerate pH of 2 for several minutes, but the pH of 1 was found destructive to all the species of lactobacilli [15, 58].

14.4.2   RESISTANCE TO BILE ACIDS

The resistance to bile acids is considered necessary to evaluate the effectiveness of probiotics. These acids are synthesized in the liver and secreted in a conjugated form (500–700 mL/d) from the gall bladder into the duodenum [56]. Antibacterial activity is exhibited by both conjugated and deconjugated form of bile acids, which inhibit the growth of various strains of *Escherichia coli*, *Enterococcus spp.* and *Klebsiella* sp. *in-vitro*, but more inhibitory effect is shown by the deconjugated forms against Gram-positive bacteria [27, 115, 141]. Among members of enteric lactobacilli of the same species, different sensitivity to bile salts was demonstrated leading to differences in the strains ability to colonize the intestinal tract of calves [40, 41].

14.4.3   PROBIOTIC ADHESION TO INTESTINAL CELL LINES

Probiotic strains, which are able to adhere to the intestinal surface of the human GI-tract strains and which persist for a long time in the intestinal tract, show better metabolic and immunomodulatory effects. Due to the adhesion of probiotic stain with mucosal layer gut-associated lymphoid tissue also come in contact, which mediates local and systemic immune effects. Further, it helps in competitive elimination of pathogenic group bacteria from the intestinal epithelium. Human intestinal cell lines, Caco-2, and HT-29 cells, have been exploited to elucidate the mechanisms of enteropathogen adhesion and also selection and subsequent assessment of LAB on the basis of their adhesion properties [6, 10, 16, 18, 130, 151].

14.4.4   ANTIMICROBIAL ACTIVITY

Probiotic strains produce several metabolic compounds, such as fatty acids, organic acids, diacetyl, and hydrogen peroxide having antagonistic activity against pathogenic bacteria. Bacteriocins are produced by probiotic strains, which are proteinaceous in nature having a specific inhibitory effect against closely related species, such as, other *Lactobacillus, Bacillus*, and *Clostridium* [57]. Low molecular weight metabolites such as lactic and acetic acid, hydrogen peroxide and other secondary metabolites have shown wide spectrum inhibition against many pathogenic bacteria like *Escherichia coli, Salmonella, Helicobacter* and *Clostridium* [50, 104]. *L. acidophilus* LA1 have shown inhibitory activity against *S. typhimurium in vivo* in a mouse model and *H. pylori in vitro* in humans [6, 94].

14.5   HEALTH BENEFITS OF PROBIOTICS

Beneficial health effects of probiotic bacteria (Figure 14.2) on the host are due to several proposed mechanisms, and broadly, these can be categorized into immunological, microbiological, and epithelial in nature [109]. All proposed health benefits of probiotic bacteria cannot be provided by a universal strain as these properties are strain-specific even vary among strains of the same species. Most investigated probiotic cultures are strains of *L. casei Shirota* (Yakult), *L. rhamnosus GG* (Valio), *B. animalis Bb-12* (Chr. Hansen) and *Saccharomyces cerevisiae Boulardii* (Biocodex) with the human health data for the management of antibiotic-associated diarrhea, lactose malabsorption, *Clostridium difficile* diarrhea, and rotavirus diarrhea.

14.5.1   ALLEVIATION OF LACTOSE INTOLERANCE

Lactose is hydrolyzed by lactase into constitutive monosaccharides, glucose, and galactose; and these are then readily absorbed into the bloodstream. Lactose intolerance is a clinical condition when Hypolactasia (decline in activity of intestinal lactase) and lactose malabsorption occurs along with abdominal pain, diarrhea, bloating, nausea, and flatulence. Due to undigested lactose in the large intestine, the increase inflow of water into the lumen occurs because of osmosis increases that causes the above-mentioned symptoms. Various clinical studies have proven that individual with hypolactasia has and better digestibility of fermented dairy products compared to an equal quantity of milk [52, 98]. The reasons for better lactose tolerance are starter cultures, which utilize lactose as a source of energy during fermentation and express intracellular β-galactosidase enzyme leading to lactose hydrolysis and oro-caecal transit time.

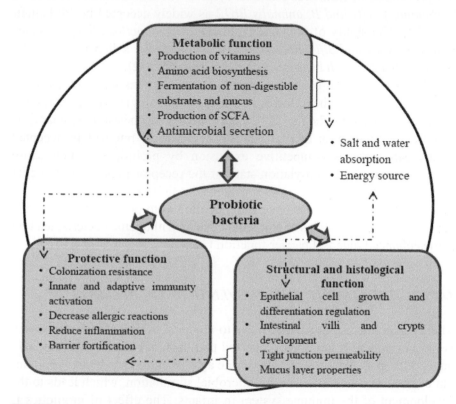

FIGURE 14.2 Summary of health benefits of probiotics.

14.5.2 TREATMENT OF DIARRHEAL SYMPTOMS

Probiotics have been used in the prevention and treatment of various intestinal diseases. Antibiotic therapy leads to a mild or severe form of diarrhea as their common side effects are due to imbalanced gut microflora. In most of the cases, it is associated with *C. difficile* colonization in the intestine, which releases exotoxins namely, toxin A and toxin B, thus causing diarrhea and colitis. Several studies have suggested that use of probiotic has reduced risk of antibiotic-associated diarrhea [93]. Treatment with probiotics, mainly *L. rhamnosus* and *S. boulardii* has been used in clinical practice. A meta-analysis on probiotics for the prevention and treatment of antibiotic-associated diarrhea has been evaluated using *L. rhamnosus, L. casei,* and *S. boulardii* and it was concluded that administration of probiotic is linked with a reduced risk of these conditions [51]. Similarly, clinical studies have shown the treatment and prevention of Rotavirus diarrhea by probiotic microorganisms such as *L. rhamnosus GG and B. animalis Bb12* as widely accepted health benefit [61, 139]. The ability of probiotics including *Lactobacillus, Streptococcus, Saccharomyces, and Bifidobacterium* to prevent traveler's diarrhea caused by enterotoxigenic *E. coli* has also been examined [53, 132].

Prophylactic effect of various probiotic preparations has been assessed against traveler's diarrhea. However, due to methodological deficiencies, the results have been conflicting [89]. There are several possible mechanisms proposed for reduction of the duration of diarrhea, but widely accepted mechanism includes competitive exclusion by probiotics, which cause modification in the glycosylation state of the receptor present in epithelial cells by excreting soluble factors thus, the adhesion of rotavirus is inhibited [30]. In addition, probiotics have also been found to prevent the disruption of the cytoskeletal proteins of epithelial cells by pathogenic bacteria, leading to the enhanced mucosal barrier function [122].

14.5.3   TREATMENT AND PREVENTION OF ALLERGY

Allergy is defined as hypersensitivity to antigens, which are harmless such as food or pollen. It has been suggested that bacterial exposure in the early stage of life may exhibit a protective role against allergy, and thus probiotics can be used as a safe alternate for microbial stimulation, which leads to the development of the immune system in infants. The effect of probiotics is mainly involved in food allergy and atopic dermatitis. The mechanism found is that bacteria and viruses in childhood would affect T-helper cells balance by stimulating the Th2 cells and insufficient stimulation of Th1 cells of the immune system due to the less exposure to pathogens and it results in a predisposition to allergy [161].

Many human trials have been conducted to prevent allergic disease by using probiotics, which include *Lactobacillus or Bifidobacterium* species or probiotic combinations, given either postnatally or prenatally. Also, it has been reported that there are quantitative as well as qualitative differences among allergic and healthy infants with the former exhibiting a more adult type of microflora colonization to support the role of intestinal microbiota in allergy [65, 112]. In a recent study, a reduction in the prevalence of atopic eczema has been reported due to the consumption of probiotic preparations containing *Lactobacillus GG* (LGG) in the early stage of life [44]. Although, immune regulation, strengthening of immune exclusion, and immune elimination have been recommended, yet the protective mechanisms of probiotics against allergic reactions are not clear [62].

14.5.4 ANTIMUTAGENIC AND ANTICARCINOGENIC PROPERTIES

In our body, bacterial or viral infections and phagocytosis will lead to the formation of mutagens and are also commonly obtained through food items. Compounds such as NO, O_2^- and H_2O_2 are released during defense mechanism by leukocytes that may cause DNA damage and mutations. Antimutagenicity can be defined as the suppression of the spontaneous and induced mutations. Studies have shown that intake of probiotics may reduce colon cancer incidence especially species of *Lactobacillus* and *bifidobacterium* lower the genotoxic activity against certain chemicals and enhance antimutagenic activity in selected media [54, 82, 146]. A significant dose-dependent anti-mutagenic activity was observed from an extract of yogurt fermented by *L. bulgaricus* 191R against mutagens such as MNNG, 4-nitroquinoline-N-oxide, 9,10-dimethyl-1,2-benz [α] anthracene, 3,2-dimethyl-4-aminobiphenyl, and Trp P2 [103].

The study conducted in human subjects has reported lesser fecal concentrations of azoreductase, β-glucuronidase and nitroreductase enzymes, which lead to carcinogen production in the gut by consumption of milk fermented with *Lactococcus lactis Lactobacillus acidophilus A1, Lactococcus cremoris* and *Bifidobacterium bifidum B1* [88]. After deconjugation of bile acids in the colon, dehydroxylation reaction of the 7 α -hydroxyl group form secondary bile acids, lithocholic, and deoxycholic acids to cause a cytotoxic effect on epithelial cells, thus increasing cell proliferation and the probability of colon cancer development [68, 79].

Feeding probiotics leads to an increased amount of net SCFA in pigs, which is a potential anti-carcinogenic agent [125]. *L. casei Shirota* has shown to reduce DNA injury in the colon of rats when exposed to the mutagens N-methyl-N-nitro, N-nitrosoguanidine. However, the heat-treated cultures of *L. acidophilus* were not able to show these anti-genotoxic effects in rat colon, which proves the importance of consumption of live organisms [118, 119]. Although the exact mechanism of anti-mutagenicity and anti-carcinogenicity by probiotic bacteria is unclear, yet a possible mechanism suggested is binding of mutagens to the microbial cell surface

[111]. Alteration of gut microflora and its metabolic activity, maintenance of intestinal permeability and immunity are other proposed mechanisms [138].

14.5.5 PREVENTION OF INFLAMMATORY BOWEL DISEASE (IBD)

Inflammatory bowel disease (IBD) refers to a group of bowel disorders, which include ulceration, inflammation, and narrowing of the intestinal tract [48]. Intestinal microbiota of IBD patients has found to be imbalanced as compared to healthy individuals with a decrease in commensal bacterial species such as *Lactobacillus* and *Bifidobacterium* and predominance of potentially harmful bacteria has been identified [19, 134]. It is of two types:

- In Crohn's disease, the immune response is regulated by Th1, an increase in interleukin (IL)-12 expression, and leads to increased tumor necrosis factor (TNF)-α and interferon (IFN)-γ (20).
- Ulcerative colitis is a Th2 driven immune response with proinflammatory cytokines, such as IL-5 amount increases.

L. reuteri has been reported to prevent colitis in IL-10 KO mice by restoring the gut microbiota balance by increasing LAB number [83]. In a placebo-controlled trial, reduction in mucosal inflammatory activity and occurrence of colon cancer has been reported by administration of *L. salivarius* UCC118 in IL-10-KO mice by reducing the *C. perfringens*, *Enterococcus*, and *Coliforms* levels in intestinal microbiota [83]. In a clinical trial, intake of the probiotic preparation VSL#3 (a combination of four lactobacilli strains, three bifidobacteria strains and one of *S. salivarius ssp. thermophilus*) by patients with UC has been shown to act by decreasing lu minal pH [154].

Another mechanism is the production of antimicrobial substances such as bacteriocins by probiotics, which inhibit pathogenic microorganisms [69]. Peptides produced by probiotic *L. salivarius* UCC118 inhibit a broad range of pathogens such as *Enterococcus, Bacillus, Salmonella, Listeria, and Staphylococcus* species [28]. During inflammation process, probiotic microorganisms, including strains of *lactobacilli, Bifidobacterium*, and *Saccharomyces,* stabilize the gut microbiota and enhance the permeability of intestinal barrier by degrading enteral antigens and altering their immunity [63].

In vivo studies have shown an initial increase in the level of IL-6 expression by *B. lactis* Bb12 and prevention in intestinal inflammation significantly caused by *B. vulgates* in rats [124]. These results indicate that commensal bacteria play a key role in initiating epithelial cell homeostasis. Some LAB has been found to prevent the local inflammatory diseases by increasing IL-10 levels and consequently decreasing inflammatory cytokines such as TNF- α and IFN- γ, which can be successfully used as an adjunct therapy along with conventional treatments.

14.5.6 IMMUNOMODULATORY PROPERTIES

The human immune system is multifaceted and is subordinate to that of the central nervous system (CNS). Adaptive and innate immunity are two main two principal components, which are especially assimilated and reciprocally dependent [55]:

- Innate immune responses are nonspecific and environmentally resistant; it is regarded as the first line of defense, which includes macrophages, eosinophils, neutrophils, epithelial cells, natural killer (NK) cells, and M cells. Among these NK cells are significant performers, which hastily respond to the infested cells at initial stages leading to the death of the target cell.
- Adaptive system is specific and attained through environmental interactions. It involves lymphocytes with receptors for a specific antigen cell, which in turns activates sub-sets of helper T cells (Th).

Memory B and T cells are the main components, which increase effective immune responses against secondary infections. Dendritic cells (DCs) were known as professional "antigen-presenting cells" (APC) as well as responsible for the correlation between the adaptive and innate immunity, similar to macrophages and monocytes. Various studies have reported the role of probiotics in boosting the immune system by regulating NK cells. The balance betweenTh1 and Th2 cytokines are important and is a determining factor for the direction and consequences of an immune response.

Production of chemokines and cytokines by epithelial cells is determined by the interaction of probiotic strains to enterocytes. Probiotic bacteria in a strain-dependent manner can inflect the down/up-regulation of pro- and anti-inflammatory cytokines, for example, *Lactobacillus sakei* induces TNF-α, IL-1βand IL-8 expression, whereas in Caco-2 cells, TGF-β production was stimulated by *L. johnsonii*. The component of innate immune system, toll-like receptors (TLRs) recognizes lipoteichoic acids groups and lipopolysaccharides (LPS) in pathogens, which trigger series of immunological defense mechanisms [5]. Probiotic strains stimulate IgA production by B-cells, which help in balancing the gut humoral immunity through binding to antigens.

The study on the administration of *L. Rhamnosus* GG along with a rotavirus vaccination showed increased number of IgA secreting cells in 2 to 5 years old children [60]. The immune response of Gram-negative *Klebsiella pneumonia* and *L. rhamnosus* were compared and induction of DC maturation was found in both cultures but with a different cytokine profile [9]. *K. pneumonia* helps in activation of T-helper (Th) 1 type cells, whereas *L. rhamnosus* decreases the production of interleukins IL-6, IL-12IL-12 and IL-18 by immature DCs and pro-inflammatory cytokines (TNF-a).

The cytokine response may also vary in response to different probiotics [49]. Despite these studies, the validity of results is still a matter of concern due to various factors such as poor study designs and lack of appropriate control *in vivo* studies. Thus, more detailed studies are still required to determine the exact mechanism of probiotics on mucosal and systemic immunity.

14.5.7 INHIBITION OF HELICOBACTER PYLORI INFECTION

Helicobacter pylori are Gram-negative bacteria that selectively colonizes the human gastric mucosa, and it increases the risk of developing a peptic ulcer, chronic gastritis, and gastric malignancies in a severe case [117]. At present, treatment of *H. pylori* infection includes combination therapy of two antibiotics (clarithromycin plus amoxicillin or metronidazole) and a proton pump inhibitor (PPI) [84]. Although this treatment is considered almost 90% effective, yet side effects, including antibiotic-associated diarrhea and antibiotic resistance have been found associated with being expensive [116].

These limitations have led to more effective alternative treatments of *H. pylori*, which include monotherapy using some component or in combination with antimicrobials [77]. Inclusion of probiotics in the diet is a potential therapy to reduce the rate of *H. pylori* infection [66]. Among several proposed mechanisms of probiotic action against *H. pylori*, the most common is the reinforcement of local microbiological homeostasis in the gut to interfere with *H. pylori*; and thus reducing inflammatory processes [3]. The inhibitory effect of probiotics is strain-dependent; for example, *L. casei* strain *Shirota* showed a significant reduction in *H. pylori* colonization level in body mucosa and antrum in a mouse model [135, 136].

Probiotics strengthen the production of several antimicrobial substances to compete with *H. pylori* for adhesion receptors, stimulate mucin production, and stabilizing the gut mucosal barrier. Metabolites, such as lactic acid in addition to its antimicrobial activity, lower the pH of intestine, which inhibits the *H. pylori* urease. These inhibitory effects are strain-dependent, e.g., *L. johnsonii* La10 produces the same amount of lactic acid as *L. johnsonii* La1 but unable to inhibit *H. pylori* infection [113]. Certain strains of lactobacilli exert anti-adhesion activity by producing various antimicrobial compounds such as bacteriocins; lactic acid such as *L. reuteri* inhibits the binding of *H. pylori* to specific glycolipid receptors [99].

In vitro studies, including *L. plantarum* and *L. rhamnosus* have shown to increase expression of MUC2 and MUC3 genes that restore the mucosal permeability; and adherence of pathogenic bacteria is inhibited [43]. A potential mechanism of probiotic efficacy is the modulation of immune response by interacting with epithelial cells and modulation of anti-inflammatory cytokines secretion; thus, gastric activity and inflammation are reduced [39]. However, various mechanisms have been studied with varying success, but their relative importance is still unclear [142].

14.5.8   HYPOCHOLESTEROLEMIC EFFECT

Cholesterol acts as a precursor to certain hormones and vitamins, and it is an important component of cell membranes and nerve cells. However, an increase in the level of serum cholesterol is one of the major risk factors for coronary heart diseases (CHDs). Both pharmacological and non-pharmacological interventions can reduce cholesterol concentrations in serum. Due to adverse effects of anti-lipid drugs, natural or alternative non-drug treatments for hyperlipidemia have been practiced.

Mann and Spoerry [85] observed the hypocholesterolemic activity of *Lactobacillus* fermented milk in the Maasai tribe of Kenya. They suggested the production of hydroxymethyl-glutarate by probiotic bacteria, which inhibit hydroxymethylglutaryl-CoA reductases (the rate-limiting enzyme in endogenous production of cholesterol). Several mechanisms have been proposed for the hypocholesterolemic effects of probiotics, which include enzymatic deconjugation of bile acids by bile salt hydrolases as one of them. After deconjugation of bile acids, they are easily absorbed in the intestine leading to their elimination through feces. *In vitro* studies using *L. plantarum* have reported that bile salt hydrolase can hydrolyze conjugated glycodeoxycholic acid and taurodeoxycholic acid leading to lowering of cholesterol level. The cholesterol-lowering effect by probiotics is also due to their ability to bind cholesterol in the small intestines [64].

BSH activity of various species and strains of LAB to hydrolyze bile salts is variable. It has been reported in almost all species and strains of bifidobacteria while only selected species of lactobacilli possesses this ability, but in *Lactococcus lactis, Leuconostoc mesenteroides,* and *S. thermophilus* BSH activity was not detected [145]. Hypocholesterolemic effect of probiotics is also attributed by its potential for the production of short-chain fatty acids from polysaccharide, protein, peptide, and oligosaccharide, mainly by anaerobic bacteria into the gastrointestinal tract (GIT) leading to alteration and reduction in serum [81]. Due to differences in conditions of *in-vitro* and *in-vivo* systems, discrepancies in the data of different effects on serum cholesterol have been observed, which is also a strain-dependent.

14.6   TECHNOLOGICAL ASPECTS OF PROBIOTICS

As there is growing scientific evidence to support a wide range of health benefits of probiotic bacteria, efforts are being made to incorporate these bacteria into our food system in order to have their functional properties. Although probiotics can be formulated into foods, drugs, and dietary supplements, yet the most common vehicle for these are dairy products.

It has also been shown that probiotic infant formulas have similar fecal levels of bifidobacteria as infants on breastfed [75]. The main challenge for food manufacturers is to deliver probiotic bacteria to the desired sites in an active and viable form due to their sensitivity to the harsh processing and chemical conditions used in the food process industry. In the world, almost 80 products have been estimated containing probiotic, and some of them are commercially available and marketed with probiotics (Table 14.2). Probiotic selection has to be following technological aspects:

- Desirable viability during processing;
- Good sensory properties;
- Resistance to phage attack;
- Stability in the product quality during storage.

14.7   PRODUCTION OF PROBIOTIC FORMULATION

Most commercial probiotic cultures are available in frozen or dry form (freeze or spray-dried) like other starter cultures for direct vat set (DVS) applications. Although freeze-drying is considered one of the best practices for reducing thermal damage, spray drying is more advantageous in terms of cost, energy, and throughput [11]. To improve the survivability and preservation of probiotics activity can be improved by the addition of protectants as compatible solutes and cryoprotectants to the growth medium or drying medium. Commercially lactobacilli and bifidobacteria manufactured as probiotic cultures consisting either a single strain or mixture of several strains. Processing conditions during manufacture may induce cellular damage to the bacteria through osmotic and thermal stresses (e.g., thermal processing, high solutes concentration) and by increasing redox potential. Also, fluctuation during storage conditions changes the physical state (glassy to rubbery), which further triggers various biochemical and enzymatic reactions that detrimentally affect the bacterial survival [32].

TABLE 14.2   Some Probiotic Strains Available Commercially

Strain	Company (Product)
B. lactis HN019 (DR10)	Fonterra
B. longum BB536	Morinaga Milk Industry Co. Ltd.
L. acidophilus LA1/LA5	Chr. Hansen
L. acidophilus LAFTI® L10	DSM Food Specialties
L. acidophilus LB	Lacteol Laboratory
L. acidophilus NCFM®	Danisco
L. acidophilus SBT-20621	Snow Brand Milk Products Co. Ltd
L. casei Immunitas	Danone
L. casei Shirota	Yakult
L. johnsonii La1	Nestle
L. paracasei F19	Medipharm
L. plantarum 299V	Probi AB
L. reuteri SD2112	Biogaia
L. rhamnosus GG1	Valio Dairy
L. rhamnosus LB21	Essum AB
L. rhamnosus R0011	Institute Rosell
L. salivarius UCC118	University College, Cork

To ensure efficient delivery of the probiotic strains in physiologically active form, technologies, such as enteric coating and microencapsulation, have been applied. The commonly used encapsulating materials in food industries are alginate, starch, and carrageenan and whey protein. Alginate is a natural polysaccharide extracted from brown seaweeds, which can be solubilized by sequestering calcium ions; thus entrapped cells are released into the system. Encapsulated probiotic bacteria have shown improved viability in various dairy products, such as frozen dairy desserts, yogurt, or freeze-dried yogurt as compared with non-encapsulated control bacteria [12]. Some of the benefits of microencapsulated bacteria are summarized in Figure 14.3.

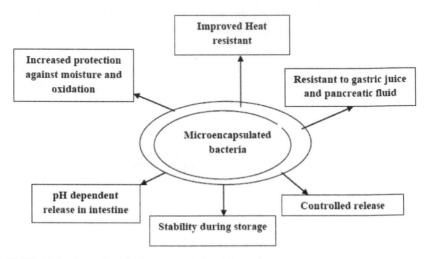

FIGURE 14.3   Benefits of micro-encapsulated bacteria.

14.8   INTERACTION BETWEEN PROBIOTIC AND STARTER

For selecting a starter culture for fermentation, the most desirable characteristic is acid-forming ability. However, health effects are taken into account while selecting probiotics beneficial. Due to the slow growth rate of the probiotic bacteria in milk, off-flavors are produced that are not desirable commercially; this could partly be overcome by using a combination of starter and probiotic organisms resulting in the product with good sensory properties and higher survivability of the probiotic bacteria [29]. By the addition of various growth factors (e.g., protein hydrolysates and yeast extract), energy sources (e.g., glucose), or nutrients such as minerals or vitamins and suitable antioxidants, the food as a substrate for the probiotic can be improved. However, sometimes it is not enough due to some differences between starter and probiotic bacteria [59].

While selecting a starter culture, the antagonistic effect on probiotic should be taken into consideration. Sometimes metabolites produced by the probiotic bacteria (lactic acid, hydrogen peroxide, and bacteriocins) may negatively affect the survival of starter culture and vice versa [156]. On the other hand, metabolites of high proteolytic or oxygen scavenging starter cultures ability have been found to promote bifidobacteria growth [59]. The interaction between probiotics and the starter organisms in the product is dependent on the stage at which probiotics are added, and it was demonstrated that when probiotic organisms are added after the fermentation process, probiotic bacteria can be consumed, e.g., the Danish product 'Cultura.'

Another approach is to increase the inoculum size of probiotics more (5–10%) than it is required for the starter culture. The physiologic state of the probiotic bacteria is also important when it is added during fermentation. Bacterial cells harvested during the stationary phase are more resistant to stress conditions and have a higher survival rate than the logarithmic phase [72]. The two-step fermentation process has been found to increase probiotic bacteria significantly in the product by primary fermentation using probiotic cultures for 2h, followed by secondary fermentation by yogurt starter bacteria for 4 h. It helps the probiotic bacteria to reach to their lag phase or early stage of log-phase resulting in higher counts of after 6 h of incubation [137].

14.8.1   INTERACTIONBETWEENPROBIOTICSANDPREBIOTICS

Prebiotics are *"non-digestible food ingredients having a beneficial effect on the host by selectively stimulating the growth and activity or both of one or more intestinal bacterial species"* [38]. It can be used in combination with probiotics to improve the viability of the product in the GIT. Foods containing a combination of probiotics and probiotics and their synergism are referred to as synbiotics.

Range of prebiotics (e.g., lactulose, galacto, and fructooligosaccharides and resistant starch) has been used as a means for enhanced activity and survivability of probiotic in fermented foods during growth and storage [80]. Probiotics adhere to the starch granule, which provides resistance and flexibility to environmental stresses during processing, storage, and transit through GIT. Oligofructose fermentation in the colon leads to increase colonic bifidobacteria, which produce inhibitory compounds for pathogenic bacteria via reduction of blood ammonia and producing digestive enzymes.

The Synbiotic product contains probiotic bacteria mainly the strains of *Lactobacillus acidophilus, Lactobacillus casei,* and *Bifidobacterium* sp. in combination with various prebiotics such as fructooligosaccharides, galactooligosaccharides, lactulose, and inulin-derived products to improve the survivability of probiotics. Table 14.3 summarizes number disorders, for which evidences has been reported from at least one clinical trial proving that oral administration of a specific probiotic strain along with prebiotics is effective against various health diseases.

14.9   STORAGE

Although the shelf-life of probiotic cultures in dried form is more suitable than the liquid form, yet the main challenge is survivability of bacteria during storage period that depends on physical parameters of the final product matrix and the storage conditions. The storage conditions include the moisture content of powders and its composition, storage temperature, relative humidity (RH), oxygen content, light exposure, and storage materials. These conditions have a significant influence on the survivability and viability of probiotic powders (freeze and spray-dried). The beneficial effect of the probiotic culture is strain-dependent, as well as the form and dose. The viability of probiotic bacteria is inversely related to storage temperature [36].

Storage of sugar-containing probiotic products at high temperature and RH has been found to decrease the viability due to their glass transition temperature [152]. Another reason for the reduction of viability during storage is the oxidation of membrane lipids of unsaturated lipids (oleic acid), which are unstable and readily oxidized to hydroperoxides in food constituents during storage. Hydroperoxides produced during lipid oxidation have been shown to cause DNA damage in bacteria [87]. In order to enhance the viability of probiotic during storage, antioxidants have been used under vacuum with controlled water activity [147]. Microencapsulation has also been shown to extend the storage life of probiotic bacteria [73]. Storage materials have a significant effect on the extension of shelf life of probiotics, and vacuum-sealed glass vials or use of nitrogen gas has shown greater viability as compared to storage in air.

14.10   REHYDRATION

For the revival of probiotic bacterial cell powders after dehydration, the final critical step is rehydration, which can be divided into four steps, i.e., wetting, submersion, dispersion, and dissolving [31]. The recovery of bacterial cells from power and its viability varies according to the rehydration media. A

TABLE 14.3 Evidence-Based Recommended Dose for Probiotics and Prebiotics in Gastroenterology

Disorder	Product	Recommended Dose	References
Adjuvant therapy for H. pylori radication	L. rhamnosus GG	6×10^9 CFU, twice daily	[150]
	B. clausii (Enterogermina strains)	2×10^9 spores, three times daily	[150]
	AB yogurt with unspecified lactobacilli and bifidobacteria	5×10^9 viable bac, twice daily	[150]
	S. cerevisiae (boulardii)	1 g or 5×10^9 CFU per day	[150]
	L. casei DN-114 001 in fermented milk with L. bulgaricus+ S. thermophilus	10^{10} CFU, twice daily	[150]
Alleviates some symptoms of irritable bowel syndrome	B. infantis 35624	10^8 CFU, once daily	[106]
	L. rhamnosus GG	6×10^9 cfu, twice daily	[37]
	VSL# 3 mixture	4.5×10^{11} CFU, twice daily	[67]
	B. animalis DN-173 1010 in fermented milk with L. bulgaricus + S. thermophilus	10^{10} CFU, twice daily	[45]
Prevention of antibiotic-associated diarrhea in children	S. cerevisiae (boulardii)	250 mg, twice daily	[131]
	L. rhamnosus GG	10^{10} CFU, once or twice daily	[131]
	B. lactis Bb12 + S. thermophilus	$10^7 + 10^6$ CFU/g of formula	[131]
Prevention of postoperative infections	Synbiotic 2000: 4 bacteria strains and fibers including the prebiotic inulin	10^{10} CFU + 10 g fibers, twice daily	[121]
Prevention of necrotizing enterocolitis in preterm infants	B. infantis, S. thermophilus, and B. bifidum	0.35×10^9 CFU each strain, once daily 10^9 CFU each, twice daily	[21]
	L. acidophilus + B. infantis		[21]

TABLE 14.3 *(Continued)*

Disorder	Product	Recommended Dose	References
Reduces symptoms associated with lactose maldigestion	Regular yogurt with	Yogurt not heat-treated after pasteurization contains suitable cultures to improve digestion of the lactose in the yogurt	[97]
	L. bulgaricus +		
	S. thermophilus		
Treatment of constipation	Lactulose	20–40 g per day	[133]
	Oligofructose	> 20 g per day	[105]
Treatment of acute infectious diarrhea in children	L. rhamnosus GG	10^{10}–10^{11}CFU, twice daily	[2]
	L. reuteri ATTC 55730	10^{10}–10^{11}CFU, twice daily	[2]
	L. acidophilus + B. infantis	10^9 CFUeach, three times daily 200 mg, three times daily	[76]
	S. cerevisiae (boulardii)		[2]
Treatment of hepatic encephalopathy	Lactulose	45–90 g per day	[133]

complex rehydration medium, which provides an environment with high osmotic pressure to reduce osmotic shock, has been recommended [20]. It has also been reported in repairing of cells damaged during the drying process by providing essential nutrients and cell components. The temperature and rate of rehydration also have a significant role in the recovery of dried and injured cells after freeze-drying, and highest number of cells during rehydration process have been recovered at 15–25°C in comparison to 35°C and 45°C.

14.10.1 NON-DAIRY PROBIOTIC DAIRY FOODS

Dairy products as a vehicle for probiotics have limitations due to the presence of some allergens and requirement of cold environment for storage. To overcome this, use of non-dairy matrices such as fruits, vegetables, legumes, and cereals are being used for delivery of probiotics as they contain essential nutrients such as minerals, vitamins, and antioxidants. Using non-dairy products as delivery agents for probiotics also encounter challenges. Survival of probiotic bacteria in food matrix is affected by various factors, such as pH, water activity, storage temperature, oxygen tension and also the presence of various competing microflora and inhibitors (e.g., NaCl). As these non-dairy products like drinks, chocolate, etc. are stored at room temperature, which creates an additional stress for cell survivability.

In products like probiotic-containing baby foods or confectionery (e.g., chocolate) where probiotic cultures are used as additives, their stability for during storage for extended periods is important as the cells do not multiply. Survivability of probiotics can be achieved to some extent by microencapsulation technologies, where the cells are entrapped into protective coating matrices. A good protective role has also been demonstrated by gelatin and vegetable gum for the strains of *Bifidobacterium* and *Lactobacillus* that are acid sensitive [13]. In vegetables, probiotic strains of *Lactobacillus* and *Leuconostoc* genera are commonly found helpful. For example, in cabbage juice, *L. plantarum, L., casei,* and *L. delbrueckii* were able to grow up to 10^8 CFU/mL at 30°C after 48 h of incubation without any nutrient supplementation [162].

Viability and bile tolerance of *L. acidophilus, L. reuteri,* and *L. plantarum* have also found to be increased in Malt, wheat, and barley extracts [95]. Cereal-based probiotic fermented products have shown beneficial health effects; for example, oat-based substrates have shown to promote the growth of *L. reuteri, L. acidophilus,* and *B. bifidum* [91]. Fermentation by probiotic bacteria reduces the non-digestible carbohydrates (poly- and oligo-saccharides) and amount of lysine and availability of the vitamin-B group increases. Also, phytates in the cereals are degraded, and minerals such as manganese, iron, zinc, and calcium are released into the medium [8].

14.11   SUMMARY

Probiotics have been widely studied and explored for commercial applications in pharmaceuticals and food sectors. Recent research and developments have shown beneficial effects of these bacteria on human health, such as immunomodulation, cholesterol-lowering, tumor prevention, lactose metabolism, etc. These health effects are strain-specific; thus, there is a demand for rapid and reliable analytical techniques for their identification and successful application. Currently, many PCR and non-PCR methods are available for identification purpose for rapid microbial analysis of complex bacterial communities. *Lactobacillus* and *Bifidobacterium* species are mainly used commercially, where their viability in the final product and stability during storage are main challenges. Microencapsulation is the most effective method used commercially for survivability of probiotic bacteria, which protects the cell from adverse conditions of processing, temperature, and pH changes. Several methods of microencapsulation of probiotic bacteria are: spray drying, extrusion, emulsion, and phase separation but microencapsulation using calcium alginate gel capsule formation is commonly used. Various factors play an important role in the efficiency of probiotic microencapsulation, including encapsulating materials and techniques of micro-encapsulation. Moreover, for large scale production of probiotic powders, freeze-drying, and spray-drying are used, but bacterial cells are exposed to a variety of stresses during these processes such as heat, cold, oxygen, and osmotic stresses which leads to loss of viability and functionality during drying and storage. Although the consumption of probiotic bacteria stimulates immunity in different ways, yet more research is needed to investigate the underlying mechanisms. Moreover, clinical trials of probiotic research have been conducted on infantile, antibiotic-related, and traveler's diarrhea; many human trails are still needed for other diseases such as cancer. One of the main unexplored areas of research is the use of non-pathogenic strains of *Bacillus*, *Entero-coccus*, and *yeasts* as probiotics.

KEYWORDS

- angiotensin-I converting enzyme
- angiotensin-II converting enzyme
- bioaccumulation
- caramelized foods
- conjugated linoleic acid
- immunoglobulins
- intestinal microbes
- oligosaccharides
- probiotics

REFERENCES

1. Achenbach, L. A., Carey, A. J., & Madigan, M. T., (2001). Photosynthesis and phylogenetic primers for detection of anoxygenic phototrophs in natural environments. *Applied Environmental Microbiology, 67*, 2922–2926.
2. Allen, S. J., Okoko, B., Martinez, E., Gregorio, G., & Dans, L. F., (2004). Probiotics for treating infectious diarrhea. *Cochrane Database of Systematic Reviews, 2*, CD-003048, PMID 15106189.
3. Alsahli, M., & Michetti, P., (2001). Lactobacilli for the management of *Helicobacter pylori. Nutrition, 17*, 268–269.
4. Amor, K. B., Vaughan, E. E., & De Vos, W. M., (2007). Advanced molecular tools for the identification of lactic acid bacteria. *The Journal of Nutrition, 137*(3), 741S–747S.
5. Akira, S., Takeda, K., & Kaisho, T., (2001). Toll-like receptors: Critical proteins linking innate and acquired immunity. *Nature Immunology, 2*(8), 675–681.
6. Bernet-Camard, M. F., Brassart, D., Neeser, J. R., & Servin, A. L., (1994). *Lactobacillus acidophilus* LA-1 binds to cultured human intestinal cell-lines and inhibits cell attachment and cell invasion by enterovirulent bacteria. *Gut, 35*, 483–489.
7. Bernet-Camard, M. F., Lievin, V., Brassart, D., Neeser, J. R., Servin, A. L., & Hudault, S., (1997). The human *Lactobacillus acidophilus* strain LA-1 secretes a nonbacteriocin antibacterial substance(s) active *in-vitro* and *in-vivo*. *Applied Environmental Microbiology, 63*, 2747–2753.
8. Blandino, A., Al-Aseeri, M. E., Pandiella, S. S., Cantero, D., & Webb, C., (2003). Cereal-based fermented foods and beverages. *Food Research International, 36*, 527–543.
9. Braat, H., De Jong, E. C., Van Den Brande, J. M., Kapsenberg, M. L., Peppelenbosch, M. P., & Van Tol, E. A., (2004). Dichotomy between *Lactobacillus rhamnosus* and *Klebsiella pneumoniae* on dendritic cell phenotype and function. *Journal of Molecular Medicine, 82*, 197–205.

10. Brassart, D., Schiffrin, E., Rochat, F., Offord, E. A., Macé, C., & Neeser, J. R., (1998). The future of functional foods: Scientific basis and future requirements. *Lebensmittel Technologie, 7/8*, 258–66.

11. Burgain, J., Gaiani, C., Linder, M., & Scher, J., (2011). Encapsulation of probiotic living cells: From laboratory scale to industrial applications. *Journal of Food Engineering, 104*, 467–483.

12. Capela, P., Hay, T. K. C., & Shah, N. P., (2006). Effect of cryoprotectants, prebiotics and microencapsulation on survival of probiotic organisms in yogurt and freeze-dried yogurt. *Food Research International, 39*, 203–211.

13. Chandramouli, V., Kailasapathy, K., Peiris, P., & Jones, M., (2004). An improved method of microencapsulation and its evaluation to protect *Lactobacillus* spp. in simulated gastric conditions. *Journal of Microbiological Methods, 56*, 27–35.

14. Chang, Y. J., Peacock, A. D., & Long, P. E., (2001). Diversity and characterization of sulfate–reducing bacteria in groundwater at a uranium mill tailings site. *Applied Environmental Microbiology, 67*, 3149–3160.

15. Charteris, W. P., Kelly, P. M., Morelli, L., & Collins, J. K., (1998). Development and application of an *in vitro* methodology to determine the transit tolerance of potentially probiotic *Lactobacillus* and *Bifidobacterium* species in the upper human gastrointestinal tract. *Journal of Applied Microbiology, 84*, 759–768.

16. Chauviere, G., Coconnier, M. H., Kerneis, S., Darfeuille-Michaud, A., Joly, B., & Servin, A. L., (1992). Competitive exclusion of diarrheagenic *Escherichia coli* (ETEC) from human enterocyte-like Caco-2 cells by heat-killed *Lactobacillus*. *FEMS Microbiology Letters, 91*, 213–218.

17. Cheah, P. Y., (1990). Hypotheses for the etiology of colorectal cancer—an overview. *Nutrition and Cancer, 14*, 5–13.

18. Coconnier, M. H., Bernet, M. F., Kerneis, S., Chauviere, G., Fourniat, J., & Servin, A. L., (1993). Inhibition of adhesion of enteroinvasive pathogens to human intestinal Caco-2 cells by *Lactobacillus acidophilus* strain LB decreases bacterial invasion. *FEMS Microbiology Letters, 110*, 299–305.

19. Conte, M. P., Schippa, S., & Zamboni, I., (2006). Gut-associated bacterial microbiota in paediatric patients with inflammatory bowel disease. *Gut, 55*(12), 1760–1767.

20. D'Haens, G., & Daperno, M., (2006). Advances in biologic therapy for ulcerative colitis and Crohn's disease. *Current Gastroenterology Reports, 8*, 506–512.

21. Deshpande, G., Rao, S., & Patole, S., (2007). Probiotics for prevention of necrotizing enterocolitis in preterm neonates with very low birth weight: A systematic review of randomized controlled trials. *The Lancet, 369*, 1614–1620.

22. Drake, M., Small, C. L., Spence, K. D., & Swanson, B. G., (1996). Rapid detection and identification of *Lactobacillus* spp. in dairy products by using the polymerase chain reaction. *Journal of Food Protection, 59*, 1031–1036.

23. Drasar, B. S., & Hill, M. J., (1974). Human intestinal flora. *London: Academic Press*, 72–102.

24. Elisha, B. G., & Courvalin, P., (1995). Analysis of genes encoding dalanine: D-alanine ligase-related enzymes in *Leuconostoc mesenteroides* and *Lactobacillus* sp. *Gene, 152*, 79–83.

25. World Health Organization (WHO), (2001). *Health and Nutritional Properties of Probiotics in Food Including Powder Milk With Live Lactic acid Bacteria*. Joint FAO/WHO expert consultation. Cordoba, Argentina. http://www.who.int/foodsafety/publications/fs_management/probiotics/en/index.html (Accessed on 20 July 2019).

26. Ferrari, A., Pacini, N., & Canzi, E., (1980). A note on bile acid transformations by strains of *Bifidobacterium*. *Journal of Applied Bacteriology, 49,* 193–197.

27. Floch, M. H., Binder, H. J., Filburn, B., & Gershengoren, W., (1972). The effect of bile acids on intestinal microflora. *American Journal of Clinical Nutrition, 25,* 1418–1426.

28. Flynn, S., Sinderen, van D., Thornton, G. M., Holo, H., Nes, I. F., & Collins, J. K., (2002). Characterization of the genetic locus responsible for the production of ABP-118, a novel bacteriocin produced by the probiotic bacterium *Lactobacillus salivarius* subsp. *salivarius* UCC118. *Microbiology, 148*(4), 973–984.

29. Fondén, R., Grenov, B., Reniero, R., Saxelin, M., & Birkeland, S. E., (2000). Industrial panel statements: Technological aspect. In: Alander, M., & Mattila-Sandholm, T., (eds.), *Functional Foods for EU-health in 2000* (Vol. 198, pp. 43–50). Report No. FAIR CT96–1028 PROBDE- MO, VTT Symposium, Rovaniemi, Finland.

30. Freitas, M., Tavan, E., Cayuela, C., Diop, L., Sapin, C., & Trugnan, G., (2003). Host–pathogens cross-talk. Indigenous bacteria and probiotics also play the game. *Biocell, 95,* 503–506.

31. Freudig, B., Hogekamp, S., & Schubert, H., (1999). Dispersion of powders in liquids in a stirred vessel. *Chemical Engineering and Processing, 38,* 525–532.

32. Fu, N., & Chen, X. D., (2011). Towards a maximal cell survival in convective thermal drying processes. *Food Research International, 44,* 1127–1149.

33. Fujii, A., & Cook, E. S., (1973). Probiotics. Antistaphylococcal and antifibrinolytic activities of omega-guanidine acids and omega-guanidinoacyl- L -histidines. *Journal of Medicinal Chemistry, 16,* 1409–1411.

34. Fuller, R., (1992). History and development of probiotics. In: Fuller, R., (ed.), *Probiotics, the Scientific Basis* (pp. 1–8). London, UK: Chapman and Hall, London.

35. Gancheva, A., Pot, B., Vanhonacker, K., Hoste, B., & Kersters, K., (1999). A polyphasic approach towards the identification of strains belonging to *Lactobacillus acidophilus* and related species. *Systemic and Applied Microbiology, 22,* 573–585.

36. Gardiner, G. E., O'Sullivan, E., Kelly, J., Auty, M. A., Fitzgerald, G. F., & Collins, J. K., (2000). Comparative survival rates of human derived probiotic *Lactobacillus paracasei* and *L. salivarius* strains during heat treatment and spray-drying. *Applied Environmental Microbiology, 66,* 2605–2612.

37. Gawronska, A., Dziechciarz, P., Horvath, A., & Szajewska, H., (2007). A randomized double-blind placebo-controlled trial of *Lactobacillus* GG for abdominal pain disorders in children. *Alimentary Pharmacology & Therapeutics, 25,* 177–184.

38. Gibson, G. R., & Robertfroid, M. B., (1995). Dietary modulation of the human colonic microbiota: Introducing the concept of prebiotics. *Journal of Nutrition, 125,* 1401–1412.

39. Gill, H. S., (2003). Probiotics to enhance anti-infective defenses in the gastrointestinal tract. *Best Practice & Research: Clinical Gastroenterology, 17,* 755–773.

40. Gilliland, S. E., & Speck, M. L., (1977). Deconjugation of bile acids by intestinal lactobacilli. *Applied Environmental Microbiology, 33,* 15–18.

41. Gilliland, S. E., Staley, T. E., & Bush, L. J., (1984). Importance of bile tolerance of *Lactobacillus acidophilus* used as dietary adjunct. *Journal of Dairy Science, 67,* 3045–3051.

42. Giraffa, G., Gatti, M., Rossetti, L., Senini, L., & Neviani, E., (2000). Molecular diversity within *Lactobacillus helveticus* as revealed by genotypic characterization. *Applied Environmental Microbiology, 66,* 1259–1265.

43. Gotteland, M., Cruchet, S., & Verbeke, S., (2001). Effect of *Lactobacillus* ingestion on the gastrointestinal mucosal barrier alterations induced by indometacin in humans. *Alimentary Pharmacology & Therapeutics, 15*, 11–17.

44. Gueimonde, M., Kalliomaki, M., Isolauri, E., & Salminen, S., (2006). Probiotic intervention in neonates-Will permanent colonization ensue? *Journal of Pediatric Gastroenterology and Nutrition, 42*, 604–606.

45. Guyonnet, D., Chassany, O., & Ducrotte, P., (2007). Effect of fermented milk containing *Bifidobacterium animalis* DN-173 010 on the health-related quality of life and symptoms in irritable bowel syndrome in adults in primary care: A multicentre, randomized, double-blind, controlled trial. *Alimentary Pharmacology & Therapeutics, 26*, 475–486.

46. Hamilton-Miller, J. M. T., Gibson, G. R., & Bruck, W., (2003). Some insight into the derivation and early uses of the word 'probiotic.' *British Journal of Nutrition, 90*, 845.

47. Hamilton-Miller, J. M. T., (2004). Probiotics and prebiotics in the elderly. *Postgraduate Medicine Journal, 80*, 447–451.

48. Hanauer, S. B., (2006). Inflammatory bowel disease: Epidemiology, pathogenesis, and therapeutic opportunities. *Inflammatory Bowel Diseases, 12*, S3–S9.

49. Hart, A. L., Lammers, K., Brigidi, P., Vitali, B., Rizzello, F., & Gionchetti, P., (2004). Modulation of human dendritic cell phenotype and function by probiotic bacteria. *Gut, 53*, 1602–1609.

50. Helander, I., Von Wright, A., & Mattila-Sandholm, T., (1997). Potential of lactic acid bacteria and novel antimicrobials against gram-negative bacteria. *Trends in Food Science and Technology, 8*, 146–150.

51. Hempel, S., Newberry, S. J., & Maher, A. R., (2012). "Probiotics for the prevention and treatment of antibiotic-associated diarrhea a systematic review and meta-analysis". *The Journal of American Medical Association, 307*(18), 1959–1969.

52. Hertzler, S. R., & Clancy, S. M., (2003). Kefir improves lactose digestion and tolerance in adults with lactose maldigestion. *Journal of the American Dietetic Association, 103*, 582–587.

53. Hilton, E., Kolakowski, P., Singer, C., & Smith, M., (1997). Efficacy of Lactobacillus GG as a diarrhea preventative. *Journal of Travel Medicine, 4*, 41–43.

54. Hirayama, K., & Rafter, J., (2000). The role of probiotic bacteria in cancer prevention. *Microbes and Infection, 2*, 681–686.

55. Hoebe, K., Janssen, E., & Beutler, B., (2004). The interface between innate and adaptive immunity. *Nature Immunology, 5*, 971–974.

56. Hoffman, A. F., Molino, G., Milanese, M., & Belforte, G., (1983). Description and stimulation of a physiological pharmokinetic model for the metabolism and enterohepatic circulation of bile acids in man. *Journal of Clinical Investigation, 71*, 1003–1022.

57. Holzapfel, W. H., Geisen, R., & Schillinger, G. U., (1995). Biological preservation of foods with reference to protective cultures, bacteriocins and food-grade enzymes. *International Journal of Food Microbiology, 24*, 343–362.

58. Hood, S. K., & Zottola, A., (1988). Effect of low pH on the ability of *Lactobacillus acidophilus* to survive and adhere to human intestinal cells. *Journal of Food Science, 53*, 1514–1516.

59. Ishibashi, N., & Shimamura, S., (1993). *Bifidobacteria*: Research and development in Japan. *Food Technology, 46*, 126–135.

60. Isolauri, E., Juntune, M., Saxilin, M., & Vesikari, T., (1995). Lactic acid bacteria in the treatment of acute rotavirus gastroenteritis. *Journal of Pediatric Gastroenterology and Nutrition, 20*, 333–336.

61. Isolauri, E., Kirjavainen, P. V., & Salminen, S., (2002). "Probiotics: A role in the treatment of intestinal infection and inflammation?" *Gut, 50*(3), 54–59.

62. Isolauri, E., Ouwehand, A. C., & Laitinen, K., (2005). Novel approaches to the nutritional management of the allergic infant. *Acta Paediatrica Supplement, 94*, 110–114.

63. Isolauri, E., Salminen, S., & Ouwehand, A. C., (2004). Microbial-gut interactions in health and disease. *Probiotics: Best Practice & Research: Clinical Gastroenterology, 18*, 299–313.

64. Jones, M. L., Martoni, C. J., Parent, M., & Prakash, S., (2011). Cholesterol-lowering efficacy of a microencapsulated bile salt hydrolase-active *Lactobacillus reuteri* NCIMB 30242 yogurt formulation in hypercholesterolaemic adults. *British Journal of Nutrition, 9*, 1–9.

65. Kalliomäki, M., Kirjavainen, P., Eerola, E., Kero, P., Salminen, S., & Isolauri, E., (2001). Distinct patterns of neonatal gut microflora in infants in whom atopy was and was not developing. *Journal of Allergy and Clinical Immunology, 107*(1), 129–134.

66. Khulusi, S., Mendall, M. A., Patel, P., Levy, J., Badve, S., & Northfield, T. C., (1995). Helicobacter pylori infection density and gastric inflammation in duodenal ulcer and non-ulcer subjects. *Gut, 37*, 319–324.

67. Kim, H. J., VazquezRoque, M. I., & Camilleri, M., (2005). A randomized controlled trial of a probiotic combination VSL# 3 and placebo in irritable bowel syndrome with bloating. *Neuro Gastroenterology & Motility, 17*, 687–696.

68. Kitahara, M., Takamine, F., Imamura, T., & Benno, Y., (2000). Assignment of *Eubacterium* sp. VPI 12708 and related strains with high bile acid 7 α- dehydroxylating activity to *Clostridium cindens* and proposal of *Clostridium hylemonae* sp. nov. isolated from human feces. *International Journal of Systematic and Evolutionary Microbiology, 50*, 971–978.

69. Klaenhammer, T. R., (1988). "Bacteriocins of lactic acid bacteria." *Biochimie, 70*(3), 337–349.

70. Kolb, H., (1955). Die *Behandlung acuter Infekte unter dem Gesichtswinkel der Prophylaxe chronischer Leiden. Ü̈ ber die Behandlung mit physiologischen bakterien* (The treatment of acute infections from the viewpoint of the prophylaxis of chronic conditions: Treatment with physiological bacteria). *Microecology and Therapy, 1*, 15–19.

71. Kollath, W., (1953). Nutrition and the tooth system, general review with special reference to vitamins. *Deutsche Zahnärztliche Zeitschrift (German Dental Journal), 8*, 7–16.

72. Kolter, R., (1993). The stationary phase of the bacterial life cycle. *Annual Review of Biochemistry, 47*, 855–874.

73. Krasaekoopt, W., Bhandari, B., & Deeth, H., (2003). Evaluation of encapsulation techniques of probiotics for yogurt. *International Dairy Journal, 13*, 3–13.

74. Kullen, M. J., & Bettler, J., (2005). The delivery of probiotics and prebiotics to infants. *Current Pharmaceutical Design, 11*, 55–74.

75. Langhendries, J. P., Detry, J., Van, H. J., Lamboray, J. M., Darimont, J., Mozin, M. J., Secretin, M. C., & Senterre, J., (1995). Effect of a fermented infant formula containing viable bifidobacteria on the fecal flora composition and pH of healthy full-term infants. *Journal of Pediatric Gastroenterology and Nutrition, 21*, 177–181.

76. Lee, M. C., Lin, L. H., Hung, K. L., & Wu, H. Y., (2001). Oral bacterial therapy promotes recovery from acute diarrhea in children. *Acta Paediatr Taiwan, 42*, 301–305.

77. Lesbros-Pantoflickova, D., Corthésy-Theulaz, I., & Blum, A. L., (2007). *Helicobacter pylori* and probiotics. *Journal of Nutrition, 137*, S812–818.

78. Lilly, D. M., & Stillwell, R. H., (1965). Probiotics: Growth-promoting factors produced by microorganisms. *Science, 147*, 747–748.

79. Ling, W. H., (1995). Diet and colonic microflora interaction in colorectal cancer. *Nutrition Research, 15*, 439–454.

80. Liong, M. T., & Shah, N. P., (2005). Optimization of cholesterol removal, growth and fermentation patterns of *Lactobacillusacidophilus* ATCC 4962 in the presence of mannitol, fructo-oligosaccharide and inulin: A response surface methodology approach. *Journal of Applied Microbiology, 98*, 1115–1126.

81. Liong, M. T., & Shah, N. P., (2006). Effects of *Lactobacillus casei* ASCC 292, fructooligosaccharide and maltodextrin on serum lipid profiles, intestinal microflora and organic acids concentration in rats. *Journal of Dairy Science, 89*, 1390–1399.

82. Lo, P. R., Yu, R. C., Chou, C. C., & Huang, E. C., (2004). Determinations of the antimutagenic activities of several probiotic bifidobacteria under acidic and bile conditions against benzo[a]pyrene by a modified Ames test. *Int. J. Food Microbiol., 93*, 249–257.

83. Madsen, K. L., Doyle, J. S., Jewell, L. D., Tavernini, M. M., & Fedorak, R. N., (1999). *Lactobacillus* species prevents colitis in interleukin 10 gene-deficient mice. *Gastroenterology, 116*(5), 1107–1114.

84. Malfertheiner, P., Megraud, F., O'Morain, C., Hungin, A. P., Jones, R., Axon, A., Graham, D. Y., & Tytgat, G., (2002). Current concepts in the management of *Helicobacter pylori* infection–the Maastricht 2–2000 Consensus Report. *Alimentary Pharmacology & Therapeutics, 16*, 167–180.

85. Mann, G. V., & Spoerry, A., (1974). Studies of a surfactant and cholesteremia in the Maasai. *The American Journal of Clinical Nutrition, 27*(5), 464–469.

86. Manzoni, P., Mostert, M., Leonessa, M. L., Priolo, C., Farina, D., Monetti, C., Latino, M. A., & Gomirato, G., (2006). Oral supplementation with *Lactobacillus casei* subspecies rhamnosus prevents enteric colonization by Candida species in preterm neonates: A randomized study. *Clin. Infectious Diseases, 42*(12), 1735–1742.

87. Marnett, L. J., Hurd, H. K., Hollstein, M. C., Levin, D. E., Esterbauer, H., & Ames, B. N., (1985). Naturally occurring carbonyl compounds are mutagens in *Salmonella tester* strain TA 104. *Mutation Research, 148*, 25–34.

88. Marteau, P., Pochart, P., Flourie, B., Pellier, P., Santos, L., Desjeux, J. F., & Rambaud, J. C., (1990). Effect of chronic ingestion of a fermented dairy product containing *Lactobacillus acidophilus* and *Bifidobacterium bifidum* on metabolic activities of the colonic flora in humans. *The American Journal of Clinical Nutrition, 52*, 685–688.

89. Marteau, P., Seksik, P., & Jian, R., (2002). Probiotics and intestinal health effects: A clinical perspective. *British Journal of Nutrition, 88*, S51–S57.

90. Marteau, P., (2001). Safety aspect of probiotic products. *Scandinavian Journal of Nutrition, 45*, 22–24.

91. Mårtenson, O., Öste, R., & Holst, O., (2001). The effect of yogurt culture on the survival of probiotic bacteria in oat-based, non-dairy products, *Food Research International, 35*, 775–784.

92. McCartney, A. L., (2002). Application of molecular biological methods for studying probiotics and the gut flora. *British Journal of Nutrition*, *88*, S29–S37.

93. McFarland, L. V., (2006). "Meta-analysis of probiotics for the prevention of antibiotic associated diarrhea and the treatment of *Clostridium difficile* disease." *The American Journal of Gastroenterology*, *101*(4), 812–822.

94. Michetti, P., Dorta, G., Wiesel, P. H., Brassart, D., Verdu, E., Herranz, M., Felley, C., Porta, N., Rouvet, M., Blum, A. L., & Corthesy-Theulaz, I., (1999). Effect of whey-based culture supernatant of *Lactobacillus acidophilus* (*johnsonii*) on *Helicobacter pylori* infection in humans. *Digestion*, *60*, 203–209.

95. Michida, H., Tamalampudi, S., Pandiella, S. S., Webb, C., Fukuda, H., & Kondo, A., (2006). Effect of cereal extracts and cereal fiber on viability of *Lactobacillus plantarum* under gastrointestinal tract conditions, *Biochemical Engineering Journal*, *28*, 73–78.

96. Mitsuoka, T., (1992). The human gastrointestinal tract. *Lactic Acid Bacteria*, *1*, 69–114.

97. Montalto, M., Curigliano, V., & Santoro, L., (2006). Management and treatment of lactose malabsorption. *World Journal of Gastroenterology*, *12*, 187–191.

98. Montalto, M., Nucera, G., Santoro, L., Curigliano, V., Vastola, M., & Covino, M., (2005). Effect of exogenous beta-galactosidase in patients with lactose malabsorption and intolerance: A crossover double-blind placebo-controlled study. *European Journal of Clinical Nutrition*, *59*, 489–493.

99. Mukai, T., Asakara, T., Sato, E., Mori, K., Matsumoto, M., & Ohori, H., (2002). Inhibition of binding of *Helicobacter pylori* to the glycolipid receptors by probiotic *Lactobacillus reuteri*. *FEMS Immunology and Medical Microbiology*, *32*, 105–110.

100. Muyanja, C., Narvhus, J. A., Treimo, J., & Langsrud, T., (2003). Isolation, characterization and identification of lactic acid bacteria from Bushera: A Ugandan traditional fermented beverage. *International Journal of Food Microbiology*, *80*, 201–210.

101. Muyzer, G., De Waal, E. C., & Uitterlinden, A. G., (1993). Profiling of complex microbial populations by denaturing gradient gel electrophoresis analysis of polymerase chain reaction-amplified genes coding for 16S rRNA. *Applied and Environmental Microbiology*, *59*, 695–700.

102. Muyzer, G., (1999). DGGE/TGGE a method for identifying genes from natural ecosystems. *Current Opinion in Microbiology*, *2*, 317–322.

103. Nadathur, S., Gould, S., & Bakalinsky, A., (1995). Antimutagenicity of an acetone extract of yogurt. *Mutation Research*, *334*, 213–224.

104. Niku-Paavola, M. L., Latva-Kala, K., Laitila, A., Mattila-Sandholm, T., & Haikara, A., (1999). New types of antimicrobial compounds produced by *Lactobacillus plantarum*. *Journal of Applied Microbiology*, *86*, 29–35.

105. Nyman, M., (2000). Fermentation and bulking capacity of indigestible carbohydrates: The case of inulin and oligofructose. *British Journal of Nutrition*, *87*, 163–168.

106. O'Mahony, L., McCarthy, J., & Kelly, P., (2005). *Lactobacillus* and *Bifidobacterium* in irritable bowel syndrome: Symptom responses and relationship to cytokine profiles. *Gastroenterology*, *128*, 541–551.

107. O'Sullivan, & Daniel, J., (2000). "Methods for analysis of the intestinal microflora." *Current Issues in Intestinal Microbiology*, *1*(2), 39–50.

108. O'Sullivan, L., Ross, R. P., & Hill, C., (2002). Potential of bacteriocins producing lactic acid bacteria for improvements in food safety and quality. *Biochimie*, *84*, 593–604.

109. Oelschlaeger, T. A., (2010). Mechanisms of probiotic actions–a review. *International Journal Medical Microbiology*, *300*, 57–62.

110. Olive, D. M., & Bean, P., (1999). Principles and applications of methods for DNA-based typing of microbial organisms. *Journal of Clinical Microbiology*, *37*, 1661–1669.

111. Orrhage, K., Sillerstrom, E., Gustafsson, J. A., Nord, C. E., & Rafter, J., (1994). Binding of mutagenic heterocyclic amines by intestinal and lactic acid bacteria. *Mutation Research*, *311*, 239–248.

112. Ouwehand, A. C., Isolauri, E., He, F., Hashimoto, H., Benno, Y., & Salminen, S., (2001). "Differences in Bifidobacterium flora composition in allergic and healthy infants." *Journal of Allergy and Clinical Immunology, 108*(1), 144–145.

113. Pantoflickova, D., Corthesy-Theulaz, I., Dorta, G., Isler, P., Rochat, F., Enslen, M., & Blum, A. L., (2003). Favorable effect of long-term intake of fermented milk containing *Lactobacillus johnsonii* on *H. pylori* associated gastritis. *Alimentary Pharmacology & Therapeutics, 18*, 805–813.

114. Parker, R. B., (1974). Probiotics, the other half of the story. *Animal Nutrition and Health, 29*, 4–8.

115. Percy-Robb, I. W., & Collee, J. G., (1972). Bile acids: A pH dependent antibacterial system in the gut? *British Medical Journal, 3*, 813–815.

116. Perri, F., Qasim, A., Marras, L., & O'Morain, C., (2003). Treatment of *Helicobacter pylori* infection. *Helicobacter, 8*, 53–60.

117. Plummer, M., Franceschi, S., & Munoz, N., (2004). Epidemiology of gastric cancer. *IARC Scientific Publications, 157*, 311–326.

118. Pool-Zobel, B., Bertram, B., Knoll, M., Lambertz, R., Neudecker, C., Schillinger, U., Schmezer, P., & Holzapfel, W. H., (1993). Antigenotoxic properties of lactic acid bacteria in vivo in the GI tract of rats. *NUT Midline Carcinoma, 20*, 271–281.

119. Pool-Zobel, B. L., Neudecker, C., Domizlaff, I., Ji, S., Schillinger, U., Rumney, C., Moretti, M., Vilarini, I., Scassellati-Sforzolini, R., & Rowland, I., (1996). *Lactobacillus* and *Bifidobacterium* mediated antigenotoxicity in the colon of rats. *NUT Midline Carcinoma, 26*, 365–380.

120. Pot, B., Ludwig, W., Kersters, K., & Schleifer, K. H., (1994). Taxonomy of lactic acid bacteria. In: De Vuyst, L., & Vandamma E. J., (eds.), *Bacteriocins of Lactic Acid Bacteria: Microbiology, Genetics and Applications* (pp. 13–90). London, UK: Chapman and Hall, London.

121. Rayes, N., Seehofer, D., & Theruvath, T., (2005). Supply of pre- and probiotics reduces bacterial infection rates after liver transplantation—a randomized, double-blind trial. *American Journal of Transplantation, 5*, 125–130.

122. Resta-Lenert, S., & Barrett, K. E., (2003). Live probiotics protect intestinal epithelial cells from the effects of infection with enteroinvasive *Escherichia coli* (EIEC). *Gut, 52*, 988–997.

123. Roy, D., (2001). Media for the isolation and enumeration of bifidobacteria in dairy products. *International Journal of Food Microbiology, 69*, 167–182.

124. Ruiz, P. A., Hoffmann, M., Szcesny, S., Blaut, M., & Haller, D., (2005). Innate mechanisms for Bifidobacterium lactis to activate transient pro-inflammatory host responses in intestinal epithelial cells after the colonization of germ-free rats. *Immunology, 115*, 441–450.

125. Sakata, T., Kojima, T., Fujieda, M., Takahashi, M., & Michbata, T., (2003). Influences of probiotic bacteria on organic acid production by pig fecal bacteria *in vitro*. *TheNutrition Society, 62*, 73–80.

126. Salminen, M. K., Rautelin, H., Tynkkynen, S., Poussa, T., Saxelin, M., Valtonen, V., & Jarvinen, A., (2006). Lactobacillus bacteremia, species identification, and antimicrobial susceptibility of 85 blood isolates. *Clinical Infectious Disease, 42*(5), 35–44.

127. Salminen, M. K., Tynkkynen, S., Rautelin, H., Saxelin, M., Vaarela, M., Ruuta, P., Sarna, S., Valtonen, V., & Jarvinen, A., (2002). *Lactobacillus bacteremia* during a rapid increase in probiotic use of *Lactobacillus rhamnosus* GG in Finland. *Clinical Infectious Disease, 35*, 1155–1160.

128. Salminen, S., Ouwehand, A., Benno, Y., & Lee, Y. K., (1999). Probiotics: How should they be defined? *Trends in Food Science and Technology, 10*, 107–110.

129. Salminen, S., Von Wright, A., Morelli, L., Marteau, P., Brassard, D., De Vos, W. M., et al., (1998). Demonstration of safety of probiotics—a review. *International Journal of Food Microbiology, 44*, 93–106.

130. Sarem, F., Sarem-Damerdji, L. O., & Nicolas, J. P., (1996). Comparison of the adherence of three *Lactobacillus* strains to Caco-2 and Int-407 human intestinal cell lines. *Letters in Applied Microbiology, 22*, 439–442.

131. Sazawal, S., Hiremath, G., Dhingra, U., Malik, P., Deb, S., & Black, R. E., (2006). Efficacy of probiotics in prevention of acute diarrhoea: A meta-analysis of masked, randomized, placebo-controlled trials. *Lancet Infectious Disease, 6*, 374–382.

132. Scarpignato, C., & Rampal, P., (1995). Prevention and treatment of traveler's diarrhea: A clinical pharmacological approach. *Chemotherapy, 41*, 48–81.

133. Schumann, C., (2002). Medical, nutritional and technological properties of lactulose. An update. *European Journal of Nutrition, 41*, 17–25.

134. Seksik, P., Rigottier-Gois, L., & Gramet, G., (2003). Alterations of the dominant faecal bacterial groups in patients with Crohn's disease of the colon. *Gut, 52*(2), 237–242.

135. Sgouras, D., Maragkoudakis, P., Petraki, K., Martinez-Gonzalez, B., Eriotou, E., & Michopoulos, S., (2004). *In vitro* and *in vivo* inhibition of *Helicobacter pylori* by *Lactobacillus casei* strain Shirota. *Applied and Environmental Microbiology, 70*, 518–526.

136. Sgouras, D. N., Panayotopoulou, E. G., Martinez-Gonzalez, B., Petraki, K., Michopoulos, S., & Mentis, A., (2005). *Lactobacillus johnsonii* La1 attenuates *Helicobacter pylori*-associated gastritis and reduces levels of pro-inflammatory chemokines in C57BL/6 mice. *Clinical and Diagnostic Laboratory Immunology, 12*, 1378–1386.

137. Shah, N. P., & Lankaputhra, W. E. V., (1997). Improving viability of *Lactobacillus acidophilus* and Bifidobacterium spp. in yogurt. *International Dairy Journal, 7*, 349–356.

138. Shah, N. P., (2006). Functional cultures and health benefits. In: *Scientific and Technological Challenges in Fermented Milk, 2nd IDF Dairy Science and Technology Week: Book of abstracts* (pp. 35–36). Sirmione–Italy.

139. Shah N. P., (2007). Functional cultures and health benefits. *International Dairy Journal, 17*(11), 1262–1277.

140. Sperti, G. S., (1971). *Probiotics* (p. 121). Avi Publishing Co., Westpoint–CT.

141. Stewart, L., Pellegrini, C. A., & Way, L. W., (1986). Antibacterial activity of bile acids against common biliary tract organisms. *Surgical Forum, 37*, 157–159.

142. Sykora, J., Valeckova, K., Amlerova, J., Siala, K., Dedek, P., & Watkins, S., (2005). Effects of a specially designed fermented milk product containing probiotic *Lactobacillus casei* DN-114 001 and the eradication of *H. pylori* in children: A prospective randomized, double-blind study. *Journal of Clinical Gastroenterology, 39*, 692–698.

143. Szajewska, H., Ruszczynski, M., & Radzikowski, A., (2006). Probiotics in the prevention of antibiotic-associated diarrhoea in children: A meta-analysis of randomized controlled trials. *Journal of Pediatrics, 149*(3), 367–372.

144. Takahashi, T., & Morotomi, M., (1994). Absence of cholic acid 7-dehydroxylase activity in the strains of *Lactobacillus* and *Bifidobacterium. Journal of Dairy Science, 77*, 3275–86.
145. Tanaka, H., Doesburg, K., Iwasaki, T., & Mierau, I., (1999). Screening of lactic acid bacteria for bile salt hydrolase activity. *Journal of Dairy Science, 82*, 2530–2535.
146. Tavan, E., Cayuela, C., Antoine, J. M., & Cassand, P., (2002). Antimutagenic activities of various lactic acid bacteria against food mutagens: Heterocyclic amines. *Journal of Dairy Research, 69*, 335–341.
147. Teixeira, P., Castro, H., Malcata, F. X., & Kirby, R. M., (1995). Survival of *Lactobacillus delbruekii* spp. bulgaricus following spray-drying. *Journal of Dairy Science, 78*, 1025–1031.
148. Temmerman, R., Scheirlinck, I., Huys, G., & Swings, J., (2003). Culture-independent analysis of probiotic products by denaturing gradient gel electrophoresis. *Applied Environmental Microbiology, 69*, 220–226.
149. Tilsala-Timisjärvi, A., & Alatossava, T., (1997). Development of oligonucleotide primers from the 16S-23S rRNA intergenic sequences for identifying different dairy and probiotic lactic acid bacteria by PCR. *International Journal of Food Microbiology, 35*, 49–56.
150. Tong, J. L., Ran, Z. H., Shen, J., Zhang, C. X., & Xiao, S. D., (2007). Meta-analysis: The effect of supplementation with probiotics on eradication rates and adverse events during Helicobacter pylori eradication therapy. *Alimentary Pharmacology & Therapeutics, 25*, 155–168.
151. Tuomola, E. M., & Salminen, S. J., (1998). Adhesion of some probiotic and dairy *Lactobacillus* strains to Caco-2 cell cultures. *International Journal of Food Microbiology, 41*, 45–51.
152. Vega, C., & Roos, Y. H., (2006). Invited review: Spray-dried dairy and dairy-like emulsions–compositional considerations. *Journal of Dairy Science, 89*, 383–401.
153. Ventura, M., & Zink, R., (2002). Specific identification and molecular typing analysis of *Lactobacillus johnsonii* by using PCR-based methods and pulsed-field gel electrophoresis. *FEMS Microbiology Letters, 217*, 141–154.
154. Venturi, A., Gionchetti, P., & Rizzello, F., (1999). Impact on the composition of the faecal flora by a new probiotic preparation: Preliminary data on maintenance treatment of patients with ulcerative colitis. *Alimentary Pharmacology & Therapeutics, 13*(8), 1103–1108.
155. Vergin, F., (1954). Anti- und Probiotika. *Hippokrates, 25*, 116–119.
156. Vinderola, C. G., Mocchiutti, P., & Reinheimer, J. A., (2002). Interactions among lactic acid starter and probiotic bacteria used for fermented dairy products. *Journal of Dairy Science, 85*, 721–729.
157. Weizman, Z., & Alsheikh, A., (2006). Safety and tolerance of a probiotic formula in early infancy comparing two probiotic agents: A pilot study. *Journal of the American College of Nutrition, 25*(5), 415–419.
158. Westerbeek, E. A., Van Den Berg, A., Lafeber, H. N., Knol, J., Fetter, W. P., & Van Elburg, R. M., (2006). The intestinal bacterial colonization in preterm infants: A review of the literature. *Clinical Nutrition, 25*(3), 361–368.
159. Wijtzes, T., Bruggeman, M. R., Nout, M. J. R., & Zwietering, M. H., (1997). A computerized system for the identification of lactic acid bacteria. *International Journal of Food Microbiology, 38*, 65–70.

160. Wolf, B. W., Wheeler, K. B., Ataya, D. G., & Garleb, K. A., (1998). Safety and tolerance of *Lactobacillus reuteri* supplementation to a population infected with human immunodeficiency virus. *Food and Chemical Toxicology, 36,* 1085–1094.

161. Yazdanbakhsh, M., Kremsner, P. G., & Van Ree, R., (2002). Allergy, parasites, and the hygiene hypothesis. *Science, 296,* 490–494.

162. Yoon, K. Y., Woodams, E. E., & Hang, Y. D., (2006). Production of probiotic cabbage juice by lactic acid bacteria, *Bioresource Technology, 97,* 1427–1430.

163. Zoetendal, E., Akkermans, A., & De Vos, W., (1998). Temperature gradient gel electrophoresis analysis of 16S rRNA from human fecal samples reveals stable and host-specific communities of active bacteria. *Applied Environmental Microbiology, 64,* 3854–3859.

APPENDIX A – LIST OF DIARY-BASED PROBIOTICS

Indian Dairy products	Origin	Description	English Name
Probiotic Lassi	Haryana, Rajasthan & Punjab	Lassi is a blend of probiotic yogurt, water and spices.	Fermented milk beverage
Probiotic Dahi	Indian subcontinent	Dahi is obtained by fermenting milk using combination of starter culture (Majorly *Lactococcus*) and probiotic culture.	Probiotic curd

Additional References

Hussain, S. A., Patil, G. R., Reddi, S., Yadav, V., Pothuraju, R., Singh, R. R. B., & Kapila, S. (2017). Aloe vera *(Aloe barbadensis Miller)* supplemented probiotic lassi prevents Shigella infiltration from epithelial barrier into systemic blood flow in mice model. *Microbial Pathogenesis, 102,* 143–147.

Yadav, H., Jain, S., & Sinha, P. R. (2007). Antidiabetic effect of probiotic dahi containing *Lactobacillus acidophilus* and *Lactobacillus casei* in high fructose fed rats. *Nutrition, 23* (1), 62–68.

CHAPTER 15

APPLICATIONS OF PROBIOTICS IN DAIRY FOOD PRODUCTS

DIVYASREE AREPALLY, SUDHARSHAN REDDY RAVULA, and
TRIDIB KUMAR GOSWAMI

ABSTRACT

Probiotics are useful viable bacteria, which confer health benefits to the host organism. The health-promoting microorganisms have been studied to show versatile benefits on our health. They aid in the stimulation of human immunity, decrease infectious, and antibiotic-mediated diarrheal incidences, decrease serum cholesterol, alleviate lactose intolerance, and repress life-threatening tumors and cancers. These benefits are mainly due to the mechanism of probiotic action on the pathogens. The useful microorganisms were included in dairy products, and a number of novel dairy foods had been developed. The nutrition, flavor, and texture depend on the type of probiotics strain incorporated for the production of probiotic dairy products.

15.1 INTRODUCTION

The natural inhabitant of the gastrointestinal (GI) system maintains our health. However, changes in our daily diet have impacted the diversity of the gut. In order to restore the microbial gut, the use of probiotics has received the attention of investigators. Research on probiotics has exploded exponentially recently in the 21st century, though the concepts have existed around for >100 years.

This chapter summarizes the history of probiotics including definitions, the characteristics that an ideal probiotic should have to apply in food products, its mode of action, health benefits accomplished by probiotics, their applications in dairy products, and challenges encountered in developing the dairy products.

15.1.1   HISTORY OF PROBIOTICS

During ancient days, fermentation was used to preserve the food. Until 6000 BC in China, cow milk and goat milk were used to prepare the cheese. Hippocrates and scientists reported that the consumption of fermented milk cured some of the disorders of the digestive system. The fermented products (like kefir, dahi, leben, and koumiss) were used before Leeuwenhoek discovered the existence of microorganisms in 1683; and lactic acid bacteria (LAB) were isolated from milk by Louis Pasteur in 1857.

Probiotics concept was hypothesized by Elie Metchnikoff in 1907. The proteolytic bacteria in the gut produce some of the toxic substances, such as indoles, phenols, and ammonia causing autointoxication [22]. Metchnikoff observed that consumption of fermented milk daily had prolonged the life of Bulgarian peasants and Russian people [43]. The fermented milk consisting of LAB was able to lower pH in the intestinal, thus suppressing the growth of proteolytic bacteria. Also, Metchnikoff was the first to characterize a specific *Lactobacillus* strain that can survive in the human intestinal tract and studied as "*the most active bacillus responsible for the souring of milk*" [23]. Later on, a series of events culminated the term "probiotic."

15.1.2   DEFINITION OF PROBIOTICS

The word probiotics ("for life") originated from the Latin word: "pro means for" and "biotic means life." The term probiotic was first expressed by Kollath in 1953 and introduced by Lilly and Stillwell in the year 1965. Since then, many definitions [12, 15, 19, 21, 34, 39] have been used to define probiotics (Table 15.1). However, the definition of probiotics must include additional explanations on the "beneficial balance," "normal population," and "stabilization of the gut microflora" [13].

Presently, the concept of probiotics alludes to the live bacteria that elevate the microbial population of the gastrointestinal tract (GIT) for the beneficial balance of our health. Consumers are familiar with the probiotic foods after EU Expert Group (in FUFOSE) defined it "as viable preparation of foods or dietary supplements to improve human health" [13]. Most frequently used probiotic bacteria belong to the genera: *Lactobacillus* and *Bifidobacterium.* However, species such as *Enterococcus*, *Bacillus,* and *Streptococcus* have also great potential.

TABLE 15.1   Definition of Probiotic

Year	Scientist	Definition of Probiotic
1953	Kollath	Food complexes, including both organic and inorganic used as supplements.
1965	Lilly and Stillwell	The microorganism produces substances that can stimulate the growth of other microorganisms.
1989	Fuller	Viable bacterial food ally that improves the intestinal bacterial equilibrium and affects the host health beneficially.
1991	Huisin't Veld and Havenaar	Single or mixed type of viable microbial cultures that can improve the indigenous microflora and benefits the host.
1996	Shaafsma	Live microorganisms on consumption in sufficient quantity exert health effects beyond the basic nutrition.
1998	Salminen	A viable microbial food ingredient that benefits the health.

15.2   CHARACTERISTICS OF PROBIOTICS

The characteristics of probiotics of a successful probiotic strain should satisfy the safety, functional, and technological criteria, and these have been defined by several reviewers [1, 11, 20, 42]. However, the probiotic strains are indicated to be of human origin and should be persistent. Research trials show that discontinuation of probiotic ingestion causes their disappearance in the GI tract. However, temporary persistence may enhance the chances of beneficial function. The main characteristics of probiotics are shown in Figure 15.1.

Some probiotic strains may have additional characteristics, such as improvement of cholesterol metabolism by lowering of serum cholesterol, lowering the pH of colonic content helps to achieve bioavailability of vitamins and minerals, fecal carcinogen and mutagen levels, increase in immune response in inflammation, intestinal response, metabolism of lactose, good technological properties, and co-aggregation with pathogens to enhance the anti-microbial activity [26].

15.3   MODE OF ACTION OF PROBIOTICS

The effect of probiotics on the human and animal health mainly depends on the mechanism of action of probiotics (Figure 15.2).

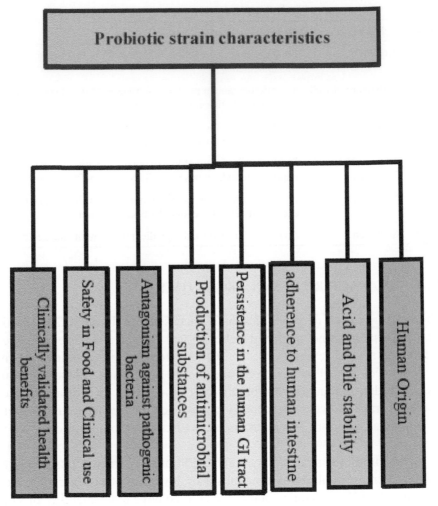

FIGURE 15.1 **(See color insert.)** Characteristics of the probiotic strain.

Probiotic action is mainly divided into three modes [25]:

- The primary mode of action is that the probiotics regulate the host's defending system, including the inherent and immune system. This action prevents occurrences of infectious diseases, provides treatment to the digestive tract, and eradicates the neoplastic host cells;
- Direct action on other organisms is its second mode of action that fights against pathogens to restore the microbial equilibrium in the gut; and

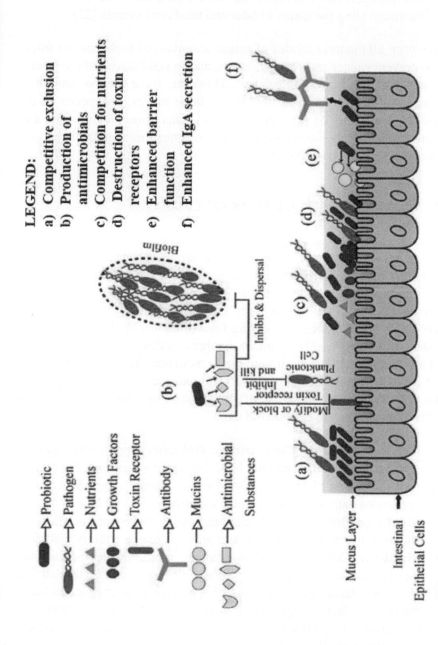

FIGURE 15.2 (See color insert.) Mechanism of probiotic action.

- The third mode affects the bacterial products like toxic substances, host products like bile salts, and food ingredients. This can be done by inactivating the toxins of host and food components [25].

However, all the three modes of action are involved to defend the infection, to prevent cancer and other diseases, and to stabilize intestinal microbiota and host physiological balance. However, there are no probiotics recognized till date that can exhibit all the three modes. It depends on the metabolic properties [45], type of strain [14]. Therefore, identification and characterization of probiotic properties had become prominent to distinguish real probiotics from non-probiotics.

15.4   HEALTH BENEFITS OF PROBIOTICS

Numerous evidences exist on the health benefits of probiotics [45]. However, the lucrative properties of probiotics are strain-dependent. The health benefits that make the probiotics more marketable are immune protection and gut comfort [22]. Some of the evidence like *L. johnsonii* La1, when consumed, accomplishes the recovery of skin immunity [28]. In turn, *L. reuteri* DSM 17938 reduced the crying time of babies by the improvement of gastric emptying [36, 37]. To attain the three modes of probiotic action, a novel mixture of probiotic strains are much more beneficial. A mixture of *L. helveticus* R0052 and *B. longum* R0175 had reduced the anxiety symptoms in the population by stimulation of the nervous system [7]. The health benefits of probiotics are categorized as:

- **Intestinal Microbial Composition:** Probiotics supports the digestive process, prevents the irritable bowel syndrome, prevents the endogenous pathogen like antibiotic-associated diarrhea, and prevents the exogenous pathogens like traveler's diarrhea.
- **Immunomodulation:** Probiotics improve the resistance to allergies, stimulate the innate immunity, and prevents the respiratory diseases.
- **Metabolic Effects:** Useful microorganisms enhance the calcium metabolism, and prevent osteoporosis, lower the toxigenic reaction in the gut, deconjugate the bile salt, and reduce the cholesterol level, improve the lactose digestion. It also includes therapeutic effects like prevention of skin diseases, Rota viral diarrhea, urogenital infection, and synthesizes the vitamins like B_2, B_6, and B_{12}.

It was found that immune modulation was achieved even with the dead probiotic bacteria [38]. The key components of bacteria, i.e., surface carbohydrates and bacterial DNA, interact with the host immune system [22]. In 1907, the products with non-viable probiotics were marketed as Lacteol™, which was an anti-diarrhea supplement [33]. Also, non-viable probiotics were found to act effectively against allergy in both children and adults [16]. The consumption of probiotic products for the better health benefits should contain a minimum of 10^6 CFU/g [30]. Although the viability of probiotics does not seem to be essential for some health benefits, yet the health benefits achieved by live probiotics and non-viable probiotics were not similar.

15.5 APPLICATIONS OF PROBIOTICS IN DAIRY PRODUCTS

There is a growing interest among consumers to consume newly developed foods that can provide health benefits beyond ample nutrition. Therefore, many industries are focusing on developing probiotic food products. Additionally, probiotics are used even in pharmaceuticals and animal feeds. In this section, the applications of probiotics in dairy foods are summarized.

Milk is a good vehicle for probiotics due to its imminent properties and milk storage at refrigerated temperatures. Various strains of probiotics are found in various dairy products (e.g., yogurt, cheese, sour, and fresh milk, etc.). These products provide a suitable habitat for probiotics supporting their growth and restoring the viability. However, several factors must be taken into consideration for applying the probiotics to dairy products. These factors include survivability of probiotics in dairy products [29, 40]; physical, chemical, and organoleptic properties of end products [2, 5], and the effect of probiotics on our health [27, 35].

During the last 20 years, probiotics have been included in dairy products, and a number of novel dairy foods have evolved including, such as sour milk, yogurt, folk-jolk, cheese, whey-based buttermilk, and fermented milks [32, 40]. The percentage of applications of probiotics to dairy products is about 49% [10]. Different kinds of probiotic strains, such as *Lactococcuslactis* ssp. *Lactic, Lactobacillus lactis* ssp. *Cremoris,* are being used in various dairy products. However, the microorganisms do not seem to be viable while passing through the GIT. Therefore, recently, dairy products are being prepared using *Lactobacillus acidophilus, Bifidobacterium* ssp., *Lb. casei, L. reuteri,* and *Lactobaciullsrhamnosus* GG. Probiotics are successfully incorporated in novel products, such as ice creams, sour cream, buttermilk, desserts, Cheddar cheese, and cottage cheese.

Fermented products date back to Persian times, and in the 8th century, Turkish nomads had given the name to the fermented milk as 'Yogurt' [24]. In 1922, the production of yogurt was firstly commercialized in Spain. Traditionally, yogurt is made by fermenting the milk at 42–45°C [4]. Metch-nikoff in the 20th century revealed the health benefits of fermented milk and yogurt. Owing to numerous health benefits, innovative variations have been developed in dairy fermented products according to nutritional and textural characteristics. The factors that affect the survivability of microorganisms in fermented milks are acidity, redox potential, pH, dissolvable oxygen content, hydrogen peroxide, starter microbes, various additives, and effects of fermented milk during its manufacturing and storage. Also, the fermented products can be used as an ingredient in cooked products, imbibed as condi-ments, snacks, and frozen desserts. To achieve good flavor and texture of the milk-based products, strains like *S. thermophiles* and *L. delbrueckii* have been used in combination [32, 40]. The examples of developed fermented milk products are illustrated in Table 15.2.

TABLE 15.2   Fermented Milk Products

Fermented Product	Equivalent English Name	Significant Raw Product
Dahi	Curd	Milk is soured by using previously soured milk as starter.
Kefir	Búlgaros	Kefir grains are used. Lactic acid, alcohol, CO_2 are responsible for sparkling characteristics.
Kishk	Cheese	Mixing of fermented milk with parboiled wheat.
Laban	Yogurt	Coagulation of milk in earthenware utensils.
Mast	Ice cream	Yogurt with cooked flavor.
Skyr	Buttermilk	Ewe milk is used along with rennet and starter.
Trahan	Yogurt cereal mix	Mixing of wheat flour.

Frozen products like ice cream are consumed by all age groups. Probi-otics are added to ice cream either prior to freezing or to milk as a substrate may be used for fermentation and mixed with the based mix. Probiotic ice cream was produced with different probiotic strains, such as *L. acidophilus, Bifidobacterium lactis, and Streptococcus thermophilus* [2]. A symbiotic chocolate mousse was produced with *Lactobacillus paracasei* sub ssp. *Para-casei* [3]. Saad et al., [31] developed coconut flan that was supplemented with *Bifidobacterium lactis* and *Lactobacillus paracasei*. They reported that the viability of *Bifidobacterium lactis* and *Lactobacillus paracasei* in coconut flan was more than acceptable level of 10^6–10^7 CFU/g. The survivability of probiotics in frozen products is higher as they are subjected to lower temperature.

Cheese due to its matrix offers a high protection to probiotic bacteria. The higher protein content of cheese provides good buffering protection with higher pH (4.8–5.6), and reduces the permeability to oxygen, thus enhances the survivability of probiotic bacteria under gastric conditions. *Lactobacillus paracasei* A13, *Bifidobacterium bifidum* A1, and *L. acidophilus* A3 were added to the Fresco cheese in Argentina, and it was found to show potential benefits [44]. The types of cheeses with different probiotics are presented in Table 15.3.

TABLE 15.3 Probiotics Used in Different Types of Cheeses

Type of Cheese	Probiotic Used	References
Cottage cheese	*Bifidobacteriuminfantis*	[6]
Fresco cheese	*Lactobacillus paracasei* A13,	[44]
	Bifidobacterium bifidum A1,	
	L. acidophilus A3	
petit-Suisse cheese	*Bifidobacterium animalis* sub ssp. *lactis*, *Lactobacillus acidophilus*	[8]
Turkish Beyaz cheese	*Lactobacillus fermentum, L. plantarum*	[18]
Turkish white cheese	*L. acidophilus* 593 N	[17]

15.6 CHALLENGES IN DEVELOPMENT OF PROBIOTIC DAIRY PRODUCTS

The viability of probiotics plays a crucial role for achieving health benefits. The recommended level of probiotics ranges from 10^6 CFU/mL to 10^7–10^8 CFU/mL, to compensate for the losses in the viability number that occurred during processing, storage, and during the passage through the GIT. Also, the viability in food matrix depends on the strain, production of hydrogen peroxide, acidity of the product, availability of nutrients, growth promoters and inhibitors, concentration of sugars, dissolved oxygen, inoculation level, and fermentation time. In general, probiotic organisms are susceptible to environmental conditions such as water activity, redox potential, temperature, and acidity and during industrial-scale production such as freeze-drying, spray drying or freeze concentration. The survivability of probiotics in frozen foods gets reduced due to acid, freeze injury, sugar concentration of product, and oxygen toxicity [9]. Therefore, cryo-protectants are added either to the growth medium or before the freezing or dehydration step in order to improve the survival of probiotic bacteria.

15.7   SUMMARY

The use of probiotics is mainly related to their ideal characteristics to incorporate in food. The list of health claims for probiotics is significant, because of the mechanism of probiotic action on the other harmful microorganisms. However, the understanding of the mechanism of probiotic effect is still fuzzy and challenging as it varies from strain to strain. Number of lactic cultures are used to make the fermented milk foods. However, these lactic cultures do not survive under GI conditions. The probiotics have received attention to add in dairy foods. However still, the limitations of dairy products (such as lactose intolerance, allergy to milk proteins, high fat and high cholesterol content and the requirement for cold storage facilities) have led to the development of several novel non-dairy products.

KEYWORDS

- **dairy products**
- **fermented milks**
- **health benefits**
- **mode of action**
- **probiotics**

REFERENCES

1. Adams, M. R., (1999). Safety of industrial lactic acid bacteria. *Journal of Biotechnology*, *68*(2/3), 171–178.
2. Akin, M. B., Akin, M. S., & Kirmaci, Z., (2007). Effects of inulin and sugar levels on the viability of yogurt and probiotic bacteria and the physical and sensory characteristics in probiotic ice-cream. *Food Chemistry*, *104*(1), 93–99.

3. Aragon-Alegro, L. C., Alegro, J. H. A., Cardarelli, H. R. C., Chiu, M. C., & Saad, S. M. I., (2007). Potentially probiotic and symbiotic chocolate mousse. *LWT—Food Science and Technology, 40*(4), 669–675.

4. Awaisheh, S. S., (2012). Probiotic food products classes, types, and processing. In: *Probiotics* (pp. 551–582). InTech Open Access.

5. Bais, H. P., Weir, T. L., Perry, L. G., Gilroy, S., & Vivanco, J. M., (2006). The role of root exudates in rhizosphere interactions with plants and other organisms. *Annual Review of Plant Biology, 57*(6), 233–266.

6. Blanchette, L., Roy, D., Belanger, G., & Gauthier, S. F., (1996). Production of cottage cheese using dressing fermented by *bifidobacteria*. *Journal of Dairy Science, 79*, 8–15.

7. Bravo, J. A., Forsythe, P., Chew, M. V., & Escaravage, E., (2011). Ingestion of *Lactobacillus* strain regulates emotional behavior and central GABA receptor expression in a mouse via the vagus nerve. *Proceedings of the National Academy of Sciences of the United States of America, 108*, 16050–16055.

8. Cardarelli, H. R., Buriti, F. C. A., Castro, I. A., & Saad, S. M. I., (2008). Inulin and oligofructose improve sensory quality and increase the probiotic viable count in potentially synbiotic petit-Suisse cheese. *LWT—Food Science and Technology, 41*, 1037–1046.

9. Davidson, R. H., Duncan, S. E., Hackney, C. R., Eigel, W. N., & Boling, J. W., (2000). Probiotic culture survival and implications in fermented frozen yogurt characteristics. *Journal of Dairy Science, 83*, 666–673.

10. De Prisco, & Gianluigi, M., (2016). Probiotication of foods: Focus on microencapsulation tool. *Trends in Food Science and Technology, 48*, 27–39.

11. Donohue, D. C., & Salminen, S. J., (1996). Safety of probiotic bacteria. *Asia Pacific Journal of Clinical Nutrition, 5*, 25–28.

12. Fuller, R., (1989). Probiotics in man and animals. *Journal of Applied Bacteriology, 66*, 365–378.

13. Holzapfel, W. H., (2006). Introduction to prebiotics and probiotics. *Probiotics in Food Safety and Human Health, 10*, 1–33.

14. Hossain, M. I., Sadekuzzaman, M., & Ha, S. D., (2017). Probiotics as potential alternative biocontrol agents in the agriculture and food industries: A review. *Food Research International, 100*, 63–73.

15. Huisint, V. J., & Havenaar, R., (1991). Probiotics in man and animal. *Journal of Chemical Technology and Biotechnology, 51*, 562–567.

16. Ishida, Y., Nakamura, F., Kanzato, H., Sawada, D., Hirata, H., Nishimura, A., Kajimoto, O., & Fujiwara, S., (2005). Clinical effects of *Lactobacillus acidophilus* strain L-92 on perennial allergic rhinitis: A double-blind, placebo-controlled study. *Journal of Dairy Science, 88*, 527–533.

17. Kasimoglu, A., Göncüoglu, M., & Akgün, S., (2004). Probiotic white cheese with *Lactobacillus acidophilus*. *International Dairy Journal, 14*, 1067–1073.

18. Kiliç, G. B., Kuleansan, H., Eralp, I., & Karahan, A. G., (2009). Manufacture of Turkish Beyaz cheese added with probiotic strains. *LWT—Food Science and Technology, 42*(5), 1003–1008.

19. Kollath, W., (1953). Nutrition: dental system. *Deutsch. Zahnaerzt. Z, 8*, 7–16.

20. Lee, Y. K., & Salminen, S., (1995). The coming of age of probiotics. *Trends in Food Science Technology, 6*, 241–245.

21. Lilly, D. M., & Stillwell, R. H., (1965). Probiotics: Growth promoting factors produced by microorganisms. *Science, 147*, 747–748.

22. Makinen, K., Berger, B., Bel-Rhlid, R., & Ananta, E., (2012). Science and technology for the mastership of probiotic applications in food products. *Journal of Biotechnology*, *162*(4), 356–365.

23. Morelli, L., & Callegari, M. L., (2005). Taxonomy and biology of probiotics. In: *Probiotics in Food Safety and Human Health* (pp. 67–83). CRC Press, Boca Raton–FL.

24. Nair, B. M., & Prajapati, J. B., (2003). The history of fermented foods. In: *Handbook of Fermented Functional Foods* (pp. 17–42), CRC Press, Boca Raton–FL.

25. Oelschlaeger, T. A., (2010). Mechanisms of probiotic actions - a review. *International Journal of Medical Microbiology*, *300*(1), 57–62.

26. Ouwehand, A. C., Pirkka, V., Kirjavainen, P., Colette, S., & Seppo, S., (1999). Probiotics: Mechanisms and established effects. *International Dairy Journal*, *9*(1), 43–52.

27. Parvez, S., Malik, K. A., Ah Kang, S., & Kim, H. Y., (2006). Probiotics and their fermented food products are beneficial for health. *Journal of Applied Microbiology*, *100*(6), 1171–1185.

28. Peguet-Navarro, J., Dezutter-Dambuyant, C., & Buetler, T., (2008). Supplementation with oral probiotic bacteria protects human cutaneous immune homeostasis after UV exposure-double blind, randomized, placebo-controlled clinical trial. *European Journal of Dermatology*, *18*, 504–511.

29. Phillips, M., Kailasapathy, K., & Tran, L., (2006). Viability of commercial probiotic cultures (*L. acidophilus, Bifidobacterium* ssp., *L. casei, L. paracasei* and *L. rhamnosus*) in cheddar cheese. *International Journal of Food Microbiology*, *108*(2), 276–280.

30. Pinto, M. G. V., Franz, C. M. A. P., Schillinger, U., & Holzapfel, W. H., (2006). *Lactobacillus* ssp. with in vitro probiotic properties from human feces and traditional fermented products. *International Journal of Food Microbiology*, *109*, 205–214.

31. Saad, S. M. I., Corrêa, S. B. M., & Castro, I. A., (2008). Probiotic potential and sensory properties of coconut flan supplemented with *Lactobacillus paracasei* and *Bifidobacterium lactis*. *International Journal of Food Science and Technology*, *43*, 1560–1568.

32. Saarela, M., Mogensen, G., Fonden, R., Mättö, J., & Mattila-Sandholm, T., (2000). Probiotic bacteria: safety, functional and technological properties. *Journal of Biotechnology*, *84*(3), 197–215.

33. Salazar-Lindo, E., Figueroa-Quintanilla, D., Caciano, M. I., Reto-Valiente, V., Chauviere, G., & Colin, P., (2007). Effectiveness and safety of *Lactobacillus* LB in the treatment of mild acute diarrhea in children. *Journal of Pediatric Gastroenterology and Nutrition*, *44*, 571–576.

34. Salminen, S., Bouley, C., & Boutron-Ruault, M. C., (1998). Functional food science and gastrointestinal physiology and function. *British Journal of Nutrition*, *80*(1), 147–171.

35. Sanders, M. E., & Klaenhammer, T. R., (2001). Invited review: The scientific basis of *Lactobacillus acidophilus* NCFM functionality as a probiotic. *Journal of Dairy Science*, *84*(2), 319–331.

36. Savino, F., Cordisco, L., Tarasco, V., & Palumeri, E., (2010). *Lactobacillus reuteri* DSM 17938 in infantile colic: A randomized, double-blind, placebo-controlled trial. *Pediatrics*, *126*, 526–533.

37. Savino, F., Pelle, E., Palumeri, E., Oggero, R., & Miniero, R., (2007). *Lactobacillusreuteri* (American type culture collection strain 55730) versus simethicone in the treatment of infantile colic: a prospective randomized study. *Pediatrics*, *119*, 124–130.

38. Schmid, K., Schlothauer, R. C., Friedrich, U., & Staudt, C., (2005). Development of probiotic food ingredients. In: *Probiotics in Food Safety and Human Health* (pp. 48–79). CRC Press, Boca Raton, FL.

39. Shaafsma, G., (1996). State of the art concerning probiotic strains in milk products. *International Dairy Federation Nutrition, News Letters, 5*, 23–24.

40. Shah, N., (2015). Novel dairy probiotic products. *Advances in Probiotic Technology, 23*, 338–355.

41. Shah, N. P., (2000). Probiotic bacteria: Selective enumeration and survival in dairy foods. *Journal of Dairy Science, 83*(4), 894–907.

42. Swain, M. R., Anandharaj, M., & Ray, R. C., (2014). Fermented fruits and vegetables of Asia: A potential source of probiotics. *Biotechnology Research International* (p. 19). Article ID 250424, https://www.hindawi.com/journals/btri/2014/250424/ (Accessed on 20 July 2019).

43. Tripathi, M. K., & Giri, S. K., (2014). Probiotic functional foods: Survival of probiotics during processing and storage. *Journal of Functional Foods, 9*, 225–241.

44. Vinderola, G., Prosello, W., Molinari, F., Ghilberto, D., & Reinheimer, J., (2009). Growth of *Lactobacillus paracasei* A13 in Argentinian probiotic cheese and its impact on the characteristics of the product. *International Journal of Food Microbiology, 135*, 171–174.

45. Wohlgemuth, S., Loh, G., & Blaut, M., (2010). Recent developments and perspectives in the investigation of probiotic effects. *International Journal of Medical Microbiology, 300*(1), 3–10.

38. Schmidt, K., Scrinivasan, R. S., Footitt, P., & Sejnif, V. (2001). Development of probiotic food ingredients: In Probiotics in Dairy, Eggs, and Plants. Saprogenic New York: Marc Press. Boca Raton, FL.

39. Shimdang, G. (1998). Role of lactose absorption problems arising in milk products intolerance. Dairy Foods on Management of Lactose. 3, 26-29.

40. Stahl, C. (2010). Novel dairy probiotic product. Advances in Nutrition Research, 24, 234-245.

41. Tamine, A. Y. (2008). Probiotic Dairy Products. Blackwell Publishing, Oxford, UK.

42. Tharah, M. R., Acharwana, S., & Ray, R. C. (2007). Fermentation and applications of lactic acid content of probiotic Bifidobacterium. Bioscience Biotechnology Biochem. 10 (25022 Lactic acid starter microbes. Food Sci. Technol. Res., V. (Accessed 28-30 July 2019).

43. Tripathi, M. K., & Giri, S. K. (2014). Probiotic functional foods: Survival of probiotics during processing and storage. Journal of Functional Foods, 9, 225-241.

44. Vinderola, G., Prosello, W., Molinari, F., Ghiberto, D., & Reinheimer, J. (2009). Growth of Lactobacillus paracasei A13 in Argentinian probiotic cheese and its impact for the characteristics of the product. International Journal of Food Microbiology, 135, 171-174.

45. Vasiljevic, T., & Shah, N. P. (2008). Research developments and probiotic foods: A perspective of probiotic efficacy. International Dairy Journal of Dairy Science, 18, 714-728.

INDEX

Printed and bound by CPI Group (UK) Ltd, Croydon, CR0 4YY

23/10/2024

01777703-0011